18th Microscopy of Semiconducting Materials Conference 2013

(MSM XVIII)

Journal of Physics: Conference Series Volume 471

Oxford, United Kingdom
7-11 April 2013

ISBN: 978-1-62993-449-5
ISSN: 1742-6588

Printed from e-media with permission by:

Curran Associates, Inc.
57 Morehouse Lane
Red Hook, NY 12571

Some format issues inherent in the e-media version may also appear in this print version.

Printed by Curran Associates, Inc. (2013)

Institute of Physics
Dirac House, Temple Back
Bristol BS1 6BE UK

Phone: 44 1 17 929 7481
Fax: 44 1 17 920 0979

techtracking@iop.org

Additional copies of this publication are available from:

Curran Associates, Inc.
57 Morehouse Lane
Red Hook, NY 12571 USA
Phone: 845-758-0400
Fax: 845-758-2634
Email: curran@proceedings.com
Web: www.proceedings.com

TABLE OF CONTENTS

Advanced semiconductor characterization with aberration corrected electron microscopes

JL Rouvière[1], E Prestat[1], P Bayle-Guillemaud[1], M Den Hertog[2], C Bougerol[1], D Cooper[3] and J Zuo[4]

[1] CEA-INAC/UJF-Grenoble 1, UMR-E, SP2M, Minatec, 17 rue des Martyrs, 38054 Grenoble, France
[2] Institut Néel, CNRS/UJF, UPR2940, 25 rue des Martyrs, 38042 Grenoble, France
[3] CEA-LETI, Minatec, 17 rue des Martyrs, 38054 Grenoble, France
[4] Dept. of Materials Science and Engineering, University of Illinois, Urbana, Illinois 61801, USA

E-mail: jean-luc.rouviere@cea.fr

Abstract. Spherical aberration (Cs) correctors were demonstrated in the last years of the twentieth century and became commercially available a few years later. In Grenoble, we received our first probe corrector on a TEM/STEM machine in 2006. Cs-correctors have allowed us to improve the spatial resolution and the contrast of high resolution images both in TEM and STEM. The aim of the article is not to give a detailed description of Cs-correctors or a thorough analysis of their pros and cons but to illustrate what the benefits of the Cs-correctors have been in four areas: (i) atomic structure determination, (ii) polarity measurement, (iii) strain determination and (iv) interface analysis. Emphasis is put on the probe corrector although some comments on image correctors are given as well.

1. Introduction

In 1936, Scherzer published a theorem that in electron microscopy, classical lenses with rotation symmetry are condemned to have positive spherical aberration (Cs>0) [1]. After the second world war, more than 10 years later he proposed several ways to build lenses without spherical aberration: non rotationally symmetric lenses, non static lenses, lenses containing space charge or a combination of lenses and mirrors. Soon after, some of his students started to work on the realisation of such Cs-corrected lenses. Later other groups in the world had also some trials (see [2] for a short history), but the real success came more than 50 years later. The first working Cs-corrector was demonstrated on a Scanning Electron Microscopy (SEM) in 1994 [3]. In Scanning Transmission Electron Microscopy (STEM), the first probe Cs-corrector was mounted on a VG microscope by Krivanek et al [4]. Soon thereafter the first image Cs-correctors was working on a Transmission Electron Microscope (TEM) [5] and commercialisation was quickly started. 15 years after the demonstration of the performance of correctors about 400 corrected microscopes are now working in the world. Of course there are great disparities between the different countries. France have 14 microscopes with Cs-correctors, half having probe correctors, half image correctors. England has about the same number, but most of them have a probe corrector, Germany has more than 60 and Italy 2 electron microscopes with Cs-correctors. My laboratory got the chance to have one of the first probe Cs-correctors of the German company CEOS. This Cs-corrector was installed by FEI on our existing TITAN in our laboratory.

The aim of this paper is not to describe in details how correctors work but, after 7 years of Cs-corrector operation, to illustrate through four topics what the benefits have been for us of having Cs-correctors. Emphasis will be put on probe Cs-correctors although a few evocations to image Cs-correctors will be made. After a short presentation of Cs-correctors in practice, illustrations and comments will concern

(i) atomic structure determination,
(ii) polarity measurement,
(iii) strain determination and
(iv) interface analysis.

Most of the presented experimental data have been taken at CEA-Grenoble on three different microscopes: a JEOL 4000EX electron microscope equipped with a LaB_6 filament, a low base FEI TITAN microscope, first without Cs-corrector and then upgraded by FEI in our laboratory with a probe Cs-corrector, and an FEI TITAN Ultimate microscope having a monochromator and both image and probe Cs-correctors. A few tests were also performed using different microscopes in Canada and in the USA.

2. Cs-correctors in practice

Cs-correctors introduce some new complexities in the microscope as several non-symmetric lenses have to be aligned and tuned in order to reduce the aberrations of the objective lens. In fact, a Cs-corrector does not only tune Cs to a desired value but also minimizes several low-order aberrations (appendix 1). The tuning is principally done by software. As aberrations are best visible when the beam is tilted out of the optical axis, the strategy of all software is to acquire series of images acquired with different tilts of the beam [5]. Typically, a table of 15 images with tilt angles up to 20 or 30 mrad in at least 8 different directions are taken and a special sample – gold particles on thin amorphous carbon films – is used for that. Hopefully, most of the aberrations are relative stable and only defocus (C1), astigmatism (A1) and coma (B2) (see appendix 1 for a list of aberrations) need a regular tuning if the microscope is kept in a given specific state, i.e. at the same high tension and same imaging technique. Tuning an image corrector is rather fast - about 20 minutes - while a probe corrector requires more time - at least 1 hour for the CEOS corrector. It is not cumbersome to have image and probe correctors on the same microscope as the tunings of both correctors are very similar and tuning the image corrector is done quickly. Up to now, only confocal experiments [6] have used the two correctors simultaneously. In normal use, only one corrector is optically active. Using a dedicated sample is not compulsory if the microscope is kept in a specific state and as there are ways to manually tune A1, C1 and B2 [7]. In our laboratory, where many users are working with different settings (for holography, TEM, STEM, diffraction), we generally prefer to start our session by introducing the 'gold on amorphous carbon' sample, which is constantly kept on a dedicated single tilt holder, and tuning accurately the aberrations.

For high resolution, either in TEM or in STEM, the benefits of the corrector are straightforward as far as image quality is concerned and the time spent in the tuning of the corrector is worth the improvements ([8], for STEM see Fig. 1). In STEM, the improvement in the transfer function is directly seen in the ronchigram which is just a diffraction pattern of an amorphous film obtained with a very high convergence angle ([9], see fig. 1e). The ronchigram gives a direct image of the aberration function $\chi(\theta,\phi)$ (see appendix for definition). One can see in fig. 1e that in the aberration corrected system, there is a large uniform region at the centre of the ronchigram (up to 20 mrad), whereas in the uncorrected system (i.e. a ronchigram obtained without the use of a Cs-corrector), the uniform region in the middle is much smaller. In STEM, the condenser aperture is introduced in order to include only the uniform region of this ronchigram. In non-Cs-corrected microscopes, typical convergence angles will be about 9 mrad, whereas they reached 15 mrad in the first generation of Cs-correctors and up to 30 mrad in the last generations of Cs-correctors that tune aberrations up to the fifth orders (i.e higher order aberration than the ones given in appendix 1). By allowing higher convergence angles, the beam intensity is also increased which is important for chemical analysis and acquisition speed [10].

In TEM, the main effects of correctors are to increase the resolution (typically from 0.14 nm to below 0.1 nm), to reduce the noise and increase the contrast and reduce the delocalisation effect at interfaces [5]. In fact, with the introduction of Cs-correctors in TEM and STEM, manufacturers had to improve the electronics and mechanical stabilities of microscopes. These stability improvements are particularly important for high resolution STEM images as STEM is particularly influenced by instabilities. From a theoretical point of view there are many analogies between TEM and STEM images. Indeed, the principle of reciprocity [11] states that similar images are obtained in TEM and STEM if the convergence angles (α) and detection angles (β) in TEM and STEM satisfy the relation given in equations (1):

$$\alpha_{STEM} = \beta_{TEM}, \quad \beta_{STEM} = \alpha_{TEM}. \tag{1}$$

In practice, however, equations (1) are rarely satisfied and in standard use, TEM and STEM images have different types of contrast.

As the first Cs-corrected microscope we got was probe-corrected, we have more practice with probe-correctors and most of the examples we will consider will be in STEM. Without starting a controversy or complete comparison of the two techniques, we have the feeling that more and more people, like us, are interested in STEM because STEM offers more possibilities in terms of chemical analysis and the image interpretation is often simpler. TEM is more and more focused on wave function reconstruction [12], low voltage electron microscopy – 80kV and lower – and small precipitates and thin membranes like graphene are more easily observed in TEM [13].

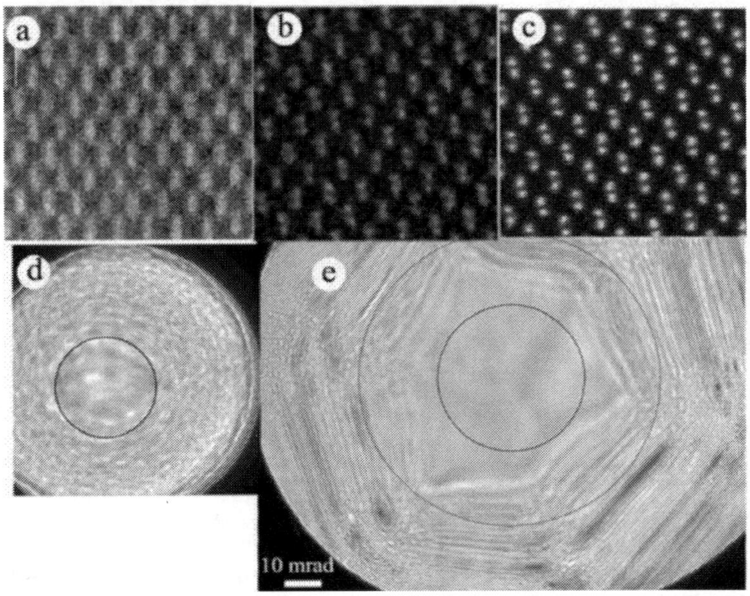

Figure 1. Comparison of STEM images of <110> silicon obtained on different microscopes: (a) FEI-TITAN 300kV without Cs-corrector, (b) the same microscope as in (a) but after the installation of a probe Cs-corrector, (c) FEI-TITAN ultimate 300kV equipped with a D-core probe Cs-corrector. d-e) Comparison of two ronchigrams, i.e. large angle diffraction patterns of amorphous carbon films, obtained on the same FEI-TITAN microscope before (d) and after (e) installation of the probe Cs-corrector. The inner circle corresponds to a suitable position for a condenser aperture for STEM.

3. Atomic structure determination: the case of an incommensurate 90° Si grain boundary

The improvement in spatial resolution enabled by Cs-correctors has allowed the solution of unknown atomic structures of interfaces or defects and several examples can be found in the literature. For instance, in [14] it is shown how the atomic structure of material around SiN masks used in GaN growth was determined thanks to the use of an image Cs-corrector that allowed clear atomic images of the interface without blurring or possible delocalisation contrast as would have been observed in non-Cs-corrected HR-TEM. In this paper, we recall the personal experience of the first author of this paper to solve the atomic structure of an incommensurate 90° grain boundary (GB) obtained by wafer bounding. More details about this specific grain boundary can be found in [15]. This structure was first observed in 2001 on a JEOL 4000EX microscope (fig. 2a). With this microscope the Si dumbbells could not be resolved and as the interface is not periodic it was impossible to build a model of the interface. As one of us was in Chicago in 2001, the sample was observed in a JEOL 200kV STEM that was able to resolve the Si dumbbells [16]. However, the stability and the resolution of this microscope was not high enough to resolve simultaneously the silicon dumbbells in the two crystals rotated by 90°. This sample was also observed in an FEI microscope equipped with an image corrector installed at McMaster University in 2007. Here, the disappointment was that nice images could be obtained only for very thin areas. In most of the areas, the sample was too thick to provide interpretable images. We had to wait for the arrival of our TITAN microscope in 2005 and the installation of the probe Cs-corrector in 2006 to solve the complex atomic structure of this grain boundary. Fig. 2 shows a part of this interface, with periods of 7/2[110] in grain 1 and 5[001] in grain 2 frequently observed in this non-periodic structure [15]. As high-angle annular dark field (HAADF) contrast does not vary much with thickness and defocus, we could observe large areas of the sample and determine at once without simulation the atomic projection of the structure (fig. 2b). The Si dumbbells are resolved in the two grains and single atomic columns can be clearly distinguished. Assuming that each Si atom has four neighbours situated approximately at the corners of a tetrahedron it was possible to build a 3D structure of the GB and minimize the energy of the configuration by using empirical potentials [15]. The projected optimized structure perfectly fits the experimental HAADF STEM image (fig. 2b). Once the structure was determined we inverted the HRTEM image of the JEOL 4000EX microscope obtained at Scherzer defocus and checked that this image was consistent with the determined structure.

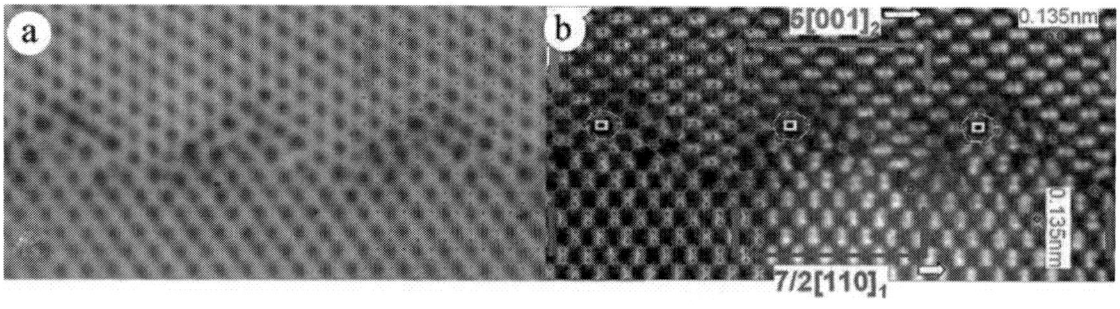

Figure 2. Comparison of two images of an incommensurate 90° grain boundary in Si. Small red spots corresponding to atomic columns have been added to the images: (a) HRTEM image obtained on a JEOL 4000EX microscope at Scherzer defocus with atomic columns in black; the contrast has been inverted in order to compare it directly to (b) the STEM image obtained in an FEI TITAN microscope equipped with a probe Cs-corrector. Some bonds between atoms have been overlaid in green in (b).

4. Polarity measurement

In non-centrosymmetric crystals, there are directions along which the positioning of atoms is not symmetric leading to some special properties of these directions: electrostatic fields can build up along these directions, planes parallel to these directions can have different atomic terminations and so different properties. Determining the ordering of atoms along these directions is determining the polarity of the structure. In [17] we proposed a mnemonic way to define the polarity in a binary compound semiconductor AB, A and B being two chemical elements, for instance A=Ga and B=N. A given surface ***n*** of the AB compound is said to have A-polarity if the ***AB*** vector defined by the A-B bonds responsible of the asymmetry of the surface has an acute angle with ***n*** (Fig. 3). A couple of years ago, the traditional way to measure the polarity in TEM was to use convergent beam electron diffraction (CBED) and look at the asymmetry of some diffraction spots [17]. The difficulty was first to calibrate the microscope as the lenses can introduce a 180° rotation between the image and the diffraction pattern, and to obtain local diffraction maps. Alternative solutions have been proposed: HRTEM image analysis [18, 19, 20], HAADF-STEM image [21, 22], or more recently Annular Bright Field (ABF) STEM images [23, 24, 25, 26]. Before the introduction of Cs-correctors, the analysis of HRTEM was not very robust unless a high voltage microscope was used [18]. Indeed, the determining of polarity relied on the shape of a profile made along a narrow inclined line [19], but the difference between the two polarities was faint and the HRTEM contrast is very sensitive to tilt, thickness and defocus. More recently Bertoni et al [20] used focal series of Cs-corrected HRTEM images. From the series, they reconstructed the exit phase function and using a channelling approximation, built a phase image where the atomic columns could be directly positioned. A more simple solution is to use Cs-corrected HAADF-STEM images and directly image the different atom columns. In AlN, this works very well because the difference between the atomic numbers of Al and N is small [21]. In [22], we showed that when the difference in atomic number is larger (for instance in GaN) it is easier not to locate directly the N-atomic columns but to locate the tunnels of the structure, which indirectly gives the N-column position and thus the polarity. Rather recently a new type of STEM image, called Annular Bright Field (ABF) [27], has been re-introduced by Findlay et al. [23]. It consists of adapting the camera length and convergence angle in order that the rim of the transmitted beam reaches an annular detector. In this way light atoms can be detected. The drawback of this method is that it requires a rather high convergence angle to resolve the GaN columns and that the ABF image suffers from similar drawbacks as TEM images, i.e. its pattern can change with defocus and thickness. Several publications have applied this ABF technique to the determination of polarity in nanowires [24, 25, 26].

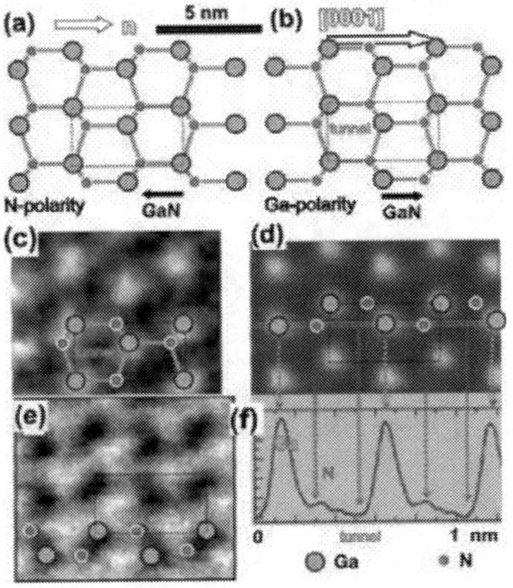

Figure 3. a-b) The GaN crystal structure projection along the [2$\underline{11}$0] direction. Bonds between atoms are in green. a) N-polarity orientation: the ***Ga-N*** vector is antiparallel to the surface normal ***n*** b) G-polarity orientation : the ***Ga-N*** vector is antiparallel to ***n***. c-d) HAADF-STEM images of GaN observed along [2$\underline{11}$0] with some atoms superimposed. c) Image obtained on TITAN 1, d) image obtained on TITAN Ultimate; the image has a better resolution and is less noisy. e) ABF-STEM image of GaN observed along [2$\underline{11}$0]. f) Average profile taken from the horizontal rectangle shown in image d). This profile shows clearly where the N columns and the tunnels are located and thus allows determination of the polarity of the layers.

5. Strain determination

Our laboratory has been involved in strain measurement at interfaces and nanostructures for a long time. It started in 1984 with the analysis of HRTEM images acquired on a JEOL 4000EX microscope. First methods used interfacial profiles or direct space methods [28]. For instance, it was possible to detect half-filled semiconductor monolayers of MnTe in CdTe [29]. Then we preferred to use the Geometrical Phase Analysis (GPA) [30, 31] because of its simplicity to provide 2D analysis and its extreme sensibility. We wrote scripts in Digital Micrographs; scripts are available through collaborations. With the arrival of our first non-corrected TITAN microscope, we presented encouraging analysis of HR-STEM images at the 16th International Microscopy Congress in Sapporo [32], but the first physical results really came after the installation of the probe Cs-corrector on our first TITAN microscope. We could partly correct for scanning errors [33], which are intrinsic to HR-STEM images, and accurately determine the strain field around edge dislocations in AlInN layers [34]. Such an analysis would not have been possible in HR-TEM due to the quality of the plan-view samples that were very thick and had a thick amorphous layer on top of them. The analysis of HR-STEM became even better with the arrival of the TITAN ultimate. With the new FEI software, 4kx4k images could be acquired instead of the previous 2kx2k scanned images. With the Gatan Digiscan option, even 8kx8k images are possible. In addition, the stability of the TITAN Ultimate has greatly reduced the scanning errors and the newly designed probe corrector gives images with higher contrast (Fig. 1). This technique is now very useful to measure strain in microelectronic devices (fig. 4, [35]). Compared to HR-TEM images, HR-STEM images have several advantages: (i) patterns do not change as much with thickness and composition, (ii) so it is possible to analyse thicker samples or samples where the sample thickness is changing smoothly, which is very important in microelectronics where most of the samples are prepared by focused ion beam (FIB), (iii) large fields of view (800nm) with 8kx8k pixels are possible and (iv) in addition to strain, some chemical information can be obtained through the intensity of the Z-contrast image [36].

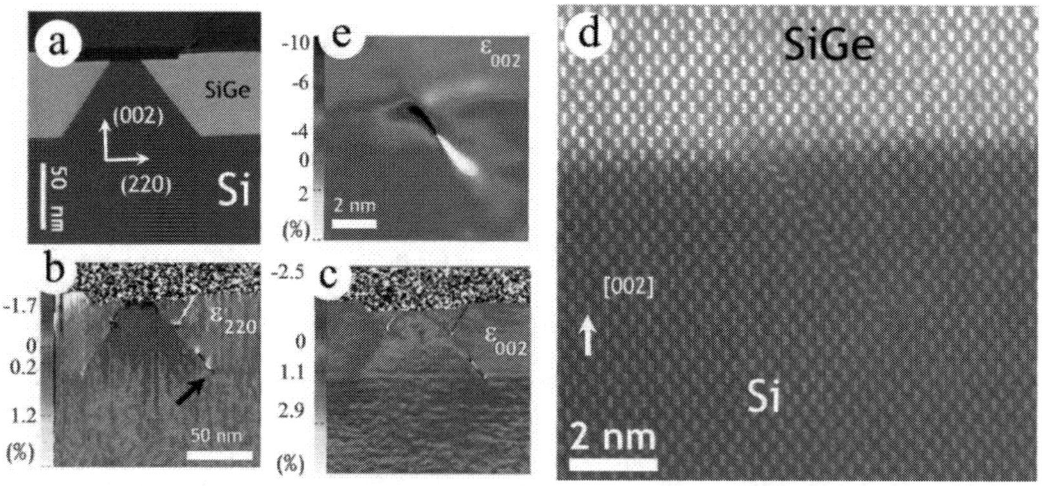

Figure 4. (a) 4kx4k HR-STEM image of a transistor with $Si_{0.64}Ge_{0.36}$ source and drain. At this scale the atomic column are not visible. (b) (220) strain map of (a). Some dislocations can be seen. A black arrow points to one of them. (c) (002) strain map of (a). (d) zoom on a dislocation present in the transistor of fig. 4a. It is situated at the SiGe/Si interface. The 60° dislocation is dissociated into two partial dislocations separated by a stacking fault. The atomic structure of the dislocation is directly determined and the step at the SiGe/Si interface introduced by the dislocation is clearly visible. (e) (002) strain map obtained from image (b). The reference of strain is here taken in the $Si_{0.64}Ge_{0.36}$ layer.

6. Interfacial profiles

As already mentioned in the previous part, the image pattern in Z-contrast images does not change much with thickness. These images can be effectively used for determining the chemical composition and the thickness at different semiconductor interfaces, by normalising these images by the incident beam intensity and by comparison with advanced simulations [36]. The method does not need a Cs-corrector and can work with low or high resolution STEM images. The advantage of the Cs-corrector will be to acquire more easily high resolution images, which have better resolution and in addition higher contrast. However, when working with high convergence angles (i.e. angles greater than 15 mrad) one should take care of the reduced depth of field of HR-STEM images [37]. On the other hand, this limited depth of field can be used for depth sectioning, that is to say, 3D reconstructions [37]. We noticed that in FIB prepared samples or in samples having some amorphous layers at the surface of the TEM lamella, the best images are not obtainable from the thinnest areas of the samples. On the contrary, for enhanced quantitative analysis, it is better to use a very thin sample without amorphous layer at its surface. In our cases, the only method that achieved this was tripod polishing [38]. In [38], we succeeded in detecting single Tm dopant atoms in an AlN matrix. More recently, single Ge atoms could be detected at SiGe interfaces (Fig. 6). Quite recently, we applied a Template Matching Algorithm (TeMA) to analyse Z-contrast interfaces [39]. Such a method was already introduced some years ago for the analysis of some special HR-TEM images [40]. Its application in HR-STEM should be more general. The method starts by defining the image unit cell size of the local patterns I_i that will be used in the analysis; here i is an integer number used to identify the local pattern. For this a small region around an atomic column is selected with the mouse and forms an initial template, T'. A cross-correlation image, or C-map, is then calculated by making all the cross-correlations between T' and the local $I_{i,j}$ patterns; here i,j run over all the pixels of the initial image. The maxima of the C-map determine the locations where the local patterns P_i in the initial image will be taken and thus determine the number n of I_i patterns. By thresholding the C-map the regions where the local patterns I_i that will be taken for the calculation of the template T are selected; T is obtained by averaging these selected patterns. The above steps (i.e. calculation of a new C-map by using the new template as starting point and new threshold of this new cross-correlation) can be repeated to refine the template T. Generally, two templates T_1 and T_2 are defined, one for either side of the interface. Then each local image I_i is decomposed into these two templates (see equation 2):

$$I_i = a_{i1}T_1 + a_{i2}T_2 + \varepsilon_i \quad (2)$$

where normally ε_i should have small values and describe noise. Amplitude images a_1 and a_2 can then be constructed and provide semi-quantitative chemical information (fig. 5). For instance, the amplitude image a_{GaSb} of fig. 5 shows the asymmetry of the interfaces GaSb/InAs: following the growth direction, the second interface in GaSb is more extended than the first one. We check that this semi-quantitative information (the amplitudes are not real chemical values of the interface) has a physical meaning by comparing this profile to chemical profiles obtained with an atom probe [39]. For instance in fig. 5, the Sb atom probe profile has been added and shows a similar asymmetry, although the asymmetry appears greater in the atom probe data.

In the case of a SiGe/Si interface, two templates, T_1 and T_2, were taken in the Si layer in order to have only one atomic column of the dumbbells (fig. 6). Each local image I_i was then expressed as function of either T_1 or T_2, depending on whether I_i was extracted at the top (T_1) or bottom (T_2) atom of the dumbbell: $I_i = a_{i1}T_1 + \varepsilon_i$ or $I_i = a_{i2}T_2 + \varepsilon_i$. In this case the template matching is just an efficient way to extract for each atomic column an intensity value, either a_{i1} or a_{i2}, that can be calibrated in terms of Ge concentration (fig. 6). In our case, this calibration was done by using EELS maps that were simultaneously acquired and measuring precisely the thickness of the sample by PACBED (Position Averaged CBED) [41]. A variation of 6 nm in thickness between the top and the bottom regions of the HR-STEM image in fig.6 was found. Details of this analysis should be published soon.

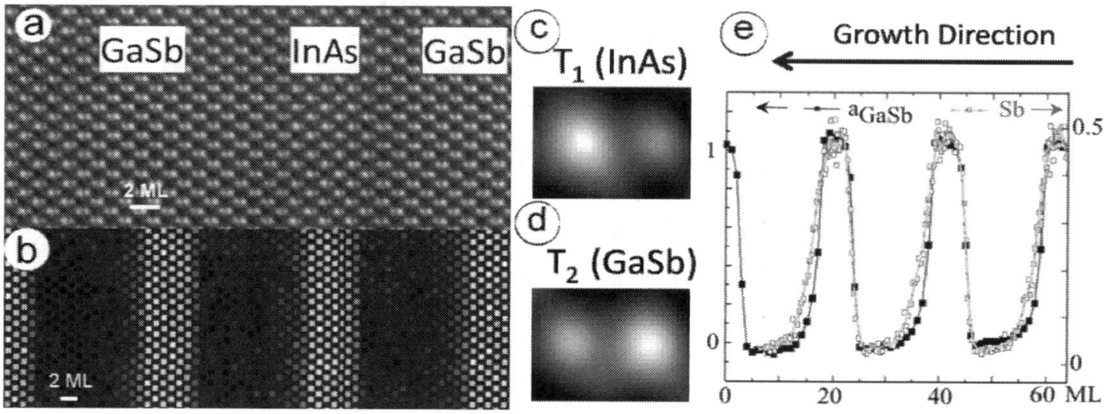

Figure 5. a) HAADF-STEM image of GaSb/InAs multilayers. b) Amplitude map of GaSb as obtained with TeMA using the templates *T1* and *T2*. c-d) Respective templates obtained in InAs and GaSb. In T_1, the As atom is on the right, In is on the left. e) Superimposition of two profiles obtained on the multilayers: in dark, profile obtained from the amplitude map of fig. 5b (left scale), in grey profile obtained from the atom probe (right scale, percentage of Sb).

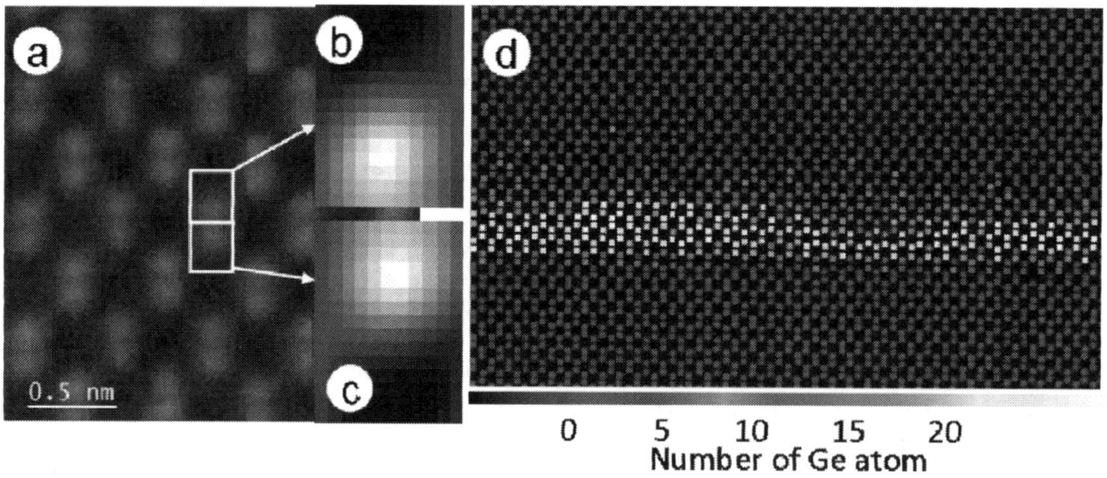

Figure 6. a) Zoom of a HAADF-STEM image showing the Si layer of SiGe interface. b-c) Templates taken at the atomic position of the silicon dumbbell. b) Template T_1. c) Template T_2. d) Amplitude image obtained at the Si/SiGe interface using the templates defined in b and c. By using EELS maps and the thickness of the sample the intensities of the amplitude map have been calibrated in multiples of integer numbers of Ge atoms.

7. Conclusion

By providing a few examples of semiconductor studies, we have tried to show what have been the benefits of Cs-correctors for our laboratory. Starting from a laboratory that was essentially doing HR-TEM, the probe Cs-corrector pushed us toward HR-STEM. We saw an important difference between HR-STEM done without a Cs-corrector (acquiring HR-STEM images demanded skills and patience) and HR-STEM done with a Cs-corrector (after its semi-automatic tuning, the Cs-corrector guarantees a good image). The introduction of Cs-correctors pushed the manufacturers to provide more stable microscopes.

Thanks to the Cs-correctors we were able to determine the structure of a complex incommensurate interface, to detect individual atoms in a column, to measure polarity and perform large field-of-view strain maps. Although not illustrated in this paper, the Cs-correctors also allow us to work at very low voltage (60kV and certainly soon below) while keeping very good spatial resolution and they permit the use of large pole piece gaps (for in-situ experiments). However, one must be aware that Cs-correctors introduce some addition aberrations (chromatic and high order aberrations) and distortions that can be annoying in some specific applications. Although Cs-correctors have boosted the electron microscopy technique and are very useful, it does not necessarily mean that Cs-correctors are and will be compulsory in a TEM. Nice experiments, for instance low resolution images, EDX chemical detection and diffraction studies can be done without Cs-correctors.

Appendix

Aberrations are introduced in the aberration function $\chi(\theta,\phi)$ which is the error in optical path length between a perfect planar wave and the actual wavefront as function of the diffraction angles θ (radial) and ϕ (azimuthal). Different notations for aberrations co-exist. For instance, in equation (3) the aberration of order n and angular multiplicity m is characterized by an amplitude $|C_{n,m}|$ and a phase $\phi_{Cn,m}$. As most of the TEM/STEM Cs-probe corrected microscopes use the correctors from CEOS company, Haider et al's notation is often used (3) although it is not very convenient as aberrations of the same order, like A2 and B2, do not always have the same pre-coefficients. In equation (3), A1, C1, A2, B2, C3, A3, S3 are respectively second-order astigmatism, defocus, third-order astigmatism, coma, spherical aberration (i.e. Cs), fourth-order astigmatism and star aberration. On a daily basis, all these aberrations are tuned. Some higher order aberrations are tuned during the build-up of the corrector in the factory.

$$\chi(\theta,\phi) = \sum_{\substack{m<n+1 \\ m+n\ odd}} \frac{1}{(n+1)} \left|C_{n,m}\right| \theta^{n+1} \cos(m\phi - \phi_{C_{n,m}}) \tag{3}$$

$$\chi(\theta,\phi) = \tfrac{1}{2} A_1\theta^2 \cos(2\phi - \phi_{A1}) + \tfrac{1}{2} C_1\theta^2 + \tfrac{1}{3} A_2\theta^3 \cos(3\phi - \phi_{A2}) + B_2\theta^3 \cos(\phi - \phi_{B2}) + \tfrac{1}{4} C_3\theta^4 + \tfrac{1}{4} A_3\theta^4 \cos(4\phi - \phi_{A3}) + S_3\theta^4 \cos(2\phi - \phi_{S3}) + \ldots \tag{4}$$

References

[1] Scherzer O, 1936 *Z. Physik* **101** 593.
[2] Hawkes P, 2001 *Biology of the Cell* **93** 432.
[3] Zach and Haider M, 1995 *Nucl. Instrum. Meth. Phys. Res.* A**363** 316.
[4] Krivanek O L et al, 1997 *Proc. EMAG'97 Inst. Phys. Conf. Ser. 153* 35.
[5] Haider M et al, 1998 *Nature* **392** 768.
[6] Cosgriff E C, D'Alfonso A J, Allen L J et al, 2008 *Ultramicroscopy* **108** 1558.
[7] Sawada H, Sannomiya T et al, 2008 *Ultramicroscopy* **108** 1467.
[8] Van Tendeloo G, Bals S et al, 2012 *Advanced materials* **24** 5655.
[9] Batson P E, 2006 *Ultramicroscopy* **106** 1104.
[10] Walther T and Stegmann H, 2006 *Microscopy and Microanalysis* **12** 498.
[11] Kirkland E J *Advanced computing in Electron Microscopy* 2010 Springer.
[12] Tillmann K, Houben L and Thust A 2006 *Phil. Mag.* **86** 4589.
[13] Lee Z, Meyer JC, Rose H and Kaiser U 2012 *Ultramicroscopy* **112** 39.
[14] Markurt T et al, 2013 *Phys. Rev. Lett.* **110** 03610.
[15] Rouviere J-L, Lançon et al, 2010 *J. of Physics : conf. Series* **209** 012041.
[16] Browning N D, James E M et al 2000 *Rev. Adv. Mat. Sci* **1** 1.
[17] Rouviere J-L, Weyher J L et al 1998 *Appl. Phys. Lett.* **73** 668.
[18] Iwamoto C, Shen X-Q et al 2002 *Materials Transactions* **43** 1542.
[19] Stirman J N, Ponce F A et al 2000 *Appl. Phys. Lett.* **76** 822.
[20] Bertoni G, Grillo V et al 2012 *ACSnano* **6** 6453.
[21] Mkhoyan K A, Batson P , Cha J et al 2007 *Science* **69** 2480.
[22] Rouviere J-L, Bougerol C, Amstatt B, et al 2008 *Appl. Phys. Lett.* **92** 201904.
[23] Findlay S D, Shibata H et al 2010 *Ultramicroscopy* **110** 903.
[24] Radtke G, Couillard M et al 2010 *Appl. Phys. Lett.* **97** 251901.
[25] De la Mata M, Magen C, Gazquez J et al 2012 *Nanoletters* **12** 2579.
[26] den Hertog M I, Gonzàlez-Posada F et al 2012 *Nanoletters* **12** 5691.
[27] Watanabe K, Kikichi Y et al 2004 *Acta Cryst.* **A60** 591.
[28] Jouneau P H, Tardot A et al 1994 *J. Appl. Phys* **75(11)** 7310.
[29] Charleux, M., J.L. Rouviere et al 1998 *J. Appl. Phys.* **84(2)** 756.
[30] Hytch M J, Snoeck E and Kilaas R 1998 *Ultramicroscopy* **74** 1998 131.
[31] Rouviere J-L and E Sarigiannidou 2005 *Ultramicroscopy* **106** 1.
[32] Rouviere J-L, Sarigiannidou E and Daudin B 2006 *Proc. of the Int. Microsc. Cong. IMC16* **2** 605.
[33] Rouviere J-L, Mouti A and Stadelmann P, 2011 *J. of Physics: Conf. Series* **326** 012022.
[34] Mouti A, Rouviere J-L, Cantoni, M et al 2011 *Phys. Rev. B* **83** 195309.
[35] Cooper D, Le Royer C, Beche A, Rouviere J-L 2012 *J. Appl. Phys.* **112** 124505.
[36] Rosenauer A, Gries K, Müller K et al 2009 *Ultramicroscopy* **109** 1171.
[37] Xin H L and Muller D A 2010 *Micros. Microanal.* **16** 445.
[38] Okuno, H., J. L. Rouviere, P. H. Jouneau et al. 2010 *Appl. Phys. Lett.* **96** 251908.
[39] Kim, Zuo J M et al 2013 *J. Appl. Phys.* **113** 103511.
[40] Ourmazd A, Baumann F H, Bode M and Kim Y 1990 *Ultramicroscopy* **34** 237.
[41] Lebeau J M, Findlay S D et al 2010 *Ultramicroscopy* **110** 118.

Dislocation dynamics in SiGe alloys

I Yonenaga
Institute for Materials Research, Tohoku University, Sendai 980-8577, Japan

E-mail: yonenaga@imr.tohoku.ac.jp

Abstract. The dislocation velocities and mechanical strength of bulk crystals of Si_xGe_{1-x} alloys grown by the Czochralski method have been investigated by the etch pit technique and compressive deformation tests, respectively. Velocity of dislocations in the SiGe alloys of the composition range $0.004 < x < 0.08$ decreases monotonically with an increase in Si content at temperature 450–700°C and under stress 3–24MPa. In contrast, velocity of dislocations in the composition range $0.92 < x < 1$ first increases, then decreases and again increases with a decrease in Si content at temperature 750–850°C and under stress 3–30MPa. The velocity of dislocations was quantitatively evaluated as functions of stress and temperature. Stress-strain behaviour in the yield region of the SiGe alloys of composition $0 < x < 0.4$ is similar to that of Ge at temperatures lower than about 600°C. However, the yield stress becomes temperature-insensitive at high temperatures and increases with increasing Si content. The stress-strain curves of the SiGe alloys of composition $0.95 < x < 1$ are similar to those of pure Si at temperatures 800–1000°C and the yield stress increases with decreasing Si content down to $x = 0.95$. The yield stress of the SiGe alloys is dependent on the composition, being proportional to $x(1-x)$, showing a maximum around $x \approx 0.5$. Built-in stress fields related to local fluctuation of the alloy composition and the dynamic development of a solute atmosphere around the dislocations, may suppress the activities of dislocations and lead to the hardening of SiGe alloys.

1. Introduction

Silicon-germanium (Si_xGe_{1-x} or Germanium-silicon $Ge_{1-x}Si_x$, where x indicates the mole fraction of silicon) is a complete solid-solution semiconductor having the diamond cubic structure. SiGe alloys have attracted great interests for both microelectronic and optoelectronic devices and various functional materials because of the potential for band-gap and strain/lattice parameter engineering they offer. That is, alloying leads to various unique effects on fundamental properties, absent in the component materials Si and Ge.

Usually these alloys are grown as thin films on Si substrates by various epitaxial techniques. The introduction of misfit dislocations is inevitable in such hetero-structures due to the stress caused by interfacial mismatch when the film thickness exceeds a critical value in the system of thin film/substrate. Although dislocations affect the electrical and optical properties of SiGe alloy and limit its application within various devices, only little is known about their dynamic properties such as the generation and motion of misfit dislocations at the film/substrate interface within stressed layers, mainly for compositions on the Si-rich side [1-5]. The large biaxial stress inherent to such hetero-structures hinders the quantitative study of the native properties of dislocations. The dynamic properties of dislocations in SiGe have often been assumed to be similar to those in Si or Ge and little attention has been paid to the unique properties which have been found in several compound alloy semiconductors [6-8]. The present author reported that the flow stresses of GaAsP and InAsP ternary alloys have an athermal component that is absent in the binary GaAs, GaP, InAs and InP compounds [9,10]. Accordingly, it is of interest to investigate the dynamic activities of dislocations in SiGe alloys using bulk crystals and to deduce their unique properties brought about by alloying. SiGe alloys are quite suitable for such a basic study. The mechanical behaviour of the element materials Si and Ge has been under investigation since the 1960's pioneering studies by groups such as Patel and Chaudhuri [11] and Bell and Bonfield [12] and at present is well understood on the basis of kinetic properties of dislocations [13].

To clarify the dynamic properties of dislocations in SiGe alloys quantitatively, it is necessary to grow bulk crystals with low dislocation densities and measure the dislocation velocities and the mechanical properties. Growth of bulk single crystals of SiGe alloys is difficult because of a large miscibility gap and the differences in the densities, lattice parameters and melting temperatures of the constituent elements. However, this type of material is also suited for a basic study on the solidification of solid solution. Here, unique features that appeared in thermo-mechanical behaviour in SiGe grown by the Czochralski (CZ) method are reviewed, based on the results by the author's group [14-19].

2. Bulk crystals

Bulk crystals of Si_xGe_{1-x} alloys in the whole composition range $0 < x < 1$ were grown by the Czochralski technique at very low pulling rates ranging from 1 to 8 mm/h in an Ar gas flowing atmosphere [14, 19-22]. Seeds prepared from a Si or Ge crystal oriented parallel to [111] or [001] were used for the growth of the crystals.

Full single crystals of large size (i.e. larger than 15 mm in diameter and longer than 40 mm) were successfully grown for the composition ranges of $0 < x < 0.15$ and $0.9 < x < 1$. Small single crystal alloys of intermediate composition were obtained in the ingot near the seeds since the transition to polycrystallinity relates to the occurrence of constitutional supercooling [22].

Striations and dislocations were observed in the grown crystals. Dislocations were generated mainly at the seed/alloy interface. The density of grown-in dislocations in the alloys was in the range 10^3–10^5 cm^{-2}. The generation process may be controlled by the magnitude of the misfit strain between the alloy and seed, the temperature, the temperature gradient and the mobility of generated dislocations in the alloy [19].

This alloy shows a variety of unique properties, such as atomistic bonding structure, electron and hole mobilities, thermal conductivity, local-vibration of oxygen impurity, muonium, etc., which have been reported separately [23-31]. Here, it should be noted that SiGe alloy is a typical disorder material and that the bond lengths and bond angles are distorted with alloy composition in SiGe, classified to be quasi-Pauling type [23,24].

3. Dislocation velocities

Velocities of dislocations were measured in Si_xGe_{1-x} alloys of composition range $0 < x < 0.08$ and $0.922 < x < 1$ with low densities of grown-in dislocations of ~ 10^3 cm^{-2}. The specimen was stressed at elevated temperature by three-point bending in a vacuum. Dislocations were generated preferentially from a scratch drawn on the surface. Displacements of dislocations caused by stressing were measured by the etch pit technique [15,17,18].

Figure 1(a) shows the velocities of 60° dislocations at 550°C plotted against the resolved shear stress in CZ-SiGe alloys of the composition range $0 < x < 0.08$ together with that in pure Ge. As seen in the figure, the logarithm of the velocity of dislocations is linear with respect to the logarithm of the stress, with approximately the same slope for all the Si contents except the following feature: the velocity of 60° dislocations in the SiGe alloys of higher Si content shows a break at stress around 8 MPa, depending on the Si content. Figure 1(b) shows the velocities of 60° dislocations in Ge-rich SiGe

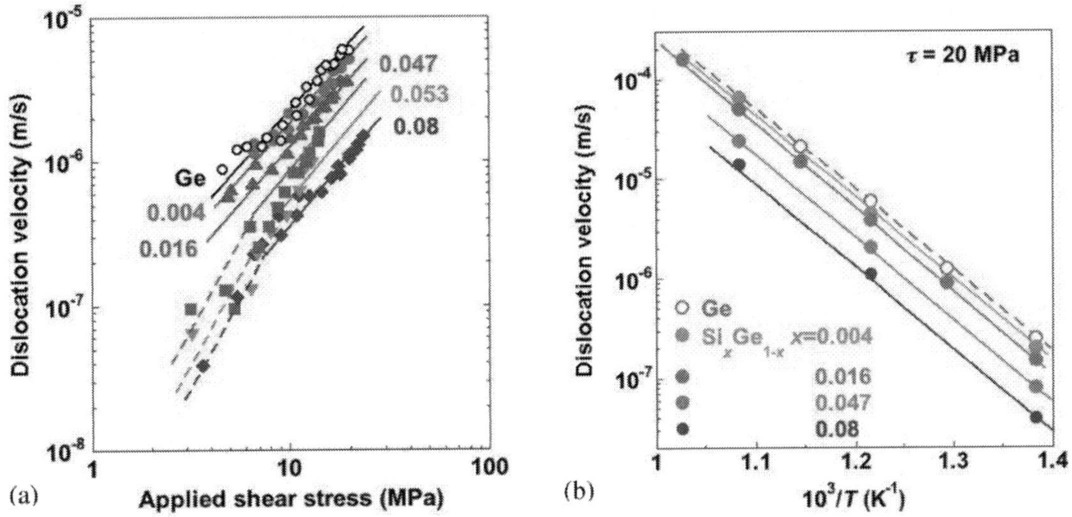

Figure 1. Velocities of 60° dislocations in the Ge-rich SiGe alloys. (a) Stress-dependence at 550°C, (b) temperature-dependence under a shear stress of 20 MPa. Numbers show the Si content.

alloys of various Si contents under a stress of 20 MPa, free from any threshold or kink, plotted against the reciprocal temperature together with that in pure Ge.

Figure 2(a) shows the velocities of 60° dislocations at 800°C plotted against the resolved shear stress up to 30 MPa in float-zone-grown (FZ-)SiGe alloys of the composition range $0.922 < x < 1$ (Si-rich side) together with that in pure Si [32]. The logarithm of the velocity of dislocations in SiGe alloys is linear with respect to the logarithm of the stress at temperatures 750–850°C. The slope of the plot of the dislocation velocity versus the stress in the SiGe alloy with $x = 0.996$ is approximately the same as that in Si. On the other hand, in SiGe alloys of $x = 0.922$–0.979 the velocity of dislocations is

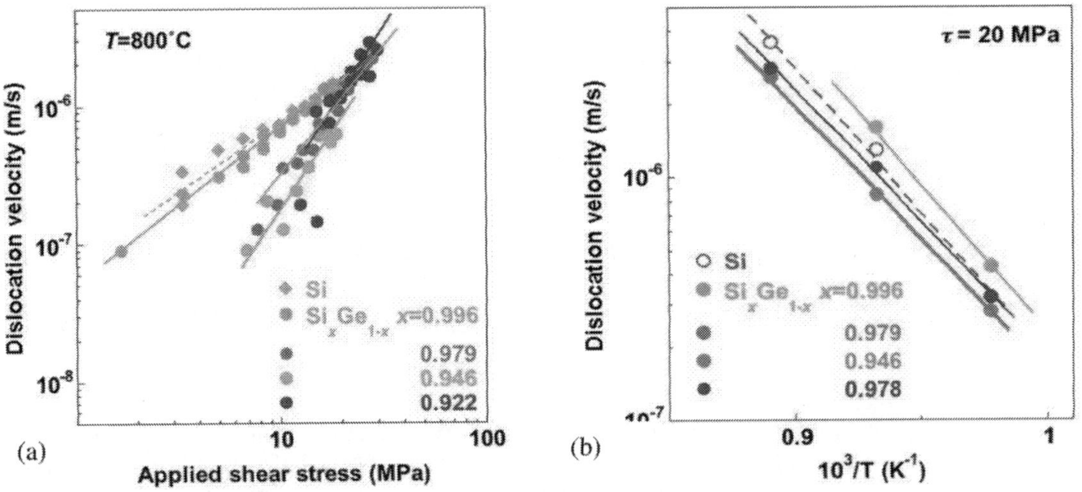

Figure 2. Velocities of 60° dislocations in the Si-rich SiGe alloys. (a) Stress-dependence at 800°C, (b) temperature-dependence under a shear stress of 20 MPa. Numbers show the Si content.

zero under a stress lower than the threshold one and then increases rapidly with an increase in stress beyond the threshold stress as reported by Iunin et al. [33]. The threshold stress for dislocation generation from a scratch increases with a decrease in the Si content. Figure 2(b) shows the velocities of 60° dislocations in Si-rich SiGe alloys of various Si contents under a stress of 20 MPa, free from an effect of the threshold, plotted against the reciprocal temperature together with that in pure Si.

In the Ge-rich SiGe alloys of composition range $0 < x < 0.08$ the dislocation velocity decreases monotonically with increasing Si content, reaching about one of seventh of that in pure Ge at $x = 0.08$ in the temperature range, as shown in figure 3. On the other hand, in the composition range $0.922 < x < 1$ (Si-rich SiGe) the dislocation velocity first increases, then decreases and again increases with increasing Ge content in the temperature range 750–850°C and the stress range 3–30 MPa, as shown in figure 3. The dislocation velocity for the Si content $x = 0.996$ is found to be higher than that in pure Si.

The velocities of dislocations in the SiGe alloys investigated in the present studies are expressed in a similar way to those in Ge, Si and other semiconductors as functions of the stress τ and temperature T by the following empirical equation [13, 34, 35]:

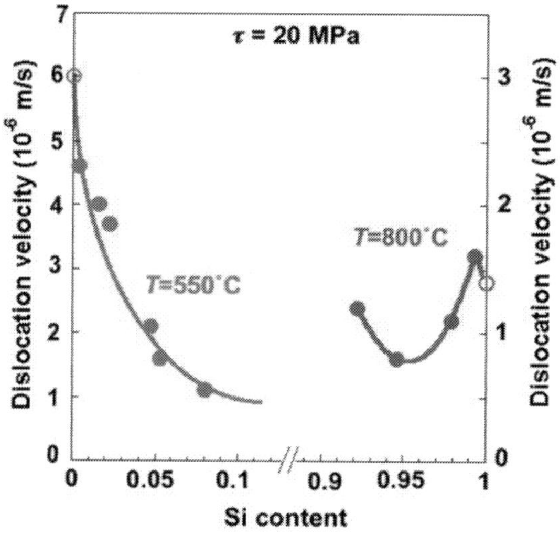

Figure 3. Velocities of 60 dislocations in the Ge-rich and Si-rich SiGe alloys at 550 and 800°C, respectively, under a shear stress 20MPa as dependent on the Si content.

$$v = v_0\,(\tau/\tau_0)^m \exp(-Q/k_B T), \quad \tau_0 = 1\ \mathrm{MPa}, \tag{1}$$

where k_B is the Boltzmann constant. The experimentally determined magnitudes of v_0, m and Q in SiGe and pure Ge and Si are listed in Table 1.

Here, it should be noted that dislocations in SiGe alloys show a typical recombination enhanced dislocation motion under an irradiation by an electron beam [36,37].

4. Mechanical strength

The mechanical strength was investigated in Si_xGe_{1-x} alloys for the composition range $0 < x < 0.4$ and $0.95 < x < 1$ with grown-in dislocation densities of 10^3–10^5 cm^{-2}. Rectangular parallelepiped specimens were compressed under a constant strain rate using an Instron-type machine at elevated temperatures [16-18].

Table 1. Magnitudes of v_0, m and Q for 60° dislocations in Si_xGe_{1-x} and pure Ge and Si.

Crystal		v_0 (m/s)	m	Q (eV)
Ge		2.9×10^2	1.7	1.62 ± 0.05
Si_xGe_{1-x}	$x = 0.016$	4.6×10^2	1.7	1.68
	$x = 0.047$	2.8×10^2	1.7	1.68
	$x = 0.080$	2.3×10^2	1.6	1.7
	$x = 0.922$	1.2×10^1	3.3	2.2
	$x = 0.946$	9.4×10^1	2.1	2.3
	$x = 0.979$	2.1×10^2	1.9	2.3
	$x = 0.996$	1.4×10^4	1.0	2.4
Si		1.0×10^4	1.0	2.4

The stress-strain curves of the Si-rich SiGe alloys of the composition range $0.95 < x < 1$ are similar to those of pure Si at temperatures 800–1000°C, characterized by a stress drop followed by an increase in the stress with strain. Such a stress drop is commonly found in other semiconductors, such as Si, Ge, GaAs, and so forth, at relatively low temperatures [34,35]. The upper and lower yield stresses and flow stress increase with a decrease in Si content. Similarly, the stress-strain curves of the Ge-rich SiGe alloys with $x = 0.01$, 0.10, 0.25 and 0.40 at low temperatures were characterised by a stress drop followed by an increase in the stress with respect to strain, but at high temperatures those show no stress drop, differing from those of the Si-rich SiGe alloys. The highly enhanced mobility of dislocations at these temperatures may contribute to the feature. It is remarkable that the SiGe alloys with x larger than 0.10 exhibit much higher levels of the yield and flow stresses than pure Ge and Si [16-18].

Figure 4 shows the temperature dependence of the upper yield stresses of various SiGe alloys and also those of Si and Ge, for comparison, under a shear strain rate of 1.8×10^{-4} s^{-1}. In a case where there is no stress drop after yielding, the yield stresses are plotted. The yield stresses in Si and Ge decrease with an increase in the temperature. Such a dependence can be described as a function of strain rate $\dot{\varepsilon}$

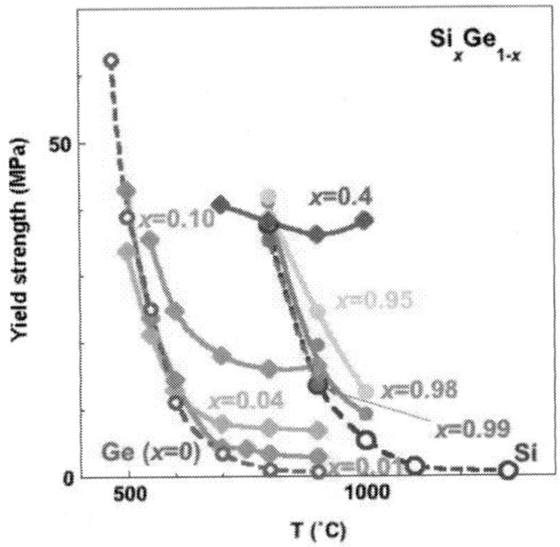

Figure 4. Yield stresses of the SiGe alloys plotted against the temperature for deformation under a shear strain rate of 1.8×10^{-4} s^{-1}.

and temperature T by the following empirical equation:

$$\tau = A\dot{\varepsilon}^{1/n}\exp(-U/k_BT), \tag{2}$$

where A, n and U are constants [34-38].

The Ge-rich SiGe alloys show similar reduction of the yield stresses with increasing the temperature in the low temperature regime. Their dependence on the temperature becomes weak in the high temperature region and finally nearly constant. The temperature-insensitive range expands towards the low temperature side with an increase of the Si content x. The magnitude of the yield stress of the alloys is higher in the high temperature region with increasing Si content up to 0.4. Typically, the yield stresses of the alloys $x = 0.4$ are temperature-insensitive in the range investigated. The magnitudes of the yield stress of the Si-rich SiGe alloy $x = 0.99$ are the same as, or slightly lower than, those of Si and the temperature dependence of the yield stress is similar to that of Si. With a decrease in the Si content to $x = 0.95$, the yield stress increases and the temperature dependence of the yield stress becomes weaker.

Figure 5 shows the composition dependence of the yield stress of the SiGe alloys under a shear strain rate of 1.8×10^{-4} s^{-1} at 900°C. The yield stress increases with increasing Si content in the composition range $x = 0$–0.4 and decreasing Si content down to $x = 0.95$ investigated. Typically, the yield stresses of the Ge-rich SiGe alloys with $x > 0.10$ are much higher than that of pure Si. Over the whole composition range of the SiGe alloys the yield stress seems to show a maximum around $x = 0.5$ and be dependent on the composition as proportional to $x(1 - x)$ [18].

Hardness of the SiGe alloys obtained with a micro indenter with a 0.5 N load for 10 s at room temperature (RT), 600°C and 900°C is shown in figure 6. The hardness at RT and 600°C increases almost linearly with the Si content x from 0 to 1, while that at 900°C shows a maximum around x = 0.5, similar to the yield stress above-mentioned. The temperature dependence of the hardness well corresponds with that of the yield stress.

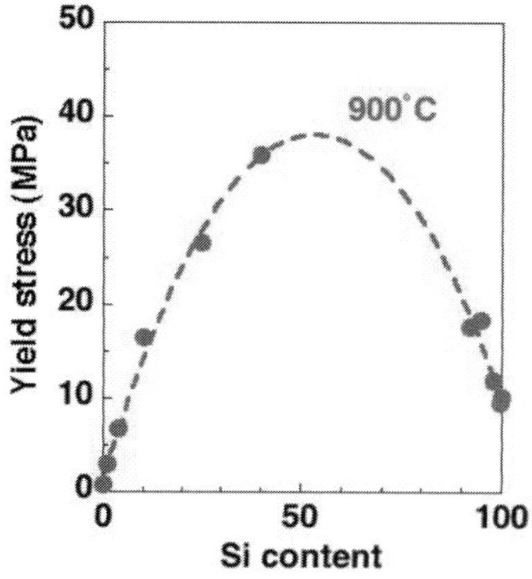

Figure 5. Yield stresses of SiGe alloys under a shear strain rate of 1.8×10^{-4} s^{-1} at 900°C.

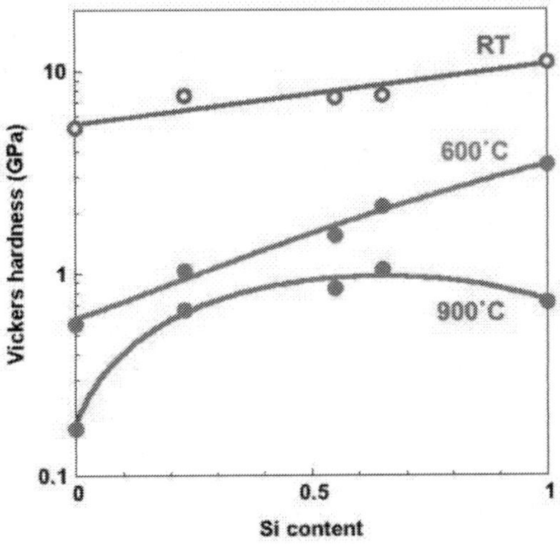

Figure 6. Micro-indentation hardness of the SiGe alloys plotted against the Si content at room temperature (RT), 600 and 900°C.

5. Origins of alloying effect

SiGe demonstrates a character as an alloy noticeably at high temperatures. This may also be understood from the fact that the hardness does not show alloying effects at room temperature but at 900°C as seen in figure 6. The velocity of isolated dislocations in SiGe alloys with compositions close to Si and Ge shows only small differences from that in Si and Ge. Thus, the difference in the dislocation mobility among various compositions of SiGe alloys may not lead to the drastic difference in the mechanical strength of the alloys observed at high temperatures. Here, it may be noted that the bulk modulus of SiGe alloys was evaluated to increase linearly with alloy composition [39]. Also, it was observed that dislocations induced by the plastic deformation are dissociated into Shockley partial dislocations bounding intrinsic stacking-faults. The intrinsic stacking-fault energy in the alloys decreases from 61 ± 10 to 55 ± 10mJ/m^2 with increasing Si content, intermediate between those of Si and Ge [40].

There are two components of the flow stress of a crystal for deformation as seen in figure 7: One is the effective stress by which dislocations move at a certain velocity against the intrinsic resistance

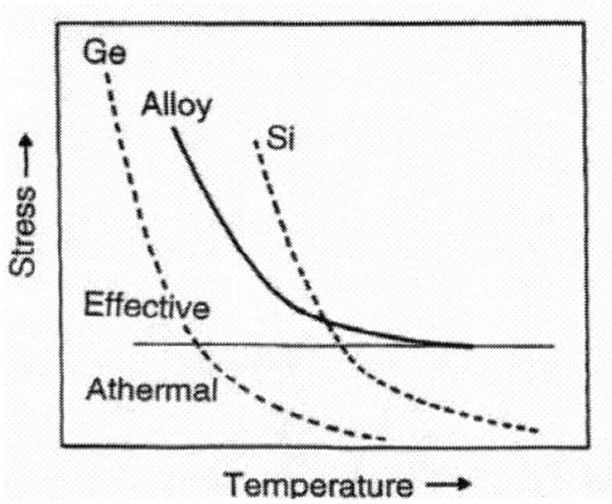

Figure 7. Schematic of the flow stress components and their dependence on temperature.

(Peierls potential) via thermal activation. The other is the athermal stress below which dislocations cannot move. The latter depends slightly on the temperature. If alloying results in a drastic increase in the Peierls potential and reduces the dislocation velocity, we may expect the stress drop at the yield point to be more remarkable in SiGe alloys than in Ge from the concept of dislocation dynamics of yielding in semiconductors. In addition, the strengthening effect caused by alloying should be less remarkable with an increase of temperature. As shown in figure 4, the yield stresses of the alloys are temperature-independent at elevated temperatures, remarkable in the alloys with the intermediate composition. Thus, the observed variation in the yield stress against the temperature in the SiGe alloys can be understood as a feature that the SiGe alloys have an athermal stress, that does not exist in other elemental and compound semiconductors. The athermal stress is in maximum in the SiGe alloy with a Si content $x \approx 0.5$. We can think reasonably that the athermal stress is related to the alloying effect.

As discussed in previous papers on GaAsP and InAsP alloys [9,10], several origins for athermal stress are plausible in alloying. First, short-range order of the $L1_1$ (CuPt-type) structure found in strained layer superlattice thin films prepared by molecular beam epitaxy [41] can lead to an extra stress of athermal nature since the motion of a dislocation destroys the short-range order along its slip plane [42]. However, there is no report detecting an ordered structure in bulk SiGe alloys [43] and also supported by XAFS study where the ordering parameter was evaluated to be around 0.22–0.29 from the coordination number [23].

Second, a long-range stress field may be developed by local fluctuation of the alloy composition in a crystal. Since the bond length of Ge is longer than that of Si by about 4%, local fluctuation of the alloy composition in the crystal, causing the development of Si or Ge enriched regions, may induce a long-range stress field that cannot be surmounted thermally by dislocations. Dislocations in SiGe alloys may move by a repeat bowing out process around the long-range stress fields.

Third, the dynamic development of a solute atmosphere around a dislocation during deformation at high temperatures leads to the additional stress for releasing the dislocation from the solute atmosphere. Indeed, many fine serrations on the stress-strain curve in the deformation under a strain rate as low as 1.8×10^{-5} s^{-1} at 900°C were observed [16]. Such a characteristic is known as the Portevin-LeChatelier phenomenon, being interpreted as repeated processes of locking and releasing of dislocations [44]. Indeed, though the width of dissociated dislocations formed by plastic deformation keeps constant, photoluminescence studies showed a variation of composition around deformation-induced dislocations by annealing [45,46]. Although the releasing process of a dislocation from its solute atmosphere is a thermally activated one, the development of a solute atmosphere around the dislocation is more enhanced at higher temperature. Thus, the contributions of these effects to the flow stress compensate each other and may give rise to a temperature-insensitive resistance to the dislocation motion, apparently looking like an athermal stress.

Either or both the local fluctuation of alloy composition and/or the dynamic development of a solute atmosphere around the dislocations are thought to suppress the dynamic activity of dislocations and result in the strengthening of bulk SiGe alloys at elevated temperatures.

6. Summary

The unique properties of Si_xGe_{1-x} crystals originating from alloying have been observed in an investigation of the mechanical strength and dislocation velocity by using bulk alloy crystals.

1. The dislocation velocity decreases monotonically with increasing Si content in the SiGe alloys of composition range $0.004 < x < 0.080$ in the temperature range 450–700°C and the stress range 3–24 MPa, while the dislocation velocity first increases, then decreases and again increases with decreasing Si content in the composition range $0.92 < x < 1$ for the temperature range 750–850°C and the stress range 3–30 MPa. The velocity of dislocations was determined as function of stress and temperature.

2. The stress-strain behaviour in the yield region of SiGe alloys of composition range $0 < x < 0.4$ is similar to that of Ge at temperatures lower than about 600°C. However, the yield stress becomes temperature-insensitive at high temperatures and increases with increasing Si content. The stress-strain behaviour of the SiGe alloys of composition range $0.95 < x < 1$ is similar to that of pure Si at

temperatures 800–1000°C and the yield stress increases with decreasing Si content down to $x = 0.95$. The composition dependence of the yield stress follows an $x(1 - x)$ relationship.

3. Built-in stress fields related to local fluctuation of the alloy composition, together with the dynamic development of a solute atmosphere around dislocations, seem to suppress the activities of dislocations and bring about a hardening of SiGe alloys.

Acknowledgments
The author expresses his gratitude to Professor K. Sumino for his encouragement and Dr. Wollweber for supplying FZ-SiGe boules.

References

[1] Tuppen G G and Gibbings C J 1990 *J. Appl. Phys.* **68** 1526
[2] Houghton D C 1990 *Appl. Phys. Lett.* **57** 2124
[3] Hull R, Bean J C, Bahnck D, Peitcolas L J, Short K and Unterwald F C 1991 *J. Appl. Phys.* **70** 2052
[4] Yamashita Y, Maeda K, Fujita K, Usami N, Suzuki K, Fukatsu S, Mera Y and Shiraki Y 1993 *Philos. Mag. Lett.* **67** 165
[5] Hull R, Bean J C, Peitcolas L J, Weir B E, Prabhakaran K and Ogino T 1994 *Appl. Phys. Lett.* **65** 327
[6] Watts D Y and Willoughby A F W 1984 *Mater. Lett.* **2** 355
[7] Cole S, Willoughby A F W and Brown M 1985 *J. Mater. Sci.* **20** 274
[8] Myles C W and Ekpenuma S N 1992 *J. Vac. Sci. Technol.* **B10** 1454
[9] Yonenaga I, Sumino K, Izawa G, Watanabe H and Matsui J 1989 *J. Mater. Res.* **4** 361
[10] Yonenaga I, Sumino K 1989 *Proc. 8th Symp. Rec. Alloy Semiconductor Physics and Electronics* (Kyoto, Organization of Special Project Research on Alloy Semiconductor Physics and Electronics) p 187
[11] Patel J R and Chaudhuri A R 196 *J. Appl. Phys.* **33** 2223
[12] Bell R L and Bonfield W 1964 *Philos. Mag.* **9** 9
[13] Sumino K 1994 *Handbook of Semiconductors* vol 3, ed S Mahajan (Amsterdam: Elsevier) p 73
[14] Yonenaga I, Matsui A, Tozawa S, Sumino K and Fukuda T 1995 *J. Crystal Growth* **154** 275
[15] Yonenaga I and Sumino K 1996 *Appl. Phys. Lett.* **69** 1264
[16] Yonenaga I and Sumino K 1996 *J. Appl. Phys.* **80** 3244
[17] Yonenaga I 1999 *Phys. Stat. Solidi* (a) **171** 41
[18] Yonenaga I 1999 *J. Mater. Sci: Mater. Electron.* **10** 329
[19] Yonenaga I 2005 *J. Crystal Growth* **275** 91
[20] Yonenaga I 2001 *J. Crystal Growth* **226** 47
[21] Yonenaga I and Ayuzawa T 2006 *J. Crystal Growth* 297 14
[22] Yonenaga I, Taishi T, Ohno Y and Tokumoto Y 2010 *J. Crystal Growth* **312** 1065
[23] Yonenaga I and Sakurai M 2001 *Phys. Rev.* B **64**, 113206
[24] Yonenaga I, Sakurai M, Sluiter M H F, Kawazoe Y and Muto S 2005 *J. Mater. Sci: Mater. Electron.* **16** 429
[25] Béraud A, Kulda J, Yonenaga I, Foret M, Salce B and Courtens E 2004 *Physica* B **350** 254
[26] King P J C, Lichti R L, Cottrell S P, Yonenaga I and Hitti B 2005 *J. Physics: Condens. Matter* **17** 4567
[27] Carroll B R, Lichti R L, King P J C, Celebi Y G, Yonenaga I and Chow K H 2010 *Phys. Rev.* B **82** 205205
[28] Yonenaga I, Nonaka M and Fukata N 2001 *Physica* B 308-310 539
[29] Yonenaga I, Li W J, Akashi T, Ayuzawa T and Goto T 2005 *J. Appl. Phys.* **98** 063702
[30] Yonenaga I 2006 *Jpn J. Appl. Phys.* **45** 2678
[31] Usami N, Nihei R, Yonenaga I, Nose Y and Nakajima K 2007 *Appl. Phys. Lett.* **90** 181914
[32] Yonenaga I and Wollweber J *unpublished work*

[33] Iunin Yu L, Orlov V I, Dyachenko-Dekov D V, Abrosimov N V, Rossolenko S N and Schröder 1997 *Solid Stat. Phenom.* **57-58** 419
[34] Yonenaga I 1997 *J. Phys. III France* **7** 1435-1450
[35] Yonenaga I 2005 *Mater. Trans.* **46** 1979
[36] Maeda K and Takeuchi S 1996 *Dislocation in Solids* vol 10, eds Nabarro F R N and Duesbery M S (Amsterdam: North-Holland Publ. Co) p 443
[37] Yonenaga I, Werner M, Bartsch M, Messerschmidt U and Weber E R 1999 *Phys. Stat. Solidi* (a) **171** 35
[38] Yonenaga I and Sumino K 1992 *Phys. Stat. Solidi* (a) **131** 663
[39] Yonenaga I and Sluiter M H F *unpublished work*
[40] Yonenaga I, Lim S -H and Shindo D 2000 *Philos. Mag. Lett.* **80** 193
[41] Ourmazd A and Bean J C 1985 *Phys. Rev. Lett.* **55** 765
[42] Fisher J C 1954 *Acta Metall.* **2** 9
[43] Stenkamp D and Jäger W 1992 *Philos. Mag.* **A 65** 1369
[44] Yonenaga I and Sumino K 1992 *J. Appl. Phys.* **71** 4249
[45] Tanaka K, Suezawa M and Yonenaga I 1996 *J. Appl. Phys.* **80** 6991
[46] Ohno Y, Tokumoto Y, Taneichi H, Yonenaga I, Togase K and Nishitani S 2012 *Physica* B **407** 3006

Boron Nitride Nanosheets: novel Syntheses and Applications in polymeric Composites

Xuebin Wang[1,3*], Chunyi Zhi[2], Qunhong Weng[1], Yoshio Bando[1,3*] and Dmitri Golberg[1]

[1] World Premier International Center for Materials Nanoarchitectonics (WPI-MANA), National Institute for Materials Science (NIMS), 3050044 Tsukuba, Japan.
[2] Department of Physics and Materials Science, City University of Hong Kong, 999077, Hong Kong.
[3] Department of Nano-Science and Nano-Engineering, Faculty of Science and Engineering, Waseda University, 1698555 Tokyo, Japan.

E-mail: Wangxb@fuji.waseda.jp; BANDO.Yoshio@nims.go.jp

Abstract. Two-dimensional (2D) boron nitride (BN) nanosheets, the rising material stars, have unique properties and amazing functionalities. The concise history of diverse synthesis routes of BN nanosheets is briefly summarized here; and the recent development towards mass production of BN nanosheets, *i.e.* "chemical blowing" technique relying on blowing molten polymer precursors into large polymeric bubbles and subsequent annealing into BN bubbles/nanosheets, is focused on. The abundant BN nanosheets enable their applications, especially in the representative polymeric composites with BN nanosheet additives, as particularly highlighted in this review. The highly thermoconductive insulating BN-filled composites are thus envisaged as high-performance packaging materials for electrical circuits.

Key words: boron nitride, nanosheet, chemical blowing, polymeric composite

1. Introduction

BN nanosheets are (002) mono-/few-layered hexagonal BN (*h*-BN). They are the structural analogues of graphenes (Figure 1). *h*-BN is a compound with equal numbers of boron and nitrogen atoms, which is widely used in skincare products, insulators, solid lubricants, pencil leads, *etc.*

BN nanosheets possess unique properties. The BN crystal lattices, similar to graphene, together with a stronger B-N bonding result in superior characteristics in mechanics and thermal applications. The bending elastic modulus of BN nanosheets is hundreds of GPa [1]. The thermal conductivity of BN layers is 1700-2000 W/mK [2]. The thermal and chemical stability of BN is better than that of

* To whom any correspondence should be addressed.

graphene [3]. On the other hand, there is a significant charge transfer from B to N atoms, so partially ionic sp^2-hybridized B-N bonds determine notably different properties from graphenes in optics and electronics, *e.g.* BN is white in color and has a large bandgap of 5.5 eV, while graphene is black and conductive.

As a result of the unusual structures and properties, BN nanosheets exhibit diverse functionalities. BN nanosheets are intrinsic insulators valuable for dielectric applications, *e.g.* as a dielectric gate layer [4,5] and as a deep ultraviolet luminescent agent [6,7]. The dielectric properties of BN (dielectric constant 3-4; breakdown voltage *ca.* 700 MV/m) are comparable to SiO_2, allowing its usage as an alternative dielectric gate in current electronics with no loss of functionality. They are also able to enhance the carrier mobility [8] and to open the bandgap of graphenes put on them [9,10], so they are envisaged as the perfect "sidekick" of graphene in the next-generation electronics. BN nanosheets show the tunable bandgap through tailoring their edge structures [11]. Besides, the weak bonding out of planes is advantageous for excellent solid-state lubricants [12]. In addition, BN nanosheets are applied in the nanomedical field due to their good biocompatibility, and in scanning probe microscopy as ultrathin and stiff tips [13]. They are also applied for filling polymeric or ceramic composites to achieve extraordinary thermal and mechanical performances.

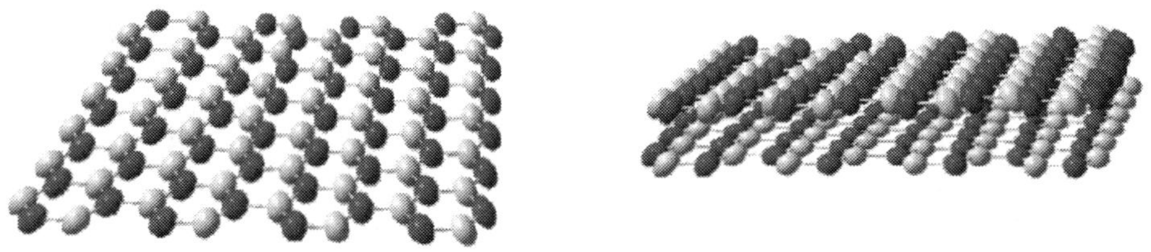

Figure 1. Mono- and double- layered BN nanosheets.

2. Diverse Syntheses of BN Nanosheets

Prior to the discovery of free-standing *h*-BN monolayers, mono-layered BN was epitaxially deposited and studied on the surfaces of transition metals, such as Ni, Pd, Ru and Pt in 1995-2004 [14,15]. A free-standing *h*-BN monolayer was firstly found in the mechanically milled residue in 2005 following the rise of graphene [16]. It is also called as "white graphene" corresponding to a bulk *h*-BN precursor which is named as "white graphite".

So far, BN nanosheets have successfully been fabricated *via* diverse methods including (see Figure 2): mechanical cleavage [17], ball milling [18], high-energy electron beam irradiation [19] and the reaction of boric acid and urea [20,21]. Ultrasonic vibration has also been available to separate the layers of bulk BN; this is used by a liquid exfoliation method. We produced ultra-thin BN nanosheets at the level of dozens of milligrams through an efficient sonication of BN precursors in a polar organic solvent of dimethylformamide [24], in which the newborn BN layers are stabilized by the solvent to prevent the reversible re-stacking [22-24]. Chemical vapor deposition (CVD) is an effective route to synthesize large-area BN nanosheets [25-28]. In 2011, a facile thermal CVD technique was developed by us to fabricate vertically-aligned BN nanosheets using cheap B, MgO and FeO precursors [29,30].

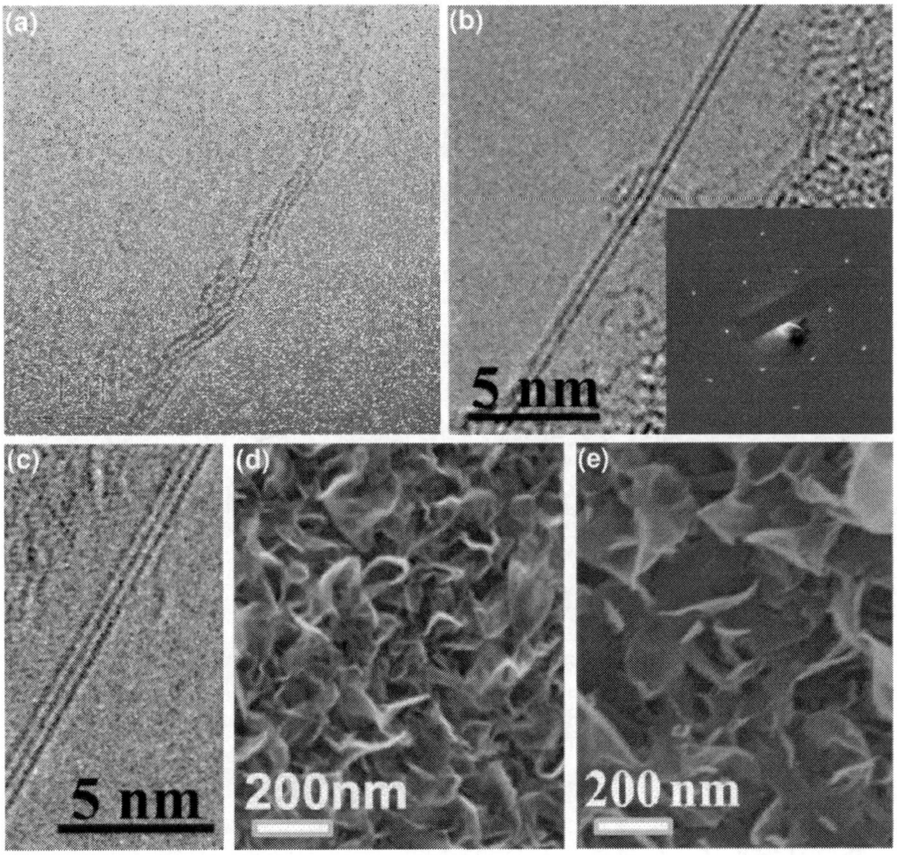

Figure 2. (a) High-resolution transmission electron microscopy (HRTEM) image of BN nanosheets with a thickness of 1.2 nm derived from liquid exfoliation method. Reprinted with permission from [24]. © 2009 WILEY-VCH Verlag GmbH & Co. KGaA, Weinheim. (b,c) HRTEM image images of BN with two/three atomic layers from the CVD method. The inset is the corresponding selected area electron diffraction (SAED). Reprinted with permission from [25]. © 2010 American Chemical Society. (d,e) Scanning electron microscopy (SEM) images of the BN nanosheets synthesized at 1100 and 1200°C respectively in CVD. Reprinted with permission from [29]. © 2011 American Chemical Society.

A smart technique of unwrapping multiwalled BN nanotubes through delicate Ar plasma etching was developed by us in 2010 to synthesize few- and mono-layered BN nanosheets (Figure 3). The typical intermediate states of the etched nanotubes were observed to confirm the unwrapping mechanism. The narrow nanosheets, *i.e.* nanoribbons terminated with N-terminated zigzag edges and vacancy defects, exhibited fantastic semiconducting properties [11]. The high conductivity of 104 S/m and carrier mobility of 58.8 cm^2/Vs were measured in such BN nanoribbons without a gate bias. First-principles calculations verified that the edge states and dangling bond states located in the zigzag edges and experimentally observed surface vacancies had been the driving sources for the semiconducting transport of BN nanoribbons, as opposed to insulating primal BN nanotubes. This study paves the way for BN nanoribbon production and usage as functional semiconductors with a wide range of fundamental issues and applications in electronics and spintronics.

Journal of Physics: Conference Series **471** (2013) 012003 doi:10.1088/1742-6596/471/1/012003

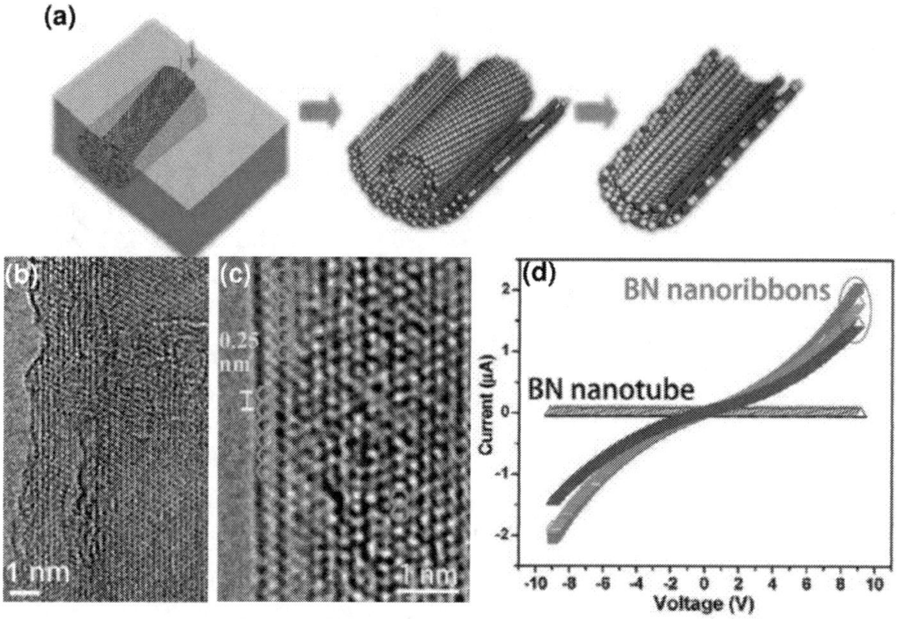

Figure 3. (a) Schematic of the unwrapping processes induced by plasma etching, the stepwise opening/unzipping, removing and exfoliating of tube walls to form BN nanoribbons. (b,c) HRTEM image of a double-layered BN nanoribbon with zigzag edge structure *via* nanotube unwrapping. (d) Electrical transport tests of several BN nanoribbons comparing with an insulating BN nanotube. Reprinted with permission from [11]. © 2010 American Chemical Society.

3. "Chemical Blowing" Synthesis for BN Nanosheets

Although many methods have been utilized to synthesize BN nanosheets, it is still a challenge to prepare BN nanosheets with a high throughput. The chemical exfoliations of graphites have realized the tons-level production of reduced graphene oxide. Unfortunately it is difficult to chemically intercalate the partially ionic BN layers because of the weaker inter-layer exchange repulsion and stronger inter-layer attraction [31]. The lack of effective mass production of BN nanosheets has hindered their relative developments. Recently, we developed a new strategy, so-called "chemical blowing" [32,33], which relies on making large bubbles with atomically thin B-N-H polymer walls starting from ammonia borane (AB) compound as a precursor (Figure 4). AB is a nontoxic, environmentally benign and stable material, which is usually used as an effective source of hydrogen [34,35]. In addition, since the cost of AB is currently decreasing, it is particularly worth utilizing this "chemical blowing" process to produce BN nanosheets with a low cost, and to finally realize their rich nanotechnology potentials. The "chemical blowing" was conducted through pre-treating AB at a low temperature followed by a controllable heating to designated temperature for dehydrogenation and chemical blowing of bubbles. The B-N-H polymer bubbles were finally annealed into BN ones at a high temperature. The BN bubbles are composed of ultra-thin BN walls, which were then crushed and converted into the BN nanosheets. This process is featured by the bubbles blown by a chemically released gas, which is thus named as "chemical blowing".

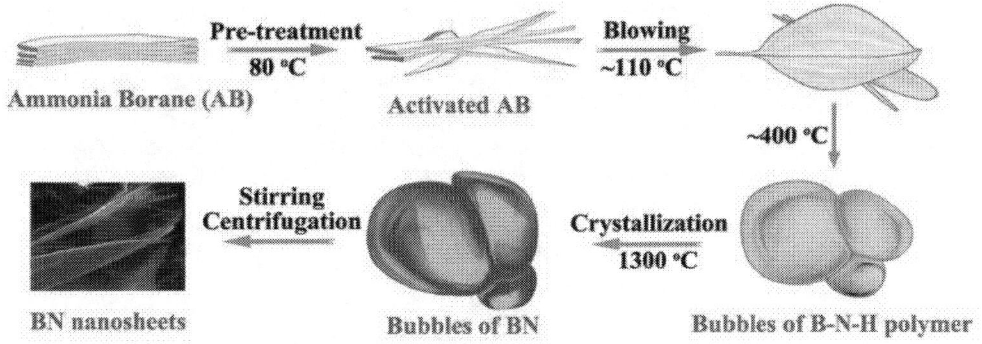

Figure 4. Synthesis scheme of BN nanosheets under "chemical blowing". Reprinted with permission from [33]. © 2012 IOP Publishing Ltd.

The products derived from "chemical blowing" are very light, with a density of 6 mg/cm^3, corresponding to a high porosity of 99.7 % (Figure 5). They have perfect B and N stoichiometry as confirmed by electron energy loss spectroscopy, except minor oxygen impurity related to B-O phases. Individual BN nanosheets have large lateral dimensions, tens of micrometers, normally larger than those obtained using liquid-exfoliation methods. BN nanosheets are uniform over a large area, and they are large enough for further assembling into electronic devices. The thickness of BN nanosheets typically ranges from 1 to 5 nm, although a few of nanosheets thicker than 5 nm may also be found. These numbers are rather small and lead to unique characters and properties as derived from the quantum effect. Few-layered nanosheets, made of as 2, 3 and 5 layers, have been demonstrated here. The layer-to-layer spacing decreases with an increase of the number of layers. It is 0.38 nm for 2-layered nanosheets, while only 0.34 nm for a 7-layered nanosheet. This change can be understood through increased inter-layer interaction with increasing number of layers. All the layer-to-layer spacing is slightly larger than those in bulk layered BN due to the relaxation of surface atoms.

The BN nanosheets obtained from "chemical blowing" are polycrystalline structures of *h*-BN with an ordered honeycomb-like hexagonal lattice, as visualized in Figure 6. The lattice distance is 0.258 nm corresponding to the B-B or N-N atom separations in *h*-BN. The locally-ordered polycrystalline structure of nanosheets is further verified by a fast-Fourier-transform (FFT) pattern over an area of hundreds of nm^2. The *h*-BN nanosheets have a strong (002) diffraction peak with a *d* value of 0.35 nm. Its broadening corresponds to the effect of downsized dimensions of *ca.* 3 nm. The *h*-BN nanosheets demonstrate the strong ultraviolet luminescence at 210 nm attributed to the intrinsic optical bandgap. As-grown BN products possess large specific surface area, 140 m^2/g, which is larger than 50 m^2/g of porous BN, or 27 m^2/g of BN nanoparticles [36]. Although it is still smaller than the theoretical value for mono-layered sheets (2600 m^2/g) owing to the overlapping and re-stacking of nanosheets, it is noticed that there is no contribution from micropores in our case. This configuration is very beneficial to create more interfaces between BN and polymer molecules, because the too small pores can not be accessed by polymer molecules to create links and effective BN-polymer interactions in the fabrication of BN-based polymeric composites.

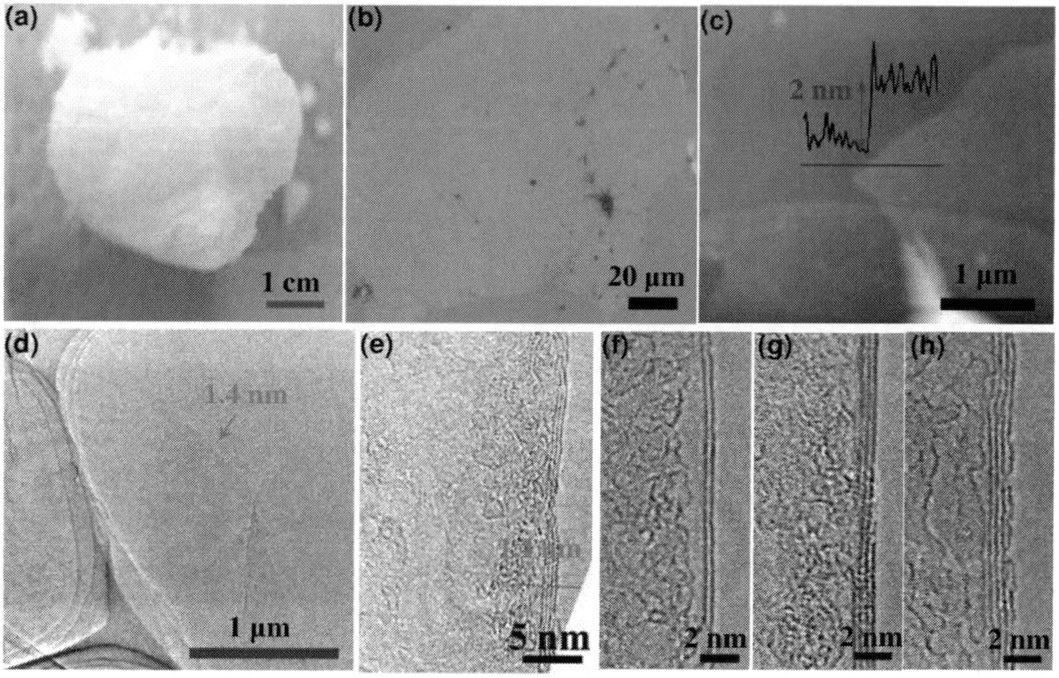

Figure 5. (a,b) Optical images of as-grown snow-white BN products and an individual ultra-thin BN nanosheet transferred on a Si substrate. (c) Atomic force microscopy (AFM) image of a BN nanosheet with the thickness of 2 nm, as pointed out by a cross-sectional profile marked in the inset. (d,e) High- and low-magnification images of a 3-4 layer BN nanosheet. (f-h) HRTEM images of two, three and four-to-five atomic-layers BN nanosheets. Reprinted with permission from [32]. © 2011 WILEY-VCH Verlag GmbH & Co. KGaA, Weinheim.

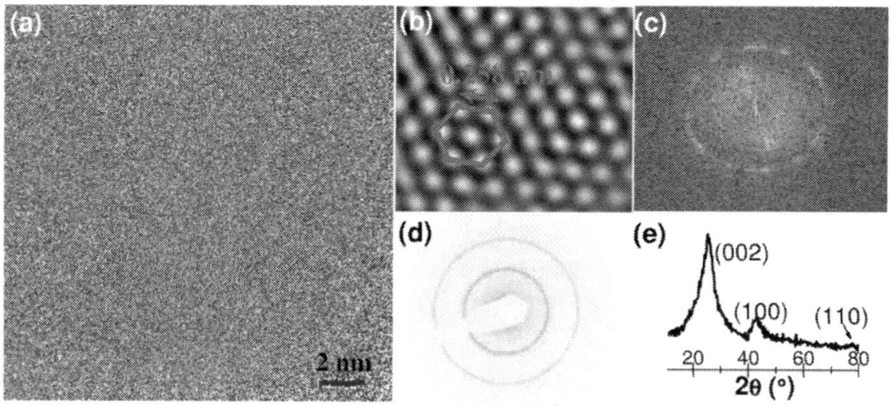

Figure 6. (a) HRTEM image of a 2-layered BN nanosheet showing polycrystalline structure. (b) Reconstructed reverse FFT image from the corresponding red framed area. (c) The FFT pattern taken from the area of 900 nm^2. (d) SAED image of a BN nanosheet taken from the area of 0.05 μm^2. (e) XRD profile of BN products confirming a hexagonal structure of the space group 194. Reprinted with permission from [32]. © 2011 WILEY-VCH Verlag GmbH & Co. KGaA, Weinheim.

The high-yield production of BN nanosheets with large lateral dimensions was optimized at a heating rate of 8°C/min following a pre-treatment under atmospheric pressure under "chemical blowing". The yield is as high as 40 wt.% with respect to AB. The "chemical blowing" thus realizes the mass production of atomically-thin (one-to-few atomic layers) free-standing BN nanosheets, which significantly prevails over the yields achievable during CVD and liquid exfoliation.

4. Polymeric Composites based on chemically blown BN Nanosheets

Comparing with 0D nanoparticles and 1D nanofibers/tubes, 2D BN nanosheet fillers can maximally expose their basal (002) crystal planes and show fast electron and phonon transport and effective force transfer in the composites. Therefore, BN nanosheets are excellent additives for polymeric composites to enhance thermal and mechanical properties.

The insulating thermoconductive polymeric composite is an advanced packaging material for quick heat-releasing and good electrical insulation in the high-speed electronics [37]. However, the traditional fillers such as AlN and Al_2O_3 powders reveal a relatively low thermal conductivity [38]. BN nanosheets are one of the best fillers to realize both high thermal conductivity and perfect insulating properties of the composites. Besides, BN can also deliver the high mechanical strength, high thermal and chemical stability. Unfortunately, research on BN-nanosheet-filled composites was rather limited due to the low-scale production and high cost of BN nanosheets. Recently, gram-level BN nanosheets have been fabricated in our laboratory under the "chemical blowing" route, which enables their utilization in polymeric composites (Figure 7).

The composites of BN/polymethyl methacrylate (PMMA) show decreased transparency with the increased filling fraction of BN nanosheets. The elastic modulus increases by 17%, whereas the yield strength increases by 32%, for a composite with 6 wt.% BN loading fraction. The mechanical reinforcement effect results from the effective transfer of external mechanical load to BN nanosheets due to the profound interfacial interactions between BN and PMMA.

The BN/PMMA composites show the improved thermal stability along with increasing a BN fraction. The two-step degradation of PMMA includes unzipping vinylidene end groups and further random scission of polymer links [39]. The surface groups of BN such as oxidative B-OH may react with the polymer end radicals. This reaction can suppress the unzipping of the polymer chain and then stabilize the composite system [40]. This interfacial stabilizing effect was amplified for the large specific surface area of BN fillers, so the onset degradation temperatures of BN/PMMA composites are much higher than that of blank PMMA. Also, the composites can withstand higher temperatures than a blank PMMA due to such stabilization provided by BN nanosheets. In a quantitative analysis, the glass transition temperature clearly increases along with an increase in the BN filling fraction. It increases from 92°C for blank PMMA to 105°C for 15 wt.% BN/PMMA, and to 110°C for 23 wt.% BN/PMMA. This indicates that the thermal mobility of polymer chains is suppressed by the strong and confined BN-PMMA interfacial interactions. The confined PMMA chains exhibit slower dynamics and thus promote an increase in glass transition temperature.

The BN/PMMA composites also demonstrate remarkable thermal conductivity. The highest thermal conductivity is 2.6 W/mK for 23 wt.% BN/PMMA. It is a 17-fold increase compared with the blank PMMA. The large lateral dimensions of the BN nanosheets result in an easily constructed thermoconductive pathway, which determines such high thermal conductivity of BN/PMMA. The good insulating properties of the PMMA matrix are basically kept in these composites. The breakdown electric fields of BN/PMMA composites are comparable with neat polymers. Although the breakdown electric field slightly decreases with increasing BN fraction due to an easier formation of conductive path, the composites still well retain their insulating abilities fully satisfying for insulating packaging.

The BN nanosheets produced with a high yield and large lateral dimensions drastically improve the mechanical, thermal and insulating performance of BN/PMMA composites. The large lateral dimensions of BN nanosheets provide spacious interfaces between BN and PMMA molecules which determines their remarkable thermal conductivity. The high yield and high throughput of such BN nanosheets further enable their large-scale practical applications in the field of functionally reinforced polymeric composites.

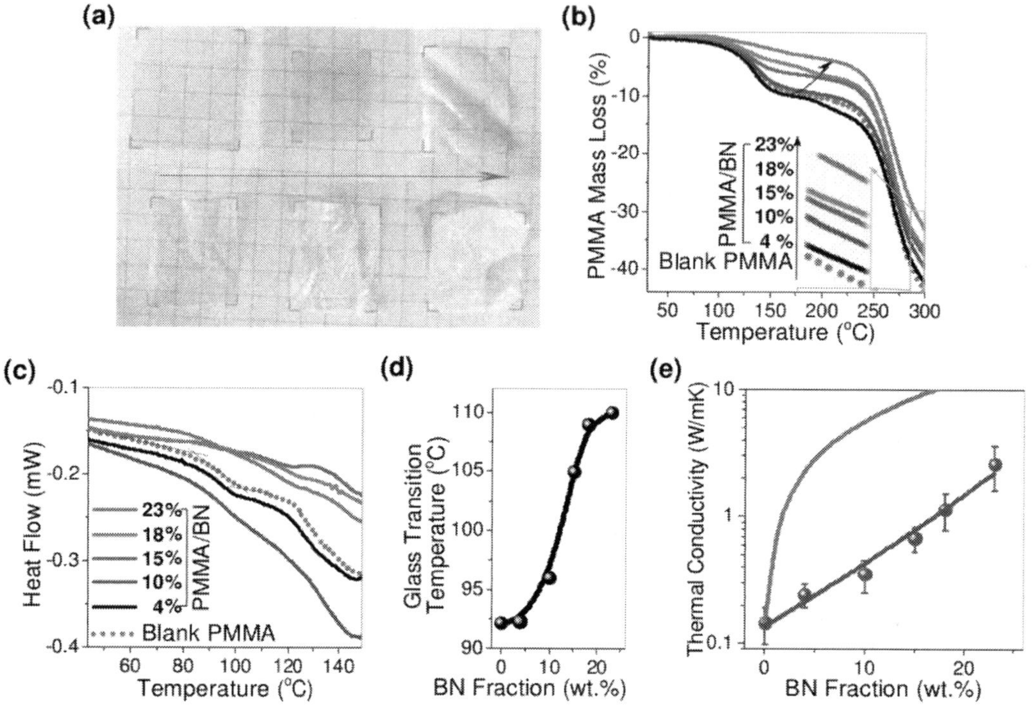

Figure 7. (a) Optical photos of blank PMMA and PMMA composites with 4, 10, 15, 18 and 23 wt.% of BN arranged from left-to-right and from top-to-bottom. (b) Normalized weight change of PMMA components in blank PMMA and in PMMA/BN composites after removal of the BN components mass loss based on their thermogravimetry curves. (c) Typical differential scanning calorimetry curves of blank PMMA and PMMA/BN composites. Glass transition temperature was determined at the mid-point in a three tangent method. (d) The increased glass transition temperature along with increasing filling fraction of BN nanosheets. (e) Increased thermal conductivity with increasing BN fraction in PMMA/BN composites. Blue curve is the fitting to experimental points using the Agari model [41]; red one is the upper limit of thermal conductivities in theory from parallel models. Reprinted with permission from [42]. © 2012 Wang *et al.*; licensee Springer.

5. Conclusion and Prospects

BN nanosheets are the representatives of 2D crystals having unique properties. So far, many synthesis routes have been developed to grow BN nanosheets. However the discussed "chemical blowing" is a specific catalyst-free and substrate-free route, which is simpler than the normal CVD method. "Chemical blowing" realizes the high product yield and laterally large areas of BN nanosheets. The obtained products normally possess larger lateral dimensions than those derived from the liquid exfoliation. These not only provide large-enough area nanosheets readily available for their electrical and mechanical performance explorations, but also give enough nanosheets' mass for fabricating ultimately strong polymeric or other composites. The developed technique opens up a wide horizon for the full realization of BN nanosheet potentials in nanotechnology, for instance growing potentially valuable BC_xN nanosheets. We believe that the cheaper BN nanosheets based on the developed economic AB or other precursors will appear in the nearest future.

The abundant BN nanosheets enable their utilization in making BN/PMMA composites. The significant 17-fold-enhanced thermal conductivity together with the 18°C increased glass transition temperature for the 23 wt% BN/PMMA was achieved. Meanwhile, BN/PMMA composites exhibited higher mechanical strength, and kept high insulating properties, decent transparency and sufficient dielectric properties. The composites are thus expected to be valuable for multi-functional applications in many fields, especially for the new-generation long-lifetime heat-release packaging materials for electrical circuits and power modules.

References

[1] Li C, Bando Y, Zhi CY, Huang Y and Golberg D 2009 *Nanotechnology* **20**, 385707
[2] Ouyang T, Chen YP, Xie YE, Yang KK, Bao ZG and Zhong JX 2010 *Nanotechnology* **21**, 245701
[3] Ghassemi HM, Lee CH, Yap YK and Yassar RS 2010 *JOM* **62**, 69
[4] Meric I, Dean C, Young AF, Hone J, Kim P and Shepard K 2010 *IEDM Tech. Dig.* **10**, 556
[5] Park CH and Louie SG 2008 *Nano Lett.* **8**, 2200
[6] Watanabe K, Taniguchi T, Niiyama T, Miya K and Taniguchi M 2009 *Nature Photon.* **3**, 591
[7] Watanabe K, Taniguchi T and Kanda H 2004 *Nature Mater.* **3**, 404
[8] Dean CR, Young AF, Meric I, Lee C, Wang L, Sorgenfrei S, Watanabe K, Taniguchi T, Kim P, Shepard KL and Hone J 2010 *Nature Nanotech.* **5**, 722
[9] Giovannetti G, Khomyakov PA, Brocks G, Kelly PJ and van den Brink J 2007 *Phys. Rev. B* **76**, 073103
[10] Slawinska J, Zasado I and Klusek Z 2010 *Phys. Rev. B* **81**, 155433
[11] Zeng HB, Zhi CY, Zhang Z, Wei X, Wang XB, Guo W, Bando Y and Golberg D 2010 *Nano Lett.* **10**, 5049
[12] Lee C, Li Q, Kalb W, Liu XZ, Berger H, Carpick RW and Hone J 2010 *Science* **328**, 76
[13] Ciofani G, Genchi GG, Liakos I, Athanassiou A, Dinucci D, Chiellini F and Mattoli V 2012 *J. Colloid. Interf. Sci.* **374**, 308
[14] Nagashima A, Tejima N, Gamou Y, Kawai T and Oshima C 1995 *Phys. Rev. Lett.* **75**, 3918
[15] Corso M, Auwarter W, Muntwiler M, Tamai A, Greber T and Osterwalder J 2004 *Science* **303**, 217
[16] Novoselov KS, Jiang D, Schedin F, J Booth T, Khotkevich VV, Morozov SV and Geim AK 2005 *Proc. Natl. Acad. Sci. USA* **102**, 10451
[17] Pacile D, Meyer JC, Girit CO and Zettl A 2008 *Appl. Phys. Lett.* **92**, 133107
[18] Li LH, Chen Y, Behan G, Zhang HZ, Petravic M and Glushenkov AM 2011 *J. Mater. Chem.* **21**, 11862
[19] Jin C, Lin F, Suenaga K and Iijima S 2009 *Phys. Rev. Lett.* **102**, 195505
[20] Nag A, Raidongia K, Hembram KPSS, Datta R, Waghmare UV and Rao CNR 2010 *ACS Nano* **4**, 1539
[21] Raidongia K, Nag A, Hembram KPSS, Waghmare UV, Datta R and Rao CNR 2010 *Chem. Eur. J.* **16**, 149
[22] Lin Y, Williams TV and Connell JW 2010 *J. Phys. Chem. Lett.* **1**, 277-283
[23] Lin Y, Williams TV, Xu TB, Cao W, Elsayed-Ali HE and Connell JW, 2011 *J. Phys. Chem. C.* **115**, 2679-2685
[24] Zhi CY, Bando Y, Tang CC, Kuwahara H and Golberg D 2009 *Adv. Mater.* **21**, 2889-2893
[25] Song L, Ci L, Lu H, Sorokin PB, Jin C, Ni J, Kvashnin AG, Kvashnin DG, Lou J, Yakobson BI, Ajayan PM 2010 *Nano Lett.* **10,** 3209

[26] Ci L, Song L, Jin C, Jariwala D, Wu D, Li Y, A rivastava S, Wang ZF, Storr K, Balicas L, Liu F, Ajayan PM 2010 *Nature Mater.* **9**, 430
[27] Shi Y, Hamsen C, Jia X, Kim KK, Reina A, Hofmann M, Hsu AL, Zhang K, Li H, Juang ZY, Dresselhaus MS, Li LJ and Kong J 2010 *Nano Lett.* **10**, 4134
[28] Gao R, Yin LW, Wang CX, Qi YX, Lun N, Zhang L, Liu YX, Kang L and Wang XF 2009 *J. Phys. Chem. C.* **113**, 15160-15165
[29] Pakdel A, Zhi CY, Bando Y, Nakayama T and Golberg D 2011 *ACS Nano* **5**, 6507
[30] Pakdel A, Wang XB, Zhi CY, Bando Y, Watanabe K, Sekiguchi T, Nakayama T and Golberg D 2012 *J. Mater. Chem.* **22**, 4818
[31] Dai BQ and Zhang GL, 2003 *Mater. Chem. Phys.* **78**, 304
[32] Wang XB, Zhi CY, Li L, Zeng HC, Li C, Mitome M, Golberg D and Bando Y 2011 *Adv. Mater.* **23**, 4072-4076
[33] Wang XB, Pakde A, Zhi CY, Watanabe K, Sekiguchi T, Golberg D and Bando Y, 2012 *J. Phys: Condens. Matter.* **24**, 314205
[34] Marder T B 2007 *Angew. Chem. Int. Ed.* **46,** 8116
[35] Fazen PJ, Beck JS, Lynch AT, Remsen EE and Sneddon LG 1990 *Chem. Mater.* **2**, 96
[36] Perdigon-Melon JA, Auroux A, Cornu D, Miele P, Toury B and Bonnetot B 2002 *J. Org. Chem.* **657**, 98
[37] Zhi CY, Bando Y, Terao T, Tang CC, Kuwahara H and Golberg D 2009 *Adv. Funct. Mater.* **19**, 1857
[38] Lee G, Park M, Kim J, Lee JI and Yoon HG 2006 *Compos: Part A* **37**, 727
[39] Grassie N and Melville HW 1947 *Faraday Soc. Disc.* **2**, 378
[40] Peterson JD, Vyazovkin S and Wight CA 1999 *Macromol. Rapid Commun.* **20**, 480
[41] Agari Y, Ueda A, Tanaka M and Nagai S 1990 *J. Appl. Polym. Sci.* **40**, 929-941
[42] Wang XB, Pakdel A, Zhang J, Weng QH, Zhai TY, Zhi CY, Golberg D and Bando Y 2012 *Nanoscale Res. Lett.* **7**, 662

Microscopic investigations of advanced thin films for photonics

S Boninelli[1,*], A Shakoor[2], K Welma[2], TF Krauss[2], L O'Faolain[2], R Lo Savio[3], S Portalupi[3], D Gerace[3], M Galli[3], P Cardile[1], G Bellocchi[1], G Franzò[1], M Miritello[1], F Iacona[1] and F Priolo[1]

[1] MATIS IMM CNR and Dipartimento di Fisica e Astronomia, Università di Catania, via Santa Sofia 64, 95121 Catania, Italy
[2] SUPA, School of Physics and Astronomy, University of St. Andrews, Fife KY16 9SS, St. Andrews, United Kingdom
[3] Dipartimento di Fisica, Università di Pavia, via Bassi 6, 27100 Pavia, Italy

E-mail: simona.boninelli@ct.infn.it

Abstract. We present the different approaches we recently followed to achieve intense room temperature photoluminescence (PL) from Si-based materials. On one side we obtained sub-bandgap PL from H-related defects induced by the H_2 plasma treatment of Si photonic crystal (PhC) nanocavities. We demonstrated that a strong and narrow PL emission can be obtained in the PhC nanocavities due to the formation of a damaged layer mainly consisting of nanometric platelets and bubbles. An overall 40000-fold enhancement of the PL signal, with respect to pure crystalline Si, has been achieved and moreover the signal can be tuned in a wide range by only changing the PhC parameters. On the other side, we focused our attention on the properties of SiO_2 and SiOC host matrices doped with Eu ions. C addition produces a strong enhancement of the Eu PL with respect to pure SiO_2 films. The chemical and structural characterization of these materials reveals an extensive Eu clustering in SiO_2-based films, while C addition induces a significant reduction of this phenomenon, enhancing the fraction of optically active Eu ions. These results can be applied for the realization of efficient Si-based light sources.

1. Introduction

The achievement of a cheap and efficient Si-based light source working at room temperature (RT) is currently the foremost objective in the field of photonics. The main performances required for this source are electrical pumping, high efficiency at room temperature, tunability, easy integration in highly compact Si microphotonic devices. The lack of an efficient Si-based light source is a result of the intrinsic inability of Si to efficiently emit light, due to its indirect bandgap. In the last decades, in our group many approaches have been followed to increase the photoluminescence (PL) efficiency of Si-based materials, such as the use of Si nanocrystals embedded in SiO_2 [1, 2], their doping with light emitting impurities,

* To whom any correspondence should be addressed.

such as Er ions [3] and, very recently, the realization of light emitting nanowires fabricated by a maskless etching technique [4].

Aim of this paper is to present an overview of the new strategies we have recently followed in our group in order to achieve intense light emission from Si-based materials.

The first approach consists of the engineering of optically active defects in crystalline Si, since it has been demonstrated in the literature that it represents a powerful method to improve the PL from this material [5, 6, 7]. Very recently the enhancement of the sub-bandgap PL via the Purcell effect in photonic crystal (PhC) cavities fabricated on bare silicon-on-insulator (SOI) has been demonstrated. The PL emission was attributed to point defects induced by the H implantation used for the production of SOI during the Smart Cut™ process [8, 9]. In those papers the authors expressed their intention to implement a suitable H_2 plasma treatment to increase the amount of optically active defects in order to further enhance the PL emission from the Si based PhC nanocavity. The purpose of this study is twofold: on one side we realized an extensive transmission electron microscopy (TEM) investigation concerning the thermal evolution of H-induced platelets in crystalline Si and their correlation with the PL properties of bulk SOI materials, on the other side, once having optimized the light emission properties in plasma treated SOI, we tried to enhance the emission from these optically active defects by introducing them in the active region of a PhC nanocavity. We explain the PL activity of H_2 plasma treated PhC nanocavities in terms of the damage induced in these structures by the process.

The second approach we want to present here regards Eu doping of Si-based materials, which has been recognized as a powerful method to provide intense PL signals spanning a wide wavelength range [10, 11, 12]. Eu ions, owing to the coexistence of both divalent (Eu^{2+}) and trivalent (Eu^{3+}) oxidation states, act as emitting centers at different wavelengths. However, the low Eu solid solubility in SiO_2 represents an important limit for the practical applications of this rare earth [13, 14]. Indeed, the main challenge for Eu application is represented by the availability of alternative host matrices, whose synthesis and processing are compatible with Si technology, which are simultaneously capable to efficiently host high concentrations of optically active Eu ions.

Here, we will compare the performances of pure SiO_2 and of SiOC as host matrices for optically active Eu ions. It will be shown that in the case of SiO_2, although both Eu^{2+} and Eu^{3+} emissions can be obtained, the efficiency of the light emission remains quite far from requirements for technological applications due to the extensive formation of Eu-containing precipitates. On the other hand, it will be shown that a SiOC matrix, owing to its chemical and structural properties, is able to ensure an intense and stable light emission centered at about 440 nm from Eu^{2+} ions, which could be of great interest for photonic applications.

2. Results and Discussion

2.1. H plasma treated SOI

For H plasma treatments, SOI films have been introduced in a parallel plate ion etching set-up where the power was fixed at 40 W and the treatment duration was 30 min. The chamber pressure was kept at 0.1 mbar and all treatments were carried at RT. In order to maximize the PL emission, the H plasma as-treated SOI samples were cut into small pieces and were subjected to thermal annealing in a conventional furnace under a controlled forming gas (FG) atmosphere at different temperatures ranging from 300 °C to 500 °C for 30 min.

Journal of Physics: Conference Series **471** (2013) 012004 doi:10.1088/1742-6596/471/1/012004

The PL was measured in a confocal μPL setup where a 640 nm cw laser was focused onto the sample with a high numerical aperture objective (NA=0.8). The emitted light from the sample was collected by the same objective and filtered by a pin-hole. The filtered light was then fed into a grating spectrometer with an InGaAs detector. Further details on the experimental apparatus can be found in references [8,15].

To observe the damage induced by H plasma on bare SOI, we performed cross-sectional TEM measurements by using a JEOL JEM 2010 instrument operating at an acceleration voltage of 200 kV. All TEM samples have been prepared using standard mechanical grinding and argon ion milling with an incidence angle of 7° and the accelerating voltage set at 5 kV.

For cavity experiments, we used modified L3 PhC cavities, which are known to have very high Q/V values [16]. The cavity was designed to have a fundamental mode at 1550 nm with a lattice constant a = 420 nm and an $r/\mathbf{a}$ ratio of 0.28. The far-field mode was optimised by enlarging alternating holes around the cavity by 15 nm, thus forming a second order grating. This folds back the $\boldsymbol{k}$-vector components near the edge of the Brillouin zone, outside the light emission cone, into the centre, resulting in an increase of the vertical emission efficiency [17, 18].

In Fig. 1a, the cross-section of TEM image of the H plasma treated SOI is reported. All the TEM images have been acquired in off Bragg defocused bright field (BF) conditions in order to avoid the background strain contrast around each defect and to emphasise the Fresnel contrast between the edge of the defect itself and the surrounding crystalline Si matrix [19, 20]. We observe that the damaged region, extending about 40 nm below the surface, presents different kind of defects: in the first 10 nm the damage consists of bubbles with a mean diameter of a few nm while, going further in depth, a buried band of platelets is found. The thickness of the band is 30 nm, the mean diameter of platelets lying on (001) is 15 nm while the mean diameter of platelets lying on {111} is 10 nm. Energy filtered TEM (not shown) reveals that the 7 nm thick amorphous layer found on the top Si surface of the H_2 plasma treated sample consists of a SiO_2 film, probably due to the hydrogenation treatment, as already speculated by Ghica et al. [21]. Moreover, we studied the thermal evolution of H-induced defects by performing annealing in FG in the temperature range 300–500 °C for 30 min in order to better passivate dangling bonds and, consequently, to decrease the efficiency of the non•radiative processes. In Fig. 1b the TEM micrograph of the sample annealed at 350 °C is reported. The thickness of the damaged layer is slightly reduced to about 30 nm with respect the as-treated sample, while the width of the band of nanobubbles is still 10 nm. The drastic reduction of the damaged layer is due to a shrinking of the platelets band to about 10 nm. Indeed, the mean diameter of {111} platelets is reduced to 5-7 nm while the (001) platelets are completely dissolved. Finally, after a thermal treatment at 500 °C for 30 min, we observe in the TEM image reported in fig. 1c a strong reduction of the damage depth at less than 15 nm and defects consisting only of nanobubbles, because both the (001) and {111} platelets are completely dissolved.

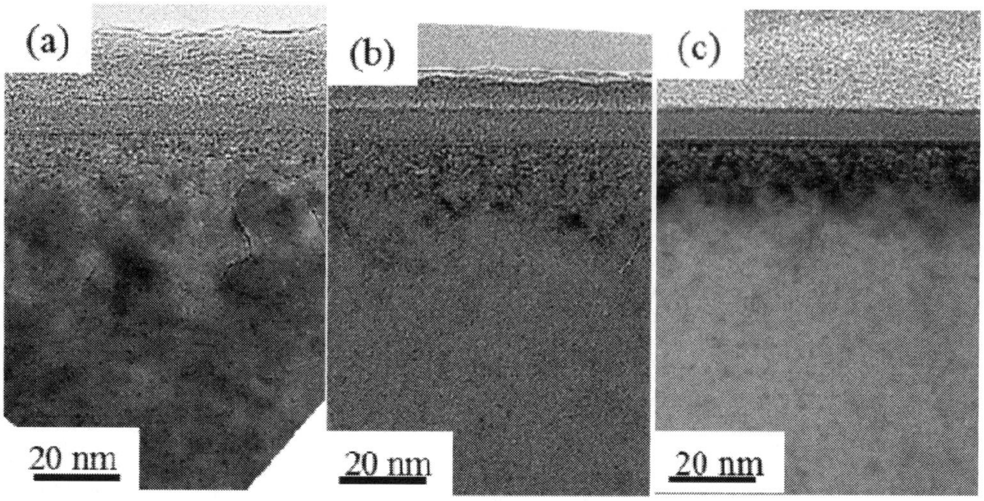

Figure 1. Defocused off-Bragg cross sectional TEM images of the H plasma treated SOI (a), after thermal annealing at 350 °C for 30 min (b) and at 500 °C for 30 min (c).

In order to correlate the structural properties with the optical ones, we have performed RT PL measurements on the H plasma treated SOI. In Fig. 2, the PL spectra of as-treated SOI (blue line and squares), of the SOI annealed at 350 °C (red line and circles) and at 500 °C (green line and triangles) are reported. The PL spectrum of the as-treated samples consists of a narrow band at ~1150 nm and a broad band with two main peaks at ~1300 nm and at ~1500 nm, in agreement with results reported in the literature [22]. After annealing at 350 °C for 30 min in FG ambient, we note an enhancement of PL with respect to the as-treated sample, while an almost full quenching is observed after the annealing at 500 °C for 30 min. These results, compared with the corresponding TEM analyses, put clearly in evidence the strong correlation between the PL intensity and the concentration of H-induced defects; indeed we note the highest PL signal in the sample annealed at 350 °C where the concentration of defects is still high and, conversely, a complete PL quenching has been observed in correspondence of the samples where there is only a residual trace of defects.

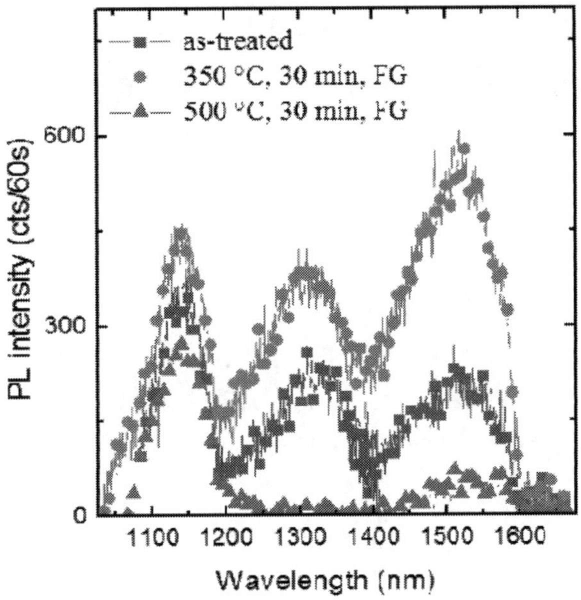

Figure 2. Room temperature PL spectra of H plasma treated SOI (blue squares), after thermal annealing at 350 °C for 30 min (red circles) and 500 °C for 30 min (green triangles).

An intense debate concerning the relationship between the broad PL band observed in H enriched samples and the H-induced damage is reported in the literature. Johnson and coworkers ascribe the PL emission observed in H plasma treated Si to the formation of the damage induced by the plasma treatment [23]. Moreover, a "strain-induced quantum-confined" model assumes the electron-hole recombination mechanism is allowed by the compressive strain in the platelets. The use of H to induce the damage further enhances the PL efficiency owing to its passivant role of non-radiative and other competitive radiative channels [22, 24].

Having optimized the light emission of H plasma-treated SOI, we have tried to enhance the emission from these optically active defects by introducing them in the active region of a L3 PhC nanocavity [25]. The scanning electron microscopy image of the cavity is shown in Fig. 3. The L3 PhC cavity was planned to have a fundamental mode at ~1.5 μm with a lattice constant a = 420 nm and an r/a ratio of 0.28, being r the radius of the PhC holes. The two holes adjacent to the cavity were reduced in size and displaced laterally to increase the cavity quality factor, Q.

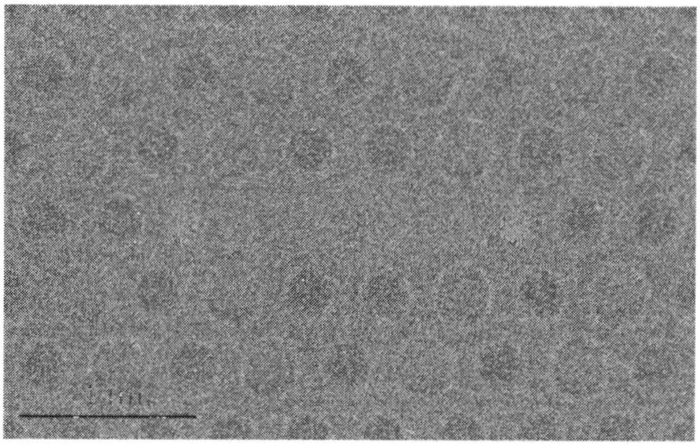

Figure 3. Scanning electron micrograph of L3 cavity

In order to control the fundamental cavity mode in the range 1300 – 1600 nm, we have also realized different PhC cavities having the period spanning from 370 to 450 nm. Indeed, in the H plasma-treated cavities, a strong and narrow PL emission can be obtained as demonstrated in Fig. 4. The sharp PL peaks, shown with different colours, correspond to the fundamental mode of each PhC cavity and can be tuned in a wide range of wavelengths by only changing the PhC parameters. We observe an overall 40000-fold increase of the PL signal at RT relative to bulk silicon.

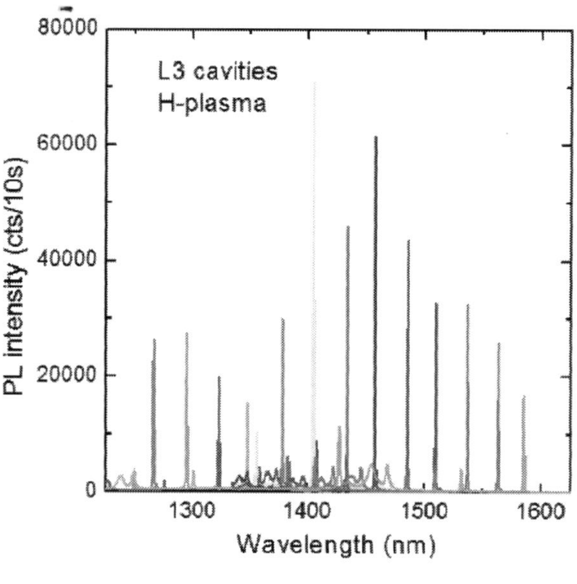

Figure 4. Room temperature PL spectra from H treated PhC cavities showing several peaks corresponding to the fundamental mode of cavities with different lattice periods.

Journal of Physics: Conference Series **471** (2013) 012004 doi:10.1088/1742-6596/471/1/012004

2.2. Eu-doped SiO_2 and SiOC films

Eu-doped SiO_2 thin films were grown on (100) Si substrates heated at 400 °C by using an ultrahigh vacuum magnetron sputtering system. The base pressure was about 1×10^{-9} mbar. The deposition has been obtained by RF co-sputtering of 4 inches diameter, water-cooled SiO_2 and Eu_2O_3 targets (99.9% purity or higher). The deposition was carried out with a sputter upwards configuration in a 5×10^{-3} mbar Ar atmosphere. Sputter rates have been set in order to obtain Eu concentrations equal to 1.5 and 5.0×10^{20} cm^{-3}. Eu-doped SiOC thin films were grown by RF co-sputtering of SiO_2, SiC and Eu_2O_3 targets (99.9% purity or higher). Sputter rates have been set in order to obtain a C concentration of about 5 at.% and an Eu concentration of 1.5×10^{20} cm^{-3}. Both SiO_2- and SiOC-based films are about 200 nm thick. After deposition, samples were thermally treated for 1 h in ultra-pure N_2 ambient by using a horizontal furnace operating at temperatures ranging from 750 to 1000 °C.
A 200 kV JEOL 2010F microscope, equipped with a Gatan Image Filter, was used to investigate the chemical composition of films by the use of energy filtered TEM (EFTEM). PL measurements were performed by pumping with the 325 nm line of a He-Cd laser. The pump power was about 3 mW and the laser beam was chopped through an acousto-optic modulator at a frequency of 55 Hz. The PL signal was analyzed by a single grating monochromator and detected with a Hamamatsu visible light photomultiplier. Spectra were recorded with a lock-in amplifier using the acousto-optic modulator frequency as a reference. All the spectra have been measured at RT and corrected for the spectral system response.

In Fig. 5a the BF cross-sectional TEM image of a SiO_2 sample doped with 5.0×10^{20} Eu/cm^3 and annealed at 750 °C in N_2 ambient is displayed. The micrograph shows that no Eu precipitation occurs in the sample. If we increase the annealing temperature the formation of clusters occurs, as clearly visible in the BFTEM images reported in Fig. 5b (sample annealed at 900 °C in N_2) and in Fig. 5c (sample annealed at 1000 °C in N_2). At 900 °C clusters are almost homogeneously distributed throughout the film but their size markedly decreases on going from the interface with the substrate (where diameters as large as about 15 nm are found) towards the surface (where the mean diameter is a few nm). At 1000 °C most of the precipitates are found close to the surface and a marked increase of their maximum diameter (about 30 nm) occurs.

The EFTEM map of elemental Eu (not shown), obtained by energy filtering at 133 eV (corresponding to the N_{IV} edge of Eu) and by using the 3-windows method [26] demonstrates that all clusters contain Eu. Moreover, the diffraction pattern reported in the inset of Fig. 5c indicates that the Eu-containing precipitates are amorphous; the only crystalline clusters are detected at the interface with the Si substrate, as shown by the high resolution TEM image shown in the inset in Fig. 5b, referring to the cluster circled by the white line. The analysis of the corresponding Fast Fourier transform (not shown) reveals that these precipitates grow epitaxially with respect to the Si substrate and that these clusters have the crystalline structure of pure Eu_2O_3. The experimental evidence on the nature of the interfacial clusters, coupled with the well-known tendency of rare earth ions, including Eu and Er, to precipitate as an oxidized phase when they are embedded in SiO_2, allow us to reasonably conclude that also the amorphous precipitates consist of Eu oxides (and/or silicates).

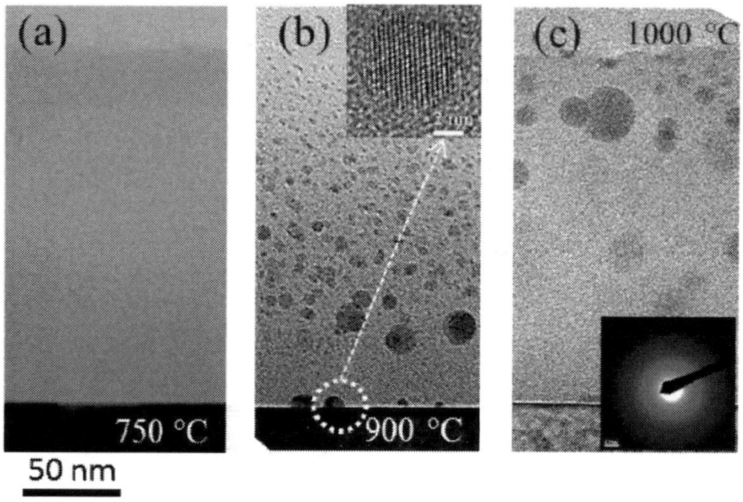

Figure 5. BF cross-sectionaol TEM images of the Eu-doped SiO_2 film annealed at 750 °C (a), at 900 °C (b) and at 1000 °C (c), for 1h in N_2 ambient. HR TEM image of the precipitate circled in white is shown in the inset of (b). The diffraction pattern of the Eu-doped SiO_2 film after the annealing at 1000 °C is reported in the inset of (c).

It is well known that Eu clustering is due to the low solid solubility of Eu in the SiO_2 matrix [12, 13, 27]. On the other hand, the peculiar temperature dependence of the precipitation phenomenon and the tendency of the clusters to move towards the surface when the annealing temperature is increased are certainly less expected. In the as-deposited sample the Eu ions are homogeneously distributed throughout the film; during the thermal process they acquire mobility but, in the absence of any concentration gradient, no relevant diffusion process is expected. Recently, we have demonstrated that the mobility acquired by the Eu atoms during the thermal annealing is due to the fact that the host matrix is O deficient and the Eu atoms, in order to be fully O-coordinated, diffuse towards the surface to trap O atoms from the contamination unavoidably present in a conventional furnace [28].

We have performed PL measurements to correlate the structural and the optical properties of Eu-doped SiO_2 films. Fig. 6 shows the PL spectra of as-deposited and annealed films, obtained at RT by exciting the samples with the 325 nm line of a He-Cd laser. The PL spectrum of the as-deposited sample consists of a broad band having its maximum at about 435 nm, corresponding to the emission of Eu^{2+} ions [29, 30], and of some narrow lines corresponding to the emission of Eu^{3+} ions [30]. After annealing the samples in N_2 ambient, we notice an increased intensity of the peak associated with Eu^{2+}, coupled with a shift to 450 nm (at 750 °C) and to 475 nm (at 900 °C); a further increase of the PL intensity and a small redshift are found at 1000 °C (spectrum not shown). The Eu^{3+} emission remains clearly visible as a peak at 750 °C and as a shoulder of the Eu^{2+} peak at 900 °C. The increased PL intensity in annealed samples accounts for a reduced density in the host matrix of defects which can constitute preferential channels for non-radiative de-excitation and for a better Eu coordination, while the peak redshift is related to the formation of regions characterized by different local Eu concentrations.

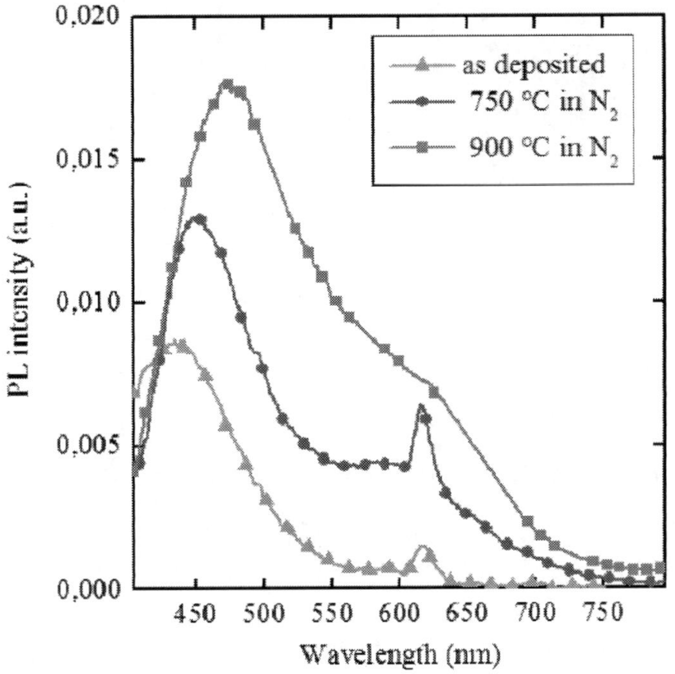

Figure 6. Room temperature PL spectra of Eu-doped SiO_2 films as-deposited and annealed at 750 and 900 °C, for 1h in N_2 ambient.

The cluster formation observed after thermal annealing, caused by the low solid solubility of Eu in silica, seriously limits the perspective of Eu application in photonics. Indeed, it is well established that the clustering occurring at high Eu concentration makes the rare-earth ions optically inactive. In this context a strong effort in the literature has been made to find strategies limiting the rare earth precipitation phenomenon [11, 12, 14, 27].

We have developed a new approach to make SiO_2 a better host for Eu ions. Indeed, it has been widely reported in literature that the Eu solubility can be remarkably increased by properly modifying the host matrix [31, 32]. We added to the Eu-doped SiO_2 matrix a C content of about 5 at. % in order to move from pure SiO_2 to a SiOC matrix, where the Eu concentration was set to 1.5×10^{20} cm^{-3}. Moreover, we deposited a SiO_2 capping layer (200 nm thick) on Eu-doped films before the annealing step, in order to avoid as much as possible additional Eu-O interactions and prevent the clustering formation. In Fig. 7, we compare the BF cross-sectional TEM images of Eu-doped SiO_2 (Figs. 7a) and SiOC films (Figs. 7b), capped with the SiO_2 film before the annealing at 900 °C in N_2 ambient. It is evident that C addition dramatically changes the structural properties of the Eu-doped SiO_2 films.

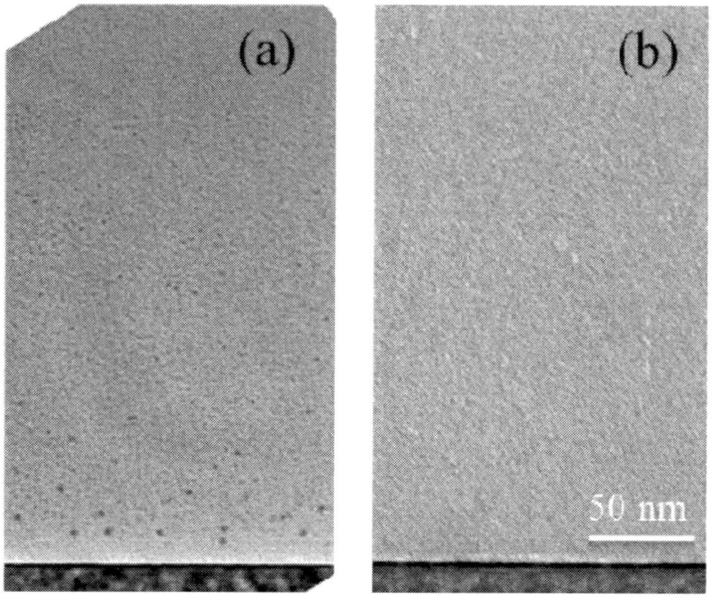

Figure 7. Cross-sectional TEM images of Eu-doped SiO_2 (a) and SiOC (b) annealed in N_2 ambient at 900 °C after the deposition of a SiO_2 capping layer.

Indeed, it is possible to note in Fig. 7a that, in the case of the SiO_2 matrix, a residual Eu clustering phenomenon is still observed in a region close to the interface with the Si substrate, while the Eu precipitation in a SiOC matrix is completely suppressed, as clearly shown by Fig. 7b. Fig. 8 reports the comparison between the RT PL spectrum of an Eu-doped SiOC sample annealed at 900 °C in N_2 (green line and triangles) and that of a SiO_2 sample containing the same Eu content and annealed under the same conditions (shown in Fig. 8 multiplied by a factor of 100 - blue line and squares). A very strong enhancement of the room temperature PL emission of about a factor of 250 has been obtained; furthermore, the peak is sharper and its maximum, detected at about 435 nm, is blue-shifted by about 60 nm. Fig. 8 also reports the PL spectra of Eu-doped SiOC samples annealed at 900 °C (red circles and line) after the SiO_2 cap deposition. In agreement with the capability of the capping layer to fully prevent Eu precipitation, demonstrated by the TEM image in Fig. 7b, a further enhancement of the PL intensity, accounting for about a factor of 1.5, has been obtained.

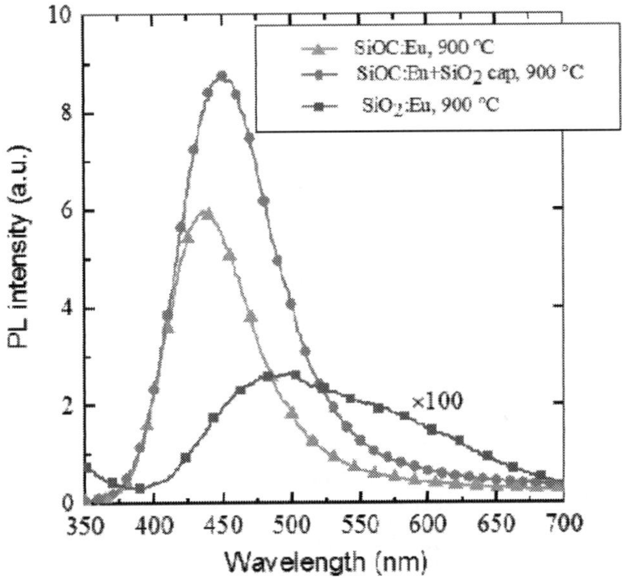

Figure 8. Room temperature PL spectra of Eu-doped SiOC and SiO_2 films annealed at 900 and in N_2 ambient. In the case of SiOC, spectra of both SiO_2-capped and uncapped samples are shown.

3. Conclusions

In this paper, we have reviewed the different approaches we recently followed for the realization and characterization of Si-based materials to be employed as efficient light emitters in Si photonics. The engineering of H-induced defects has been exploited in order to obtain light emission from crystalline Si. By comparing the structural and optical properties we clearly demonstrate that the luminescence is strictly correlated with the damage produced by the treatment, in agreement with the "strain-induced quantum-confined" model proposed in the literature. Moreover, an overall 40000–fold improvement of the emission from these optically active defects with respect to pure Si has been obtained by placing them in the active region of a PhC nanocavity. In addition, we demonstrated that the emission can be tuned over a wide range of wavelengths by only changing the PhC parameters.

On the other side, limits and perspectives of pure SiO_2 and SiOC matrices as a host for optically active Eu ions have been presented and discussed. In the case of SiO_2 the occurrence of an extensive Eu precipitation in thermally annealed films seems to be unavoidable. We have therefore proposed the addition of C to the Eu-doped SiO_2 matrix and a capping with SiO_2 as a strategy to minimize clustering phenomena and to improve the efficiency of RT light emission. Indeed, TEM data have demonstrated that Eu ions in SiOC are characterized by an enhanced mobility and solubility. This peculiarity leads to a strongly reduced Eu precipitation, and, as a consequence, to a very intense and stable light emission, centered at about 440 nm, from Eu^{2+} ions. Since SiOC can be considered a material fully compatible with Si technology and an appreciable electrical conduction is expected in this matrix, we believe that the Eu-doped SiOC films may open new and interesting perspectives for photonic applications of Eu-containing materials.

References

[1] Iacona F et al 2004 *J. Appl. Phys.* **95** 3723-3732.
[2] Boninelli S et al 2005 *Nanotechnology* **16** 3012–3016
[3] Franzò G et al 2003 *Appl. Phys. Lett.* **82** 3871-3873
[4] Irrera A et al 2012 *Nanotechnology* **23** 075204
[5] Henry A et al 1991 *J. Appl. Phys.* **70** 5597–5603
[6] Weber J 1991 *Phys. B, Cond. Mat.* **170** 201–217
[7] Canham L T et al 1989 *Mat. Sci. Eng. B* **4** 41–45
[8] Lo Savio et al 2011 *Appl. Phys. Lett.* **98** 201106
[9] Shakoor A et al 2012 *Phys. B* **407** 4027–4031
[10] Li D et al 2010 *Opt. Express* 18 27191-27196
[11] Rebohle L et al 2008 *Appl. Phys. Lett.* **93** 071908
[12] Rebohle L et al 2009 *J. Appl. Phys.* **106** 123103
[13] Afify N D et al 2009 *Phys. Rev. B* **79** 024202
[14] Lægsgaard J et al 2002 *Phys. Rev. B* 65, 174114
[15] Galli M et al 2009 *Appl. Phys. Lett.* **94** 071101
[16] Akahane Y et al 2003 *Nature* **425** 944-947
[17] Tran N V Q et al 2009 *Phys. Rev. B* **79** 041101
[18] Portalupi S L et al 2010 *Opt. Express* **18** 16064
[19] Grisolia J et al 2000 *Appl. Phys. Lett.* **76** 852-854
[20] Boninelli S et al 2006 *Appl. Phys. Lett.* **89** 171916
[21] Ghica C et al 2011 *J. Phys. D : Appl. Phys.* **44** 295401
[22] Weman H et al 1990 *J. Appl. Phys.* **87** 1013-1021
[23] Johnson N M et al 1987 *Phys. Rev. B* **35** 4166-4169
[24] Weman H et al 1990 *Phys. Rev. B* **42** 3109-3112
[25] Shakoor A et al 2012 *Laser Photonics Rev.* **7** 1 114
[26] Krivanek O L et al 1994 *Mater. Res. Soc. Symp. Proc.* **332** 341
[27] Nazarov A N et al 2010 *J. Appl. Phys.* **107** 123112
[28] Boninelli S et al 2013 *J. Appl. Phys* **113** 143503
[29] Tian Y et al 2011 *J. Appl. Phys.* **109** 053511
[30] Ye X et al 2009 *J. Appl. Phys.* **105** 064302
[31] G Bellocchi et al 2012 *Opt. Express* **20** 5501
[32] Q Zhang et al 2010 *Opt. Mater.* **32** 427

Atomic-scale characterization of (mostly zincblende) compound semiconductor heterostructures

David J Smith[1,2], T Aoki[3], J K Furdyna[4], X Liu[4], M R McCartney[1,2] and Y-H Zhang[2,5]

[1] Department of Physics, Arizona State University, Tempe, AZ 85287, USA
[2] Center for Photonics Innovation, Arizona State University, Tempe, AZ 85287, USA
[3] LeRoy Eyring Center for Solid State Science, Arizona State University, Tempe, AZ 85287, USA
[4] Department of Physics, University of Notre Dame, Notre Dame, IN 46556, USA
[5] School of Electrical, Computer and Energy Engineering, Arizona State University, Tempe, AZ 85287, USA

E-mail: david.smith@asu.edu

Abstract. This paper provides an overview of our recent atomic-scale studies of semiconductor heterostructures, based primarily on combinations of zincblende compound materials grown by molecular beam epitaxy. Interfacial strain due to lattice mismatch inevitably causes growth defects to be introduced. Analysis of defect type and distribution using image filtering allows residual strain to be estimated. Exploratory investigations using aberration-corrected electron microscopy, which enables individual atomic columns to be resolved, are also described.

1. Introduction

Compound semiconductors offer a multitude of optoelectronic and photonic device opportunities based on their wide range of band gaps that cover wavelengths from far-infrared to near-ultraviolet. These possibilities are illustrated in Fig. 1, which depicts the lattice constants and band-gap energies of many common semiconductors [1]. Ternary (and quaternary) alloys offer additional flexibility in terms of band-gap engineering, as well as potentially allowing for avoidance of any lattice-mismatch issues. Nevertheless, the epitaxial growth of compound semiconductor heterostructures with two (or more) dissimilar materials presents many challenges. Successful growth requires suitable preparation of the substrate surface and careful attention to the growth conditions. For materials with differing lattice parameters, there are several additional problems. As well as lattice mismatch, which inevitably leads to strain and probable defect formation, valence mismatch and differences in thermal expansion are further factors that can seriously impact whether or not high quality materials can be grown.

The transmission electron microscope (TEM) provides a wide range of techniques for structural characterization of compound semiconductors including high-resolution electron microscopy (defect identification and strain field analysis), Z-contrast imaging (cation distribution), convergent-beam electron diffraction (local lattice parameter), and electron holography (internal electric field). These TEM methods provide powerful complementary approaches for characterizing and understanding the often-competing effects of growth conditions and compositional differences. In the following, we briefly review our recent atomic-scale investigations of heterostructures consisting of II-VI compound semiconductors grown epitaxially on several common III-V substrates.

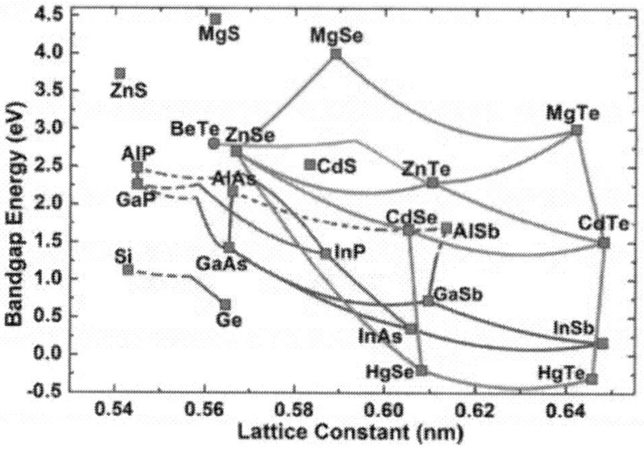

Figure 1. Schematic showing band gaps of common semiconductors vs. corresponding lattice parameter.

2. Experimental details

Most of the heterostructures described here were grown by molecular beam epitaxy (MBE), as described elsewhere [2], using a Riber 32 system consisting of two separate III-V and II-VI chambers connected via an ultrahigh-vacuum transfer module. The different III-V substrates were normally deoxidized, followed by growth of a thin buffer layer of the same material (except in the case of InP), before being cooled down to room temperature and then transferred to the II-VI chamber. Monitoring of the II-VI surface reconstruction during growth using reflection-high-energy electron diffraction (RHEED) was used to maintain optimal deposition conditions.

Samples were prepared for cross-sectional TEM observation using standard mechanical polishing and dimpling to reduce sample thicknesses to ~10μm, followed by argon-ion-milling at low energy (~2.0-3.0keV) with the sample held at liquid-nitrogen temperature to minimize ion-beam damage [3]. All samples were prepared for observation along [110]-type zone axes so that the interface normal would be perpendicular to the electron-beam direction [4]. The observations described here were made either with a JEM-4000EX high-resolution electron microscope (HREM) operated at 400keV (structural resolution of ~1.7Å), or a probe-corrected JEM-ARM200F scanning TEM (STEM) operated at 200keV (probe size ~0.8Å). Images from the latter instrument were primarily recorded using high-angle annular-dark-field (HAADF) and/or bright-field (BF) imaging modes, with corresponding detector collection angles of ~90-170mrad and ~0-22mrad, respectively.

3. Results

3.1. High-resolution lattice-fringe imaging

The characterization of zincblende compound semiconductors by electron microscopy methods has attracted much attention over several decades. Here we illustrate some recent developments by comparing changes in the microstructure of epitaxial ZnTe layers as a function of lattice mismatch with the underlying substrate material, focusing primarily on the stress-relieving defects that are present at the hetero-interfaces. Projections of individual atomic columns are clearly resolved in probe-corrected HAADF and BF images, as will be shown below.

In the case of the ZnTe(001)/GaSb(001) heterostructure, where the lattice mismatch was ~0.13%, misfit dislocations were occasionally visible at the interface but they were usually highly separated (>0.1μm apart). Moreover, the exact position of the interface was otherwise difficult to determine except in some diffraction contrast images, which was ascribed to the close similarity in the combined atomic numbers of the two elements in each material [3]. Similar results were also obtained in related studies of ZnTe(211)/GaSb(211)B heterostructures, as shown by the example in Fig. 2 [5].

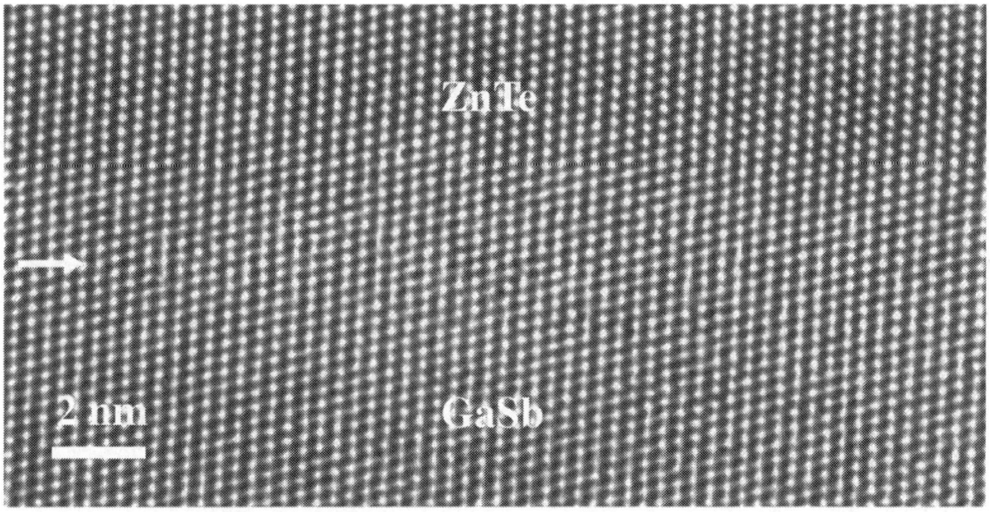

Figure 2. High-resolution electron micrograph establishing the highly coherent nature of the ZnTe(211)/GaSb(211)B interface (arrowed). Black spots correspond to closely-spaced pairs of atomic columns that are not separately resolved [5].

For the ZnTe(001)/InAs(001) sample, where the lattice mismatch between the two materials was slightly larger (~0.74%), threading dislocations were observed in the thick ZnTe epilayer and some strain-related contrast at the interface was also visible in diffraction-contrast images [6]. However, as shown by the lattice-fringe image in Fig. 3, the ZnTe/GaAs interface was invariably abrupt and coherent although its exact position was difficult to recognize.

Figure 3. High-resolution lattice-fringe image of ZnTe/InAs(001) interface. The location of the interface (arrowed) is almost impossible to identify in the absence of misfit dislocations.

The interfacial defect density increased substantially as the difference in lattice parameters was increased. In the case of the ZnTe(001)/InP(001) sample (lattice mismatch ~3.85%), Burgers' circuit analysis of HREM images revealed that the most common defects present at the interface were either perfect Lomer edge dislocations with a Burgers' vector of (1/2)***a***<110> along the interface, or perfect 60° dislocations with Burgers' vector of the same length but inclined at an angle of 45° to the interface [6]. Aberration-corrected images of these two types of defects are shown below.

The separation between interfacial defects was found to decrease as the lattice mismatch was further increased. For the ZnTe(001)/GaAs(001) heterostructures (mismatch ~7.38%), the interfacial defects were observed to be pseudo-periodic with average separations of ~5.5nm. Figure 4 shows a representative region of the ZnTe/GaAs interface, with the positions of misfit dislocations indicated by arrows. Burgers' circuit analysis of individual defects was again used to identify the dislocation type, although digital image processing provided a far more efficient approach for analyzing larger fields of view. Thus, similar lattice-fringe images were digitized, processed by Fast Fourier Transform (FFT), and then filtered by selecting specific (111) diffraction spots for inverse FFT. The location *and* type of the interfacial misfit dislocations could then be easily determined: perfect Lomer edge dislocations were identified when two corresponding (111) planes terminated at the same location, whereas 60° dislocations were present when a single (111) plane terminated at the interface [5]. Measurements showed that the ratio of Lomer dislocations to the total number of dislocations for the ZnTe/GaAs sample was about 39%. Further analysis taking the type of defect into account also indicated that the residual interfacial strain was ~0.1% for this sample, so that the interface could thus be considered as being relaxed to within experimental error [6].

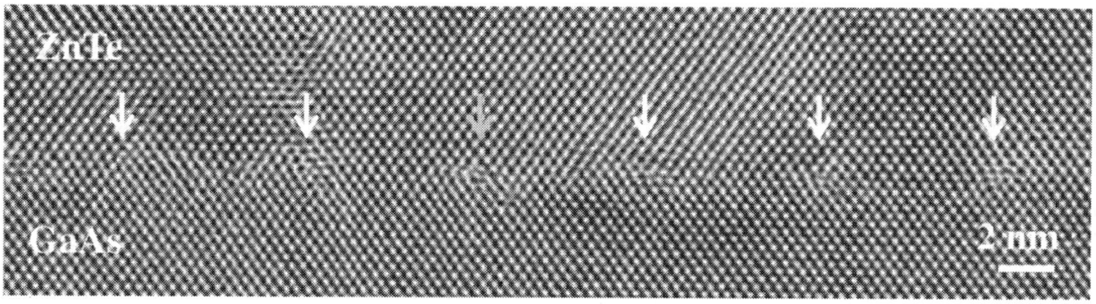

Figure 4. High-resolution lattice-fringe image showing array of dislocations (pseudo-periodic) accommodating the misfit at ZnTe(001)/GaAs(001) interface [6].

3.2. Aberration-corrected electron microscopy

The recent emergence of multiple techniques for aberration correction using either (on-line) hardware or (off-line) software approaches, augmented by improved beam coherence and improved mechanical and electrical stabilities, has enabled microscope information limits to be pushed routinely to beyond the 1 Å resolution barrier [7]. This revolution in aberration correction has dramatically altered the landscape for advanced materials research by making it straightforward to achieve atomic-resolution imaging on a regular basis from many different types of crystalline materials, including metals, oxides and semiconductors. For the specific cases of elemental and compound semiconductors observed in the most commonly used <110> orientation, it then becomes possible to resolve the projections of individual atomic columns, which are often referred to as 'dumbbells' [8]. As an illustrative example, Fig. 5 compares HAADF and BF images, recorded simultaneously with a probe-corrected STEM, which show the interface region of the ZnTe(001)/InP(001) heterostructure mentioned earlier. Individual atomic columns appear with white contrast in the ("Z-contrast") HAADF image, while they are visible with dark contrast in the BF image. The insets show line traces across several In-P dumbbells. From inspection, it is clear that the separate In and P atomic columns which are separated by 0.146nm are well-resolved in both detector geometries despite the considerable difference in their atomic numbers.

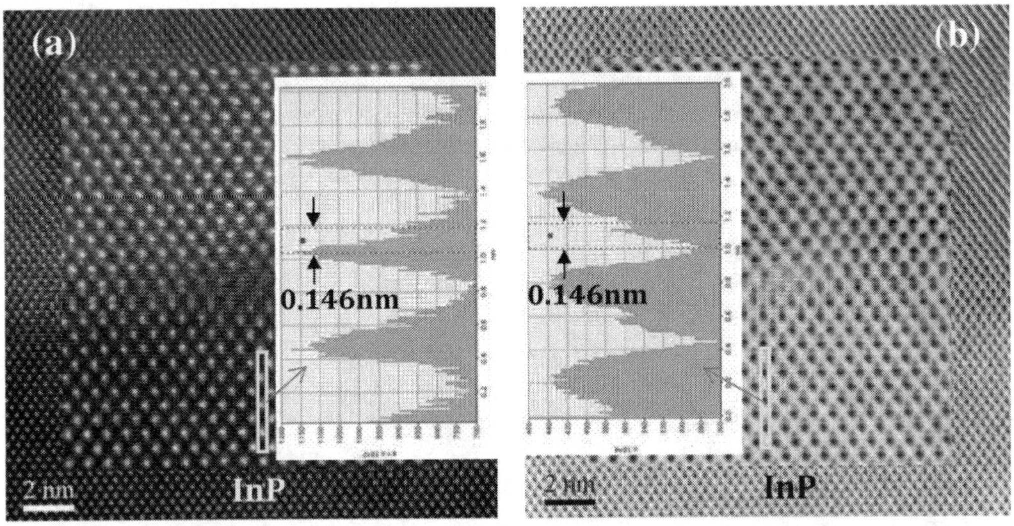

Figure 5. Aberration-corrected STEM images of ZnTe/InP(001) interface: (a) HAADF image (90~170mrad); (b) BF image (0~22mrad). Inset profiles show resolved In-P dumbbells in both cases.

Aberration-corrected imaging is already being widely used although compound semiconductors have so far attracted relatively less attention [8]. However, it should be abundantly clear that the enhanced capability for imaging individual atomic columns should enable improved insights about defect microstructure and structural properties to be obtained, as recently demonstrated, for example, in AC-STEM studies of CdTe solar cells [9]. As another illustration of these possibilities, Fig. 6 shows aberration-corrected STEM images of the two most common types of interfacial defects observed at the ZnTe/InP(001) interface, together with the corresponding Burgers' circuit analysis. By inspection, it seems that the BF image displays clearer 'dumbbell' contrast but neither image shows well-resolved structure at the dislocation cores, which is perhaps attributable to e-beam damage during observation.

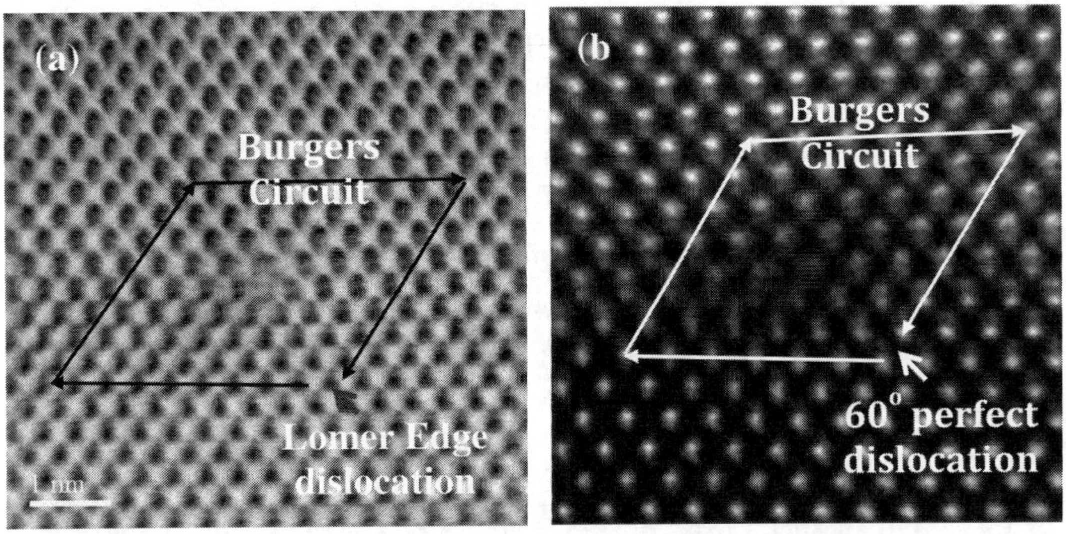

Figure 6. Aberration-corrected STEM images of ZnTe/InP(001) interface: (a) BF image showing Burgers' circuit analysis for Lomer edge dislocation; (b) HAADF image showing Burgers' circuit for perfect 60° misfit dislocation. Individual atomic columns have better visibility in BF image.

Figure 7 compares AC-STEM HAADF and BF images of a perfect Lomer edge dislocation as observed at the ZnTe/GaAs(001) interface. Both images clearly show the 'dumbbell' structure for each material, while line traces of the GaAs region (not shown here) enable the separate Ga and As atomic columns to be identified on the basis of their intensity. The atomic structure of the dislocation core is, however, unclear and further observations are needed to establish whether this apparent disorder has been caused by sample preparation artifacts and/or electron-beam irradiation or whether misfit dislocations at such heterovalent semiconductor interfaces are intrinsically likely to be dissociated.

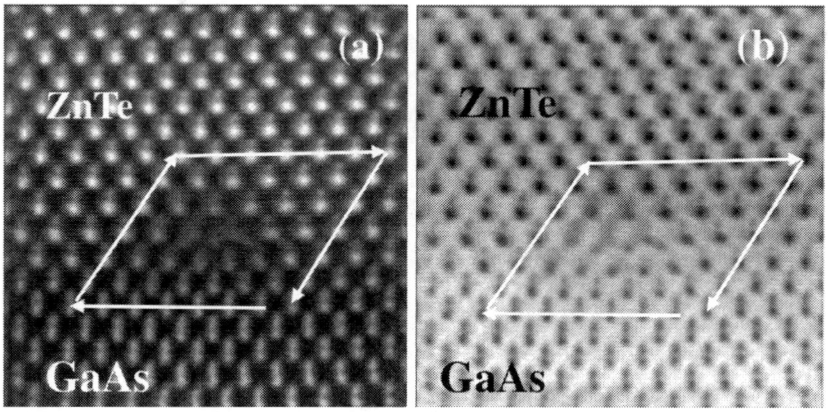

Figure 7. Aberration-corrected STEM images showing Burgers' circuit analysis identifying perfect Lomer edge dislocation at ZnTe/GaAs(001) interface. Individual atomic columns are clearly visible in both images and both materials except close to the dislocation core.

4. Conclusions

Atomic-resolution imaging with the electron microscope has been demonstrated to play a valuable role in characterizing the defect microstructure of zincblende semiconductors. The increased resolving power of aberration-corrected instruments offers further scope for detailed structural investigations at the level of individual atomic columns. Complementary theoretical modeling is still sorely needed to determine the equilibrium structure for heterovalent semiconductor interfaces [9].

References

[1] Zhang YH, Wu SN, Ding D, Yu SQ and Johnson SR 2008 *33rd. IEEE PVSC* pp.1-5
[2] Fan J, Ouyang L, Liu X, Furdyna JK, Smith DJ and Zhang YH 2011 *J. Cryst. Growth* **323** 127
[3] Wang CZ, Smith DJ, Tobin S, Parados T, Zhao J, Chang Y and Sivananthan S 2006 *J. Vac. Sci. Technol.* A **24** 995
[4] Zhang X, Wang S, Ding D, Liu X, Tan JH, Furdyna JK, Zhang YH and Smith DJ 2009 *J. Electron. Mater.* 38 1558
[5] Chai J, Noriega OC, Dinan JH, Smith C, Chau N, Pena J, Dedigama A, Kim JJ, Smith DJ and Myers TH 2013 *J. Electron. Maters.* (in press).
[6] Ouyang L, Fan J, Wang S, Lu X, Zhang YH, Liu X, Furdyna JK and Smith DJ 2011 *J. Cryst. Growth* **330** 30
[7] Smith DJ 2008 *Microsc. Microanal.* **14** 2
[8] Smith DJ, Aoki T, Mardinly J, Zhou L and McCartney MR 2013 *Microscopy* **62** S65
[9] Li C, Poplawsky J, Wu Y, Lupini AR, Mouti A, Leonard DN, Paudel N, Jones K, Yin W, Al-Jassim M, Yan Y and Pennycook SJ 2013 *Ultramicroscopy* (in press)
[10] We acknowledge use of facilities in the John M. Cowley Center for High Resolution Electron Microscopy at Arizona State University. Acquisition of the JEM-ARM200F was supported by NSF DMR-0820196, and the work at Notre Dame was supported by NSF ECC1002072.

The influence of atomic ordering on strain relaxation during the growth of metamorphic solar cells

A G Norman, R M France, W E McMahon, J F Geisz and M J Romero

National Renewable Energy Laboratory, 15013 Denver West Parkway, Golden CO 80401, USA

E-mail: andrew.norman@nrel.gov

Abstract. The occurrence of single variant $CuPt_B$ ordering during growth of InGaP graded buffer layer structures on offcut (001) GaAs substrates for inverted metamorphic solar cells is found to have a strong influence on strain relaxation mechanisms. Since the surface-induced $CuPt_B$ ordering is metastable in the bulk of the material, a strong preference is observed for the nucleation and glide of 60° type misfit dislocations with Burgers vectors that introduce an antiphase boundary into the ordered structure. This results in an overall epitaxial layer tilt in the opposite sense to that normally observed for the direction of substrate offcut. Furthermore, in InGaP buffer layers graded to InP, a switch in the dislocation glide plane preference back to that normally observed for the direction of substrate offcut is observed as the degree of atomic ordering falls below a critical value. This results in the nucleation and glide of new misfit dislocations resulting in an increase in the threading dislocation density that is found to have a deleterious effect on device efficiency. Understanding the materials science behind this behavior will enable the engineering of more effective, lower threading dislocation density strain relief buffer layers resulting in improved performance of subsequently grown devices.

1. Introduction

The growth of lattice-mismatched semiconductor heterostructures introduces increased degrees of freedom in the design of optoelectronic devices for desired properties such as emission wavelength of light emitting diodes and optimal band gap combinations for ultra-high efficiency multijunction solar cells. Fundamental to the achievement of the highest performance devices is an understanding of the factors that affect strain relaxation processes that occur during the epitaxial growth of these lattice-mismatched layer structures, such as dislocation nucleation and glide, and epitaxial layer tilt.

The highest efficiency solar cells so far produced are multijunction solar cells that are typically operated under concentrated light. There are presently two main approaches to producing this type of cell: (i) an all lattice-matched approach using a dilute III-nitride alloy as an $\approx$1 eV junction [1], or (ii) a metamorphic approach where one or more of the junctions are lattice-mismatched with respect to the substrate used for growth [2–7]. Both of these methods have resulted in multijunction solar cells with efficiencies greater than 40 % under concentrated light. Multijunction solar cells are able to reach such high efficiencies because their design reduces the thermalization and transmission losses that are typically associated with single junction devices and hence enables a fuller utilization of the solar spectrum [e.g., 8]. However, to produce such high efficiency devices requires extremely high quality, epitaxial III-V semiconductor materials. These can only be grown on very expensive single crystal substrates using costly epitaxial growth techniques. In order to reduce costs, multijunction solar cells

Journal of Physics: Conference Series **471** (2013) 012006 doi:10.1088/1742-6596/471/1/012006

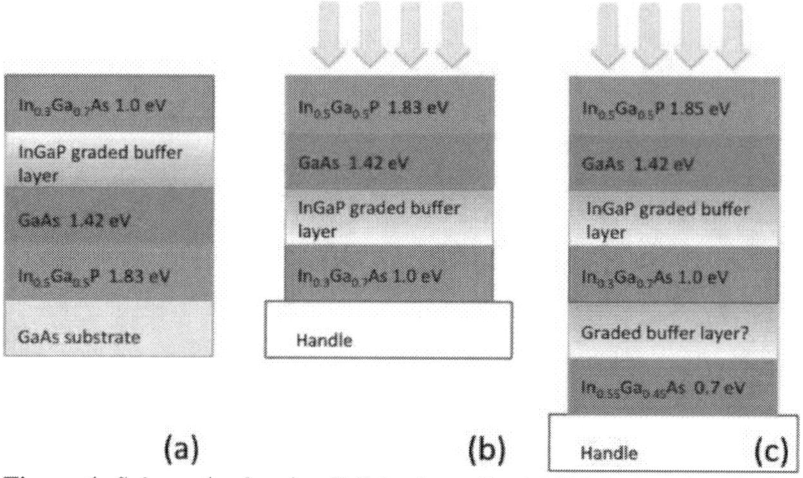

Figure 1. Schematic showing IMM solar cells. (a) Triple-junction device as grown on GaAs substrate. (b) Triple-junction device after flipping the structure over, bonding to a suitable handle, and removal of the substrate. (c) Four-junction device.

for terrestrial applications are generally used in a concentrator system where the light is gathered from a large area and then focused using a relatively cheap optical element onto a very small area multijunction solar cell. An added bonus is that the efficiency of the multijunction solar cells generally increases under concentrated light [e.g., 9].

The particular structure being investigated at NREL is the inverted metamorphic multijunction (IMM) solar cell [2, 4, 5]. A schematic showing this structure is shown in figure 1. An efficiency of 42.6 % has recently been reported for this triple-junction device structure and >44% efficiency is expected after further optimization [10]. However, further significant efficiency gains beyond this are only likely to be obtained by the addition of a fourth junction with a band gap of ≈ 0.7 eV, figure 1(c) [e.g., 11]. The IMM solar cell has several advantages over the upright lattice-matched and upright metamorphic multijunction (MM) solar cells. The first is that a wider range of band gaps is potentially available for the metamorphic approach than for the lattice-matched approach, figure 2. In the IMM design, the two most power producing high band gap junctions are lattice-matched and hence a higher efficiency may be expected due to a lower sensitivity to the presence of mismatched-induced defects than in the case of the upright MM solar cell where the high band gap junctions are lattice-mismatched. The IMM design also has the potential for a large cost reduction by enabling substrate reuse after removal of the cell active layers that may also be bonded to a handle material with more favorable properties than the original substrate, such as a lighter weight or flexibility. However, to obtain the highest efficiency IMM solar cells require the growth of highly perfect, transparent, graded buffer layers to minimize optical losses and recombination losses associated with the introduction of structural defects such as dislocations resulting from the lattice-mismatch.

To achieve the four-junction IMM device illustrated in figure 1, we are developing a graded InGaP buffer layer structure to step the lattice parameter from that of 1.0 eV band gap InGaAs to that of 0.7 eV InGaAs, figure 2. For this application the InGaP graded buffer layer must be transparent to photons with energies between 0.7 eV and 1.0 eV and also be electrically conducting. It is also essential to promote dislocation glide during growth of this graded buffer layer to minimize the density of threading dislocations that would have a deleterious effect on the efficiency of the final device. Strain relaxation in lattice-mismatched III-V semiconductor epitaxial growth normally occurs by the nucleation and glide of 60° type misfit dislocations once a certain critical thickness has been exceeded. These misfit dislocations generally lie along the orthogonal <110> directions at the interfaces between layers in samples grown on (001) substrates and their *a*/2<011> Burgers vectors are at an angle of 60°

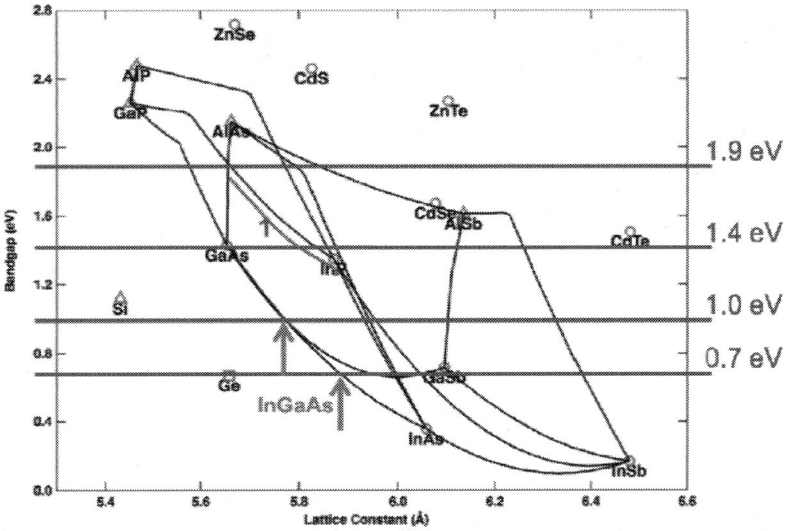

Figure 2. Band gap versus lattice parameter plot for selected group IV, III-V, and II-VI semiconductors showing lattice-mismatched InGaAs junction materials of band gaps 1.0 and 0.7 eV and a potential graded buffer layer path, 1, using InGaP alloys between the lattice parameters of GaAs and these two InGaAs alloys.

to the <110> line direction of the dislocations, and at 45° to the interface, as shown in figure 3 [12]. These dislocations possess screw, misfit, and tilt components and glide on {111} planes. Depending on the sign of the strain and the <110>-line direction, these dislocations can either have group V atoms at the core, α-dislocations, or group III atoms at the core, β-dislocations. For the InGaP compressively strained graded buffer layer of interest in this work, α-dislocations have a line vector of [1-10], whilst β-dislocations have a line vector of [110]. It should also be noted that these 60° type glide dislocations are frequently observed to dissociate into two Shockley partial dislocations separated by an intrinsic stacking fault [13, 14]. The possible slip systems for 60° dislocations in the compressively strained InGaP semiconductor alloy are listed in table I together with the dislocation line vectors, Burgers vectors and components, slip planes and dislocation type [15].

Under the growth conditions normally used, the InGaP graded buffer layers are found to atomically order on {111}B planes [16–18]. This $CuPt_B$ form of ordering is predicted to be energetically unstable in the bulk of the material but is found to be energetically favorable during growth on a reconstructed surface containing [110] rows of [-110]-oriented P dimers [19–21]. On exact (001) oriented substrates, a mixture of the two {111}B variants is observed but a preferential selection of a single variant can be obtained by growth on a substrate offcut a few degrees from (001) towards one of the {111}B planes [21]. Since the ordered planes are {111}, the same as the preferred glide planes for 60° misfit dislocations in zincblende III-V semiconductors, this has important consequences on the dislocation nucleation and glide behavior and strain relaxation mechanisms in these graded buffer layer structures. Depending on the Burgers vector, certain of the 60° misfit dislocations will leave behind an antiphase boundary (APB) in the ordered structure as they glide [16, 17, 22], see table 1. Since the ordered structure is metastable in the bulk, this results in a net reduction in the energy of the layer giving an energetic preference for the nucleation and glide of misfit dislocations that produce APBs. We present experimental results that illustrate this behavior and discuss the consequences it has for strain relaxation in these graded buffer layer systems and what effect it might have on the resulting threading dislocation density.

2. Experimental Procedures

InGaP layers and compositionally graded buffer layers were grown on (001) GaAs substrates offcut by 2° towards one of the {111}B planes by atmospheric pressure metal organic vapor phase epitaxy (MOVPE). The InGaP layers were grown at 675° C. Transmission electron microscope (TEM) cross-section samples were typically prepared by standard mechanical polishing and dimpling techniques followed by argon ion milling with the sample rotated and cooled by liquid nitrogen to preserve the atomic ordering that was normally present. A few samples were prepared in an FEI dual beam FIB workstation using a lift out technique. Cross-section TEM and transmission electron diffraction (TED) experiments were performed in an FEI ST 30 TEM operated at 300 kV. Threading dislocation density was measured by cathodoluminescence (CL) imaging in a JEOL 5800 scanning electron microscope (SEM) equipped with Ge and InGaAs detectors. Epitaxial layer tilt, misfit, and strain relaxation of the samples were measured by high-resolution x-ray diffraction (XRD) reciprocal space mapping (RSM) using a Bede D1 diffractometer.

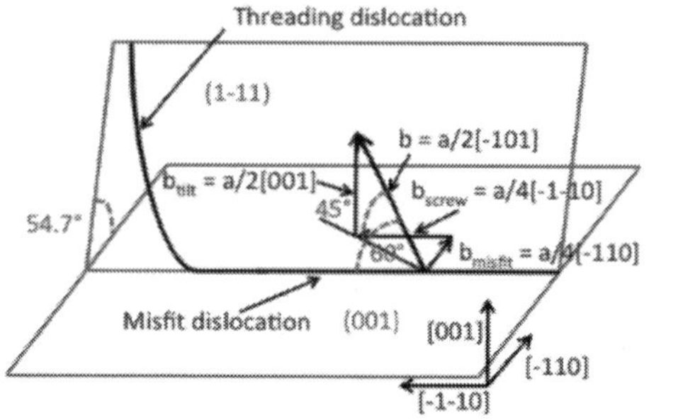

Figure 3. Diagram illustrating the nature of the typically observed 60° type misfit dislocation and associated threading dislocation in zincblende III-V semiconductors.

3. Experimental Results

Figure 4 illustrates TEM, TED, and XRD RSM results obtained from an InGaP step-graded buffer layer, grown on an (001) GaAs substrate offcut 2° towards (-111), where the lattice parameter was graded from that of GaAs to that of InP. In the 220 dark field (DF) image of figure 4(a) the array of 60° misfit dislocations that are introduced to relieve the strain during the growth of this structure can clearly be seen. The InP layer grown on top of the graded buffer layer contains a threading dislocation density ≈ 2.2 x 10^7 cm^{-2}, showing the potential of the graded buffer layer technique for reducing the density of harmful threading defects in the device structure that is typically grown above. In figure 4(b), selected area TED patterns, obtained from < 1μm areas of the sample, are shown of the InGaP graded buffer layer at various positions through the structure. Strong ½{-111} superlattice spots are observed in the first few InGaP layers whose composition is close to 50% Ga, indicating the presence of strong single variant $CuPt_B$ ordering in the alloy. The single variant nature of the ordering is a consequence of the growth on an offcut substrate, 2° towards (111)B, that is typically used in MOVPE to obtain improved quality material. As the InGaP alloy composition becomes more In-rich the maximum degree of atomic ordering possible decreases linearly to zero at InP and this is visible in the TED patterns as a gradual disappearance of the ½{111} superlattice spots as the alloy approaches the InP rich region. The XRD RSM of this sample shown in figure 4(c) shows several interesting features. First, the growth of the InGaP graded buffer layer results in the generation of a significant epitaxial layer tilt indicating an imbalance in the density of misfit dislocations that have nucleated and glided into the material on the (-111) and (1-11) planes. The second interesting point is that the sense of the layer tilt is in the opposite direction to that expected for the substrate offcut. The third interesting feature is that as the InGaP graded buffer layer composition approaches InP, eventually the sense of the overall epitaxial layer tilt begins to reverse back to that normally expected for the substrate offcut and this seems to correlate with the disappearance of the atomic ordering. This change in the epitaxial layer tilt is associated with a switch in the preferred {111} glide plane of the misfit dislocations and results in a sudden increase in the threading dislocation density above this point that is harmful to any device grown on such a buffer layer structure [22].

Table 1. Details of possible slip systems and 60° degree misfit dislocations possible in compressively strained InGaP layers grown on a (001) substrate.

Slip System	Line Direction	Burgers vector	b_{misfit}	b_{screw}	b_{tilt}	Slip plane	Type	APB in (-111) ordering
1	[110]	a/2[011]	a/4[-110]	a/4[110]	a/2[001]	(1-11)	β	Yes
2	[110]	a/2[-101]	a/4[-110]	a/4[-1-10]	a/2[001]	(1-11)	β	Yes
3	[110]	a/2[-10-1]	a/4[-110]	a/4[-1-10]	a/2[00-1]	(-111)	β	No
4	[110]	a/2[01-1]	a/4[-110]	a/4[110]	a/2[00-1]	(-111)	β	No
5	[1-10]	a/2[01-1]	a/4[110]	a/4[-110]	a/2[00-1]	(111)	α	No
6	[1-10]	a/2[10-1]	a/4[110]	a/4[1-10]	a/2[00-1]	(111)	α	Yes
7	[1-10]	a/2[101]	a/4[110]	a/4[1-10]	a/2[001]	(-1-11)	α	No
8	[1-10]	a/2[011]	a/4[110]	a/4[-110]	a/2[001]	(-1-11)	α	Yes

Introducing an offcut to the (001) substrate used for growth introduces an asymmetry in the resolved shear stress on different {111} glide planes and to the misfit component of the Burgers vector of various 60° misfit dislocations making nucleation and glide of some of the misfit dislocations more favorable. (It also introduces an asymmetry in the glide path length for the dislocations on the different {111} glide planes from the growth surface to the interface that is also important). The effect of offcut on the misfit component of the Burgers vector is illustrated in figure 5 that shows for an (001) substrate, an offcut towards (-111) results in the dislocations gliding on (-111) having a larger misfit component to their Burgers vector and hence would be expected to nucleate and glide preferentially with respect to the misfit dislocations that glide on (1-11). As these two types of dislocation have tilt components in opposite directions, an imbalance in the numbers of misfit dislocations gliding on these

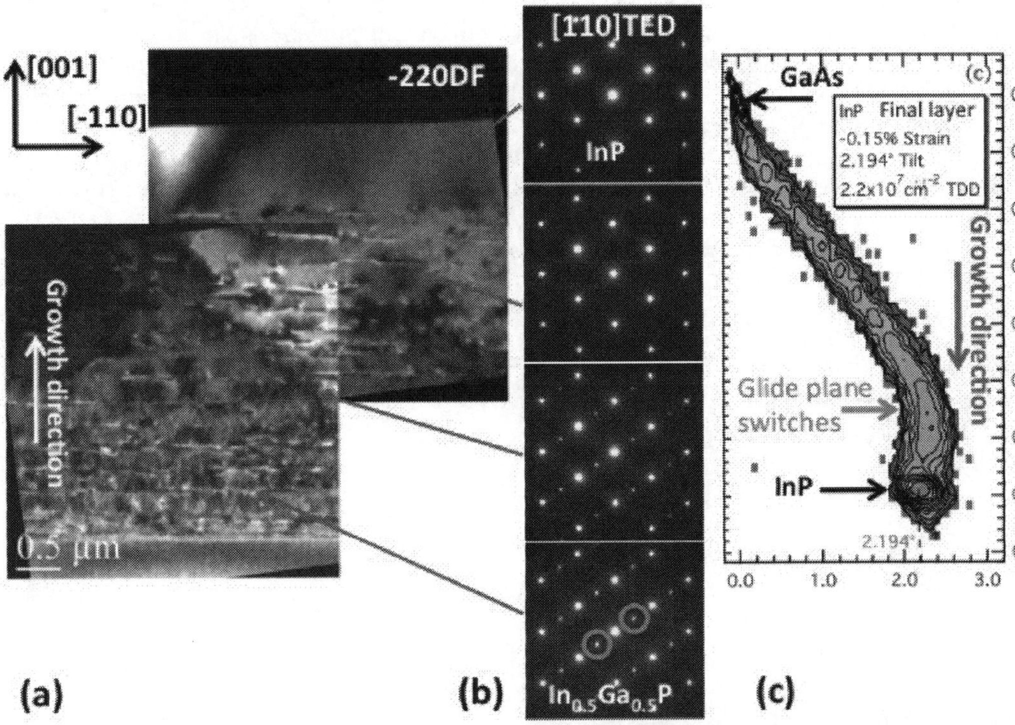

Figure 4. (a) -220 dark field (DF) image of InGaP graded buffer layer structure showing the misfit dislocation network. (b) [110] TED patterns from various parts of the graded buffer showing the absence of $CuPt_B$ atomic ordering at InGaP alloy compositions close to InP. (c) XRD RSM showing presence of unexpected direction of epitaxial layer tilt and a dislocation glide plane preference switch at an alloy composition close to InP.

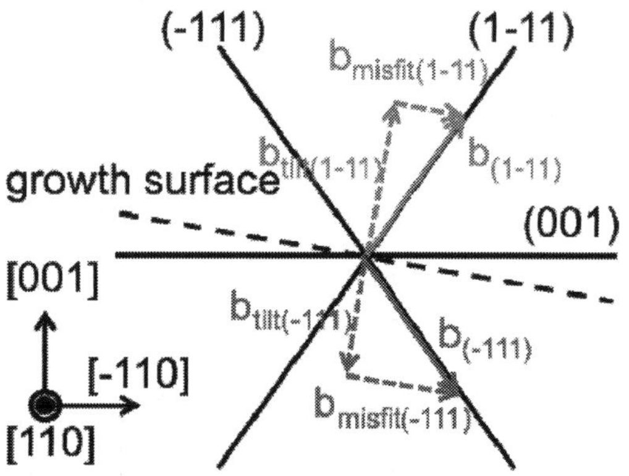

Figure 5. Diagram showing the effect of substrate offcut on the misfit component of the Burgers vector of two different 60° type dislocations gliding on (-111) and (1-11).

two different {111} planes will result in an overall epitaxial layer tilt that can be measured after growth using XRD. For the case of a compressively strained alloy layer grown on GaAs with the offcut illustrated in figure 5, the preferential nucleation and glide of misfit dislocations typically occurs on (-111) resulting in rotation of the lattice planes in the epitaxial layer towards the surface normal such that the average surface offcut becomes smaller [23–25].

However, in the case of the InGaP graded layers examined here, the substrate offcut introduces another asymmetry during the growth by preferentially selecting single variant $CuPt_B$ atomic ordering on (-111) planes, figure 6. A consequence of this is that glide of misfit dislocations on (1-11) planes leaves behind an antiphase boundary (APB) in the ordered structure whilst glide of misfit dislocations on (-111) does not, figure 6. As the $CuPt_B$ type ordering is metastable in the bulk, the formation of an APB in the single variant ordering in the crystal results in a reduction in energy and an energetic preference for nucleation and glide of misfit dislocations on (1-11) that outweighs the normal effect of the offcut. This we believe is the cause of the unusual epitaxial layer tilt behavior we observe in these graded InGaP buffer layers, figure 4. The glide plane switch occurs when the degree of ordering in the InGaP drops below a critical value as the alloy composition approaches InP resulting in a loss of the energetic preference for the nucleation and glide of misfit dislocations on (1-11) and a return to the more normally observed preference for nucleation and glide of misfit dislocations on (-111) planes for the sense of substrate offcut used. The introduction of new misfit dislocations gliding on (-111) planes naturally causes new threading dislocations resulting in the observed jump in the threading dislocation density for graded buffer layer structures grown beyond the point that the glide plane switch occurs.

To investigate this behavior further we have performed detailed transmission electron microscopy (TEM) and diffraction (TED) studies on a number of samples. Figure 7 shows ½{113} bright field (BF) and dark field (DF) images of the same area of the [110] cross-section of a single $In_{0.6}Ga_{0.4}P$ layer grown on a (001) GaAs substrate offcut 2° towards (-111). The inset TED pattern in figure 7(b) shows that this sample contained mostly domains of the (-111) variant of the atomic ordering as expected for the substrate offcut used. The superlattice dark field image of figure 7(b) exhibits contrast arising from two different types of APBs as reported previously [17, 18]. The wavy dark features rotated clockwise from the InGaP/GaAs interface are the naturally occurring

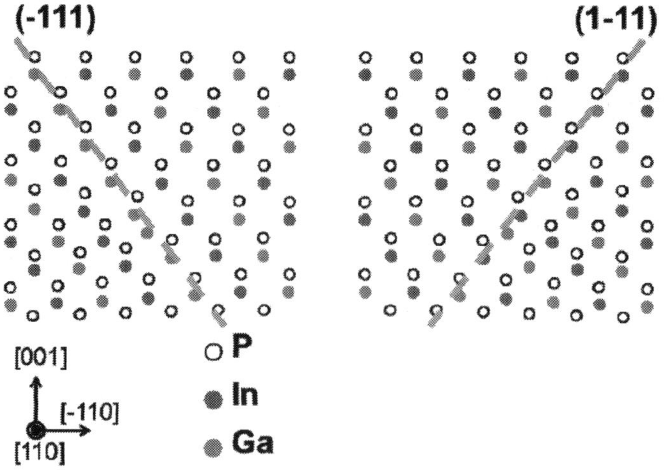

Figure 6. Schematic showing glide of 60° misfit dislocations on (-111) and (1-11) planes in (-111) ordered $In_{0.5}Ga_{0.5}P$. Glide on (1-11) planes results in an APB whilst glide on (-111) does not.

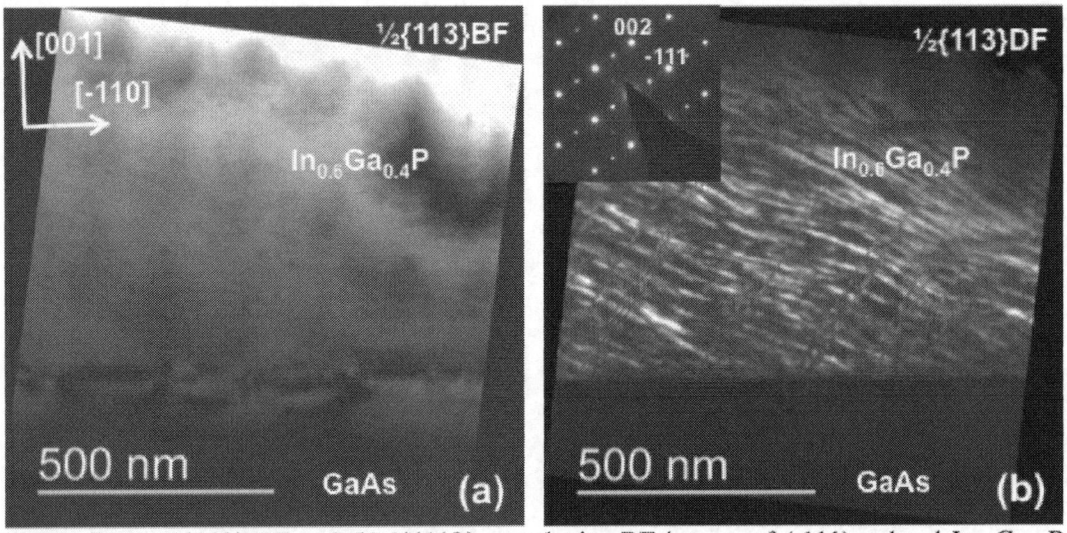

Figure 7. (a) ½{113} BF and (b) ½{113} superlattice DF images of (-111) ordered $In_{0.6}Ga_{0.4}P$ layer grown on (001) GaAs substrate offcut 2° towards (-111). Inset in (b) is the (110) TED pattern showing strong ½[111] superlattice spots arising from the ordering on (-111) planes. The dark field image of (b) shows two types of APBs as described in the text.

APDs in the ordered structure. The sharp, flat dark lines lying on (1-11) planes are APBs left behind by the glide of 60°, β-type misfit dislocations on this plane to the interface. Comparison with the bright field image of figure 7(a) reveals that all of the interfacial misfit dislocations left behind an APB in the ordered structure as observed previously by Spiecker et al in a similar sample [18]. No misfit dislocations were observed to have glided to the interface on (-111) planes consistent with the unusual direction of tilt observed in XRD studies of single variant atomically $CuPt_B$ ordered InGaP single layer and graded buffer layer samples. Further evidence of this was provided by high-resolution TEM studies of the misfit dislocations at the InGaP/GaAs interface. The vast majority of the dislocations were found to be dissociated into two partial dislocations with the stacking fault between them lying on the (1-11) glide plane, figure 8. The occurrence of dissociation of the misfit dislocations is useful as it enables identification of the glide plane for misfit dislocations even in a disordered sample. The above results are consistent with the XRD RSM obtained from this sample that revealed an epitaxial layer tilt of +0.3° indicating that glide on (1-11) planes was dominant.

In order to test whether the single variant $CuPt_B$ ordering is indeed responsible for the unusual dislocation glide plane preference and epitaxial tilt behavior observed in these samples, we have performed an experiment to remove the atomic ordering and see whether the dislocation glide plane preference switches back to that expected for the substrate offcut used. In this experiment, an Sb surfactant, known to disrupt or entirely prevent the $CuPt_B$ type atomic ordering [26, 27], was used during growth of the InGaP. The layer structure was the same as the sample illustrated in figures 7 and 8. [110] cross-section, high-resolution TEM studies of the misfit dislocations at this interface revealed the presence of misfit dislocations that had mainly glided in on (-111) planes, figure 9, indicating that after removing the single variant $CuPt_B$ atomic ordering the glide plane preference and epitaxial layer tilt returned to that normally expected for the direction of substrate offcut. Some residual ordering is visible at the InGaP/GaAs interface in figure 9 since on initiating growth of the InGaP it takes a while for the Sb surfactant surface concentration to build up to that required to eliminate the atomic ordering. The above results are consistent with the XRD RSM obtained from this sample that revealed an epitaxial layer tilt of -0.33° indicating that glide on (-111) planes was dominant.

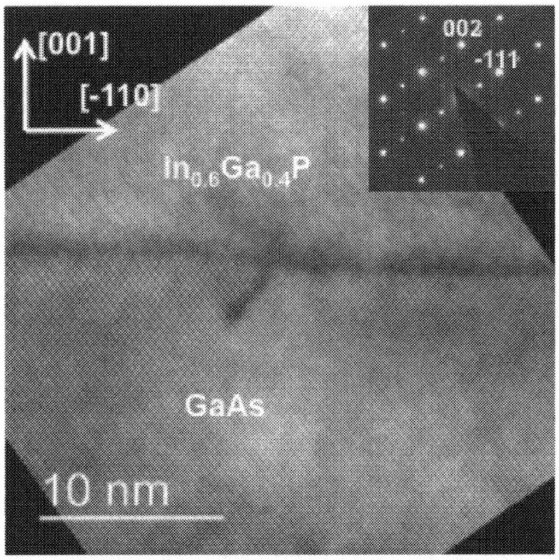

Figure 8. [110] cross-section, high-resolution TEM image showing interfacial misfit dislocation resulting from glide on (1-11) plane in a (-111) ordered InGaP layer grown on (001) GaAs substrate offcut 2° towards (-111) plane. Inset is the [110] TED pattern from the ordered InGaP layer.

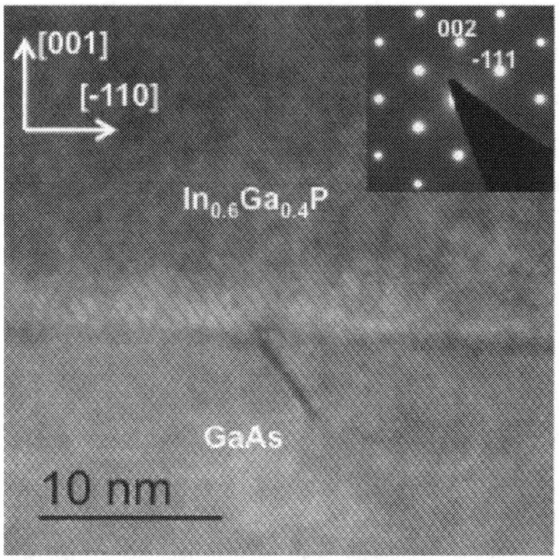

Figure 9. [110] cross-section, high-resolution TEM image showing interfacial misfit dislocation resulting from glide on (-111) plane in a random InGaP layer grown in the presence of Sb surfactant on (001) GaAs substrate offcut 2° towards (-111) plane. Inset is the [110] TED pattern from the random InGaP layer.

4. Discussion

Recent calculations indicate a substantial lowering in energy of the $CuPt_B$ InGaP ordered crystal by the introduction of an APB, giving rise to a strong preference for the glide of misfit dislocations with Burgers vectors that create an APB in this material [28]. The experimentally observed glide plane switch in the InGaP graded buffer layers occurs when the degree of atomic ordering falls below a critical value as the alloy composition approaches InP and nucleation and glide of misfit dislocations on the more normally reported (-111) plane for the substrate offcut used becomes energetically favorable and results in a substantial increase in the threading dislocation density. Although not presented in this paper, the presence of the single variant $CuPt_B$ atomic ordering also influences the glide of the α-type 60° misfit dislocations with line vectors along [1-10].

As well as affecting dislocation glide, the presence of atomic ordering will also influence the initial nucleation of misfit dislocations during growth of the graded buffer layers. As described by Marée et al [29], this is normally thought to occur at the layer surface by the nucleation of a dislocation half-loop, probably at the site of a stress concentrator such as a step bunch. The energy of a perfect dislocation half-loop E_{hl} of radius r is given by [29]

$$E_{hl} = E_l + E_s - E_\tau, \qquad (1)$$

with

$$E_l = [\mu b^2 r(1-\nu/2)/4(1-\nu)]\ln(\alpha r/b),\ E_s = r\sigma_0 b\sqrt{3},\ \text{and}\ E_\tau = \tfrac{1}{2}\pi r^2 \boldsymbol{\tau}\bullet\mathbf{b}.$$

where E_l is the energy of a line dislocation of length πr, E_s is the energy associated with the creation of a surface step, and E_τ is the strain energy released by the loop. μ is the shear modulus, ν is Poisson's ratio, σ_0 the surface tension, b the magnitude of the Burgers vector, α is the dislocation core parameter, and τ is the magnitude of the resolved shear stress in the glide plane. In the case of nucleation of a dislocation half-loop on a {111} glide plane in the ordered InGaP with a Burgers vector such that it would create an APB, this introduces an extra energy lowering term in equation (1), $E_{APB} = \tfrac{1}{2}\pi r^2 \xi$, where ξ is the energy per unit area released by the formation of the APB such that equation (1) now becomes

$$E_{hl} = E_l + E_s - E_\tau - E_{APB} \quad (2)$$

This results in an energetic preference for the nucleation at the surface for misfit dislocation loops that generate an APB in the (-111) atomic ordering by reducing the critical radius and maximum energy required for nucleation of such half-loops over half-loops that do not generate an APB in the ordered structure.

In order to avoid a harmful increase in threading dislocation density associated with the observed glide plane switch during growth of the InGaP graded buffer layer, re-engineering of the buffer layer structure is required [30]. Further understanding of misfit dislocation nucleation and glide in these complex materials may enable control of the Burgers vector of the misfit dislocations to be achieved so as to enhance dislocation glide and promote favorable dislocation interactions to occur, leading to a reduced threading dislocation density in the final device.

5. Conclusions

The occurrence of single variant $CuPt_B$ atomic ordering during MOVPE growth of InGaP graded buffer layer structures on offcut (001) substrates has a profound influence on strain relaxation mechanisms in IMM solar cell structures. A strong preference is observed for the nucleation and glide of misfit dislocations that generate an APB in the ordered crystal, resulting in an epitaxial layer tilt in the opposite sense to that typically observed for the direction of substrate offcut used. As the alloy composition approaches InP, the reduction of atomic ordering leads to a switch in the glide plane preference back to that normally observed and the generation of new threading dislocations that harms device performance. The knowledge gained from these studies will enable the engineering of new graded buffer layer architectures resulting in higher efficiency devices.

Acknowledgements

The authors are pleased to thank W J Olavarria for growth of samples; K M Jones for FIB preparation of TEM samples; J M Olson, J Simon and M A Steiner for valuable discussions; and D J Friedman and M M Al-Jassim for technical leadership. This work is supported by the U.S. Department of Energy under Contract No. DE-AC36-08-GO28308 with the National Renewable Energy Laboratory.

References

[1] Sabnis V, Yuen H and Wiemer M 2012 *AIP Conf. Proc.* **1477** 14

[2] Wanlass M W, Ahrenkiel S P, Ahrenkiel R K, Albin D S, Carapella J J, Duda A, Geisz J F, Kurtz S, Moriarty T, Wehrer R J and Wernsman B 2005 *Proc. of the 31st IEEE PV Specialists Conference (Lake Buena Vista FL)* (Piscataway NJ: IEEE) p530

[3] King R R, Law D C, Edmondson K M, Fetzer C M, Kinsey G S, Yoon H, Sherif R A and Karam N H 2007 *Appl. Phys. Lett.* **90** 183516

[4] Geisz J F, Kurz S, Wanlass M W, Ward J S, Duda A, Friedman D J, Olson J M, McMahon W E, Moriarty T E and Kiehl J T 2007 *Appl. Phys. Lett.* **91** 023502

[5] Geisz J F, Friedman D J, Ward J S, Duda A, Olavarria W J, Moriarty T E, Kiehl J T, Romero M J, Norman A G and Jones K M 2008 *Appl. Phys. Lett.* **93** 123505

[6] Guter W, Schöne J, Philipps S P, Steiner M, Siefer G, Wekkeli A, Weiser E, Oliva E, Bett A W and Dimroth F 2009 *Appl. Phys. Lett.* **94** 223504

[7] Chiu P, Wojtczuk S, Zhang X, Harris C, Pulver D and Timmons M *Proc. of the 37th IEEE PV Specialists Conference (Seattle WA)* (Piscataway NJ: IEEE) p771

[8] Dimroth F and Kurtz S 2007 *MRS Bulletin* **32** 230

[9] Bremner S P, Levy M Y and Honsberg C B 2008 *Prog. Photovolt: Res. Appl.* **16** 225

[10] Geisz J F, Duda A, France R M, Friedman D J, Garcia I, Olavarria W, Olson J M, Steiner M A, Ward J S and Young M 2012 *AIP Conf. Proc.* **1477** 44

[11] Friedman D J 2010 *Current Opinion in Solid State and Mater. Sci.* **14** 131

[12] Beanland R, Dunstan D J and Goodhew P J 1996 *Advances in Physics* **45** 87

[13] Hirsch P B 1985 *Mats. Sci. Tech.* **1** 666

[14] Yonenaga I 1997 *J. de Physique III* **7** 1435
[15] Olsen J A, Hu E L, Lee S R, Fritz I J, Howard A J, Hammons B E and Tsao J Y 1995 *J. Appl. Phys.* **79** 3578
[16] Gomyo A, Suzuki T and Iijima S 1988 *Phys. Rev. Lett.* **60** 2645
[17] Baxter C S, Stobbs W M and Gibbings C J, 1993 *Phil. Mag. Lett.* **67** 59
[18] Spiecker E, Seibt M, Schröter W, Winterhoff R and Scholz F 2002 *Appl. Surf. Sci.* **188** 61
[19] Froyen S and Zunger A 1991 *Phys. Rev. Lett.* **66** 2132
[20] Bernard J E, Froyen S and Zunger A 1991 *Phys. Rev. B* **44** 11178
[21] Philips B A, Norman A G, Seong T-Y, Mahajan S, Booker G R, Skowronski M, Harbison J P and Keramidas V G 1994 *J. Cryst. Growth* **140** 249
[22] France R M, McMahon W E, Norman A G, Geisz J F and Romero M J 2012 *J. Appl. Phys.* **112** 023520
[23] Ayers J E, Ghandi S K and Schowalter LJ 1991 *J. Cryst. Growth* **113** 430
[24] Mooney P M, LeGoues F K, Tersoff J and Chu J O 1993 *J. Appl. Phys.* **75** 3968
[25] LeGoues F K, Mooney P M and Chu J O 1993 *Appl. Phys. Lett.* **62** 140
[26] Shurtleff J K, Lee R T, Fetzer C M and Stringfellow G B 1999 *Appl. Phys. Lett.* **75** 1914
[27] Stringfellow G B, Lee R T, Fetzer C M, Shurtleff J K, Hsu Y U, Jun S W, Lee S and Seong T-Y 2000 *J. Electron. Mater.* **29** 134
[28] McMahon W E, Kang J, France R M, Norman A G, Friedman D J and Wei S H 2013 *Appl. Phys. Lett.*, submitted for publication
[29] Marée P M J, Barbour J C, van der Veen J F, Kavanagh K L, Bulle-Lieuwma C W T and Viegers M P A 1987 *J. Appl. Phys.* **62** 4413
[30] France R M, García I, McMahon W E, Norman A G, Simon J, Geisz J F, Friedman D J and Romero M J 2013 *IEEE J. Photovoltaics*, submitted for publication

Near-infrared nano-spectroscopy and emission energy control of semiconductor quantum dots using a phase-change material

Nobuhiro Tsumori, [1] Motoki Takahashi, [1] Nurrul Syafawati Humam, [1] Phillipe Regreny, [2] Michel Gendry[2] and Toshiharu Saiki [1]

[1] Graduate School of Science and Technology, Keio University, 3-14-1 Hiyoshi, Kohoku-ku, Yokohama-shi, Kanagawa 223-8522, Japan
[2] Université de Lyon, Institut des Nanotechnologies de Lyon, UMR-CNRS 5270, Ecole Centrale de Lyon, 69134 Ecully, France

Email: saiki@elec.keio.ac.jp

Abstract. We have proposed a method to achieve near-field imaging spectroscopy of single semiconductor quantum dots with high sensitivity by using an optical mask layer of a phase-change material. Sequential formation and elimination of an amorphous aperture allows imaging spectroscopy with high spatial resolution and high collection efficiency. We present numerical simulation and experimental result that show the effectiveness of this technique. Inspired by this optical mask effect, a new approach which can precisely control the emission energy of semiconductor quantum dots has been proposed. This method uses the volume expansion of a phase change material upon amorphization, which allows reversible emission energy tuning of quantum dots. A photoluminescence spectroscopy of single quantum dots and simulation were conducted to demonstrate and further explore the feasibility of this method.

1. Introduction

Semiconductor quantum dots (QDs) have shown great promise as efficient single photon emitters and entangled photon sources, making them attractive for quantum communication and quantum information processing applications. Self-assembled InAs QDs on InP substrates are promising as near-infrared (NIR) single photon and entangled photon emitters [1-3]. In order to clarify and control the optical properties of QDs for telecommunication devices, photoluminescence (PL) spectroscopy studies of single QDs with high spatial resolution at NIR wavelength is necessary. The most useful technique to attain this is by using near-field scanning optical miscroscopy (NSOM) [4-6]. The advantage of NSOM is its ability to perform imaging spectroscopy of single QDs with a spatial resolution determined by the aperture size without the need to process the sample. On the other hand, NSOM has a lower PL collection efficiency at NIR wavelength than at visible wavelength. The transmission efficiency of NIR radiation through the aperture decreases inversely with about the sixth power of the wavelength [7]. This problem inhibits NIR-PL spectroscopy based on NSOM to be practically realized. Therefore, we proposed a method to overcome the low NIR-PL spectroscopy by using a nanoaperture on an optical mask layer of phase-change material (PCM).

PCM mask effect has also the potential to be applied in emission energy control of QDs. One of the main problems for realization of quantum communication applications is the precise control of the emission energy of QDs. The three main existing approaches to tune emission energy include externally applied magnetic field [8], electric field [9] and mechanical strain [10, 11]. The magnetic field method is believed to be the most promising approach in tuning energy over a wide range; however, large equipment is required. Electric field method allows one to tune emission energy over a large range, however, significant decrease in excitonic emission intensity was observed with a high electric field. Mechanical stress method can in principle be used to tune energy; however, the energy tuning range is still limited. In addition, another drawback of the current techniques is they have not shown precise energy control of individual QDs. Scanning probe miscroscopy (SPM) allows a spatially local perturbation with nanoscale resolution [11-13]; however, this technique requires a feedback control mechanism, making it mechanically unstable and impractical. As a modified method of nanoindentation with a probe tip, we proposed a new approach to precisely control the emission energy of QDs by applying local strain using volume expansion of phase-change material. In this study, we experimentally demonstrate and further explore the feasibility of our proposed method.

This paper is organized as follows. Section 2 presents the principles, mainly focusing on the properties of phase-change material. Section 3 shows the sample and measurement setup used in experimental studies. Section 4 demonstrates the numerical simulation and experimental results for near-infrared nano-spectroscopy of QDs using a phase-change mask layer. We show that an amorphous nanoaperture allows imaging spectroscopy with high spatial resolution and high collection efficiency. Section 5 focuses on the experimental and simulation results of emission energy control of QDs using volume expansion of a phase-change material. We show that this approach can serve as a method which allows reversible energy tuning of individual QDs.

2. Principles

A phase-change material (PCM) is a substance that is widely used for rewritable data storage applications. A PCM can be transformed from crystalline phase to amorphous phase, and vice versa, by laser pulse irradiation. When a short pulse of a high intensity laser beam heats the crystalline region above the melting point, it is melted and rapidly quenched into the amorphous phase. Longer and weaker pulse heating to a temperature above the glass transition temperature leads to the transition of amorphous phase to crystalline state. The large optical contrast between amorphous state and crystalline state is advantageous for recording media applications. One of the most prominent examples of a PCM is GeSbTe (GST).

The idea of PCM mask layer has been implemented by Tominaga *et al.* in super-resolution near field structure disk for high-density optical data storage [14]. According to reference [15], $Ge_2Sb_2Te_5$ has high optical contrast at excitation wavelengths of 600nm-900nm which allows aperture-size dependent resolution, and also has low extinction coefficient at telecommunication wavelength of 1550nm which results in high collection efficiency. These features are essential in measuring photoluminescence of quantum dots. Figure 1 illustrates the schematic diagram of nanospectroscopy using a PCM mask.

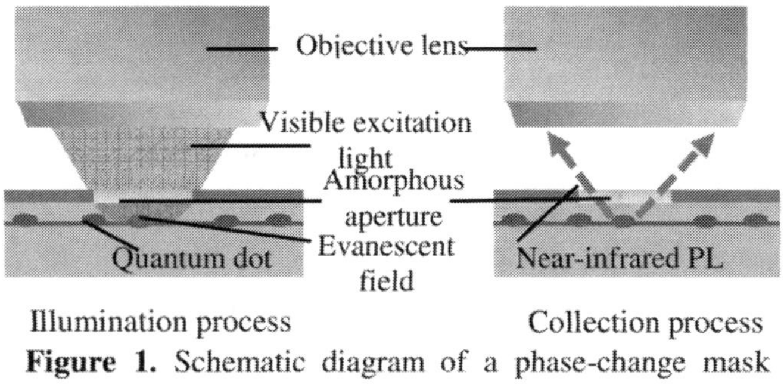

Figure 1. Schematic diagram of a phase-change mask layer above QDs.

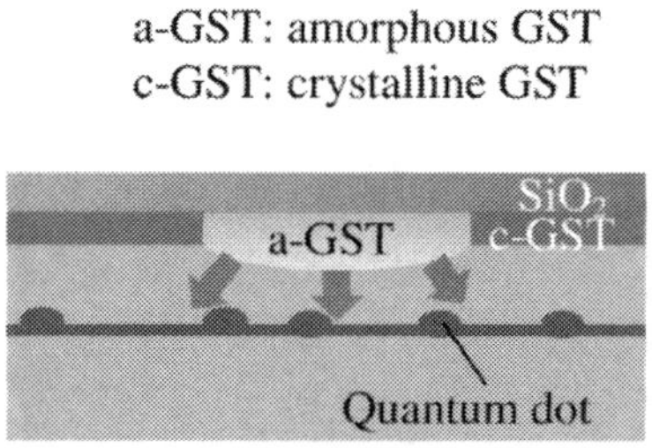

Figure 2. Schematic illustration of strain induced QDs using volume expansion of a GST layer.

A focused laser beam forms a local amorphous region in the crystalline GST film. Approximately 10% volume expansion accompanies the transformation from crystalline phase to amorphous phase. Following the theoretical prediction that volume of PCM should increase upon amorphization, stress can be induced on QDs and cause energy shift. This stress is released when amorphous phase is transformed into crystalline phase using laser annealing. The magnitude of energy shift can be controlled because the amorphized volume is proportional to the number of laser pulses during irradiation. Figure 2 shows the schematic diagram of compressive strain to individual QDs by local amorphization of a GST layer.

3. Experimental details

3.1 Sample growth

Low density self-assembled double-capped InAs/InP QDs grown by solid-source molecular beam epitaxy (SSMBE) were used as sample for our experimental studies. 250nm thick InP buffer layer was deposited on semi-insulating InP(001) substrate at 600°C. The temperature was then reduced to 500°C. Self-assembled InAs QDs were formed on InP surface by using double capping method. 1.4ML of InAs was deposited on InP surface. Then, a thin InP layer (2nm), also known as first cap layer, was grown on the QDs. After that, 120s growth interruption occurred under P_2 flow, where As/P exchange occurred. Temperature was increased to 600°C and the structure was then capped with another InP layer (100nm), also known as InP second cap. Finally, a $Ge_2Sb_2Te_5$ (GST) film layer, was deposited directly on the whole structure to a thickness of 40nm, followed by a 50nm thick SiO_2 layer by sputtering. The SiO_2 layer acts as a protective layer.

3.2 Micro-photoluminescence measurement setup

We used micro-photoluminescence (μPL) measurement to investigate the emission properties of QDs in the experiments. All measurements were conducted at low temperature. The sample was first placed on the cold finger of a helium flow cryostat, ensuring temperature below 10K. During the PL measurement, QDs were excited using a He-Ne laser (λ=632.8nm) or a CW laser (λ=662nm) through a spot, and PL signal was detected by a spectrometer and a liquid-nitrogen cooled InGaAs diode array.

4. Near-infrared nano-spectroscopy of QDs using a phase-change mask layer

4.1 Numerical simulation

The performance of the PCM mask method was evaluated by conducting simulation focusing on the spatial resolution aspect. We calculated three-dimensional electromagnetic field distributions in the vicinity of the amorphous aperture by employing finite-difference time-domain (FDTD) method. We simulated the illumination profile obtained using a focused visible light beam (λ=633 nm). Figure 3(a) presents a schematic diagram of the model. In this model, a crystalline $Ge_{10}Sb_2Te_{13}$ film (c-GST) with 50nm thickness was coated as a mask layer on a bulk InP substrate (refractive index n=3.2). $Ge_{10}Sb_2Te_{13}$ has a refractive index of n=3.16 + 4.24i at wavelength, λ=633 nm. At the centre of the mask layer, an amorphous aperture (a-GST) with a diameter of 300 nm was formed and the aperture was illuminated using x-polarized focused light. The simulation volume was 4×4×4 μm^3.

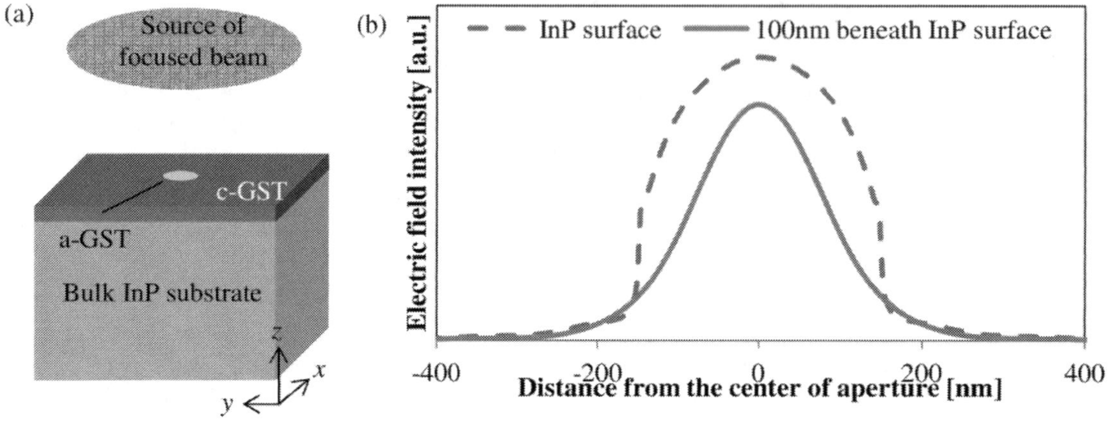

Figure 3.(a) Schematic of calculation model for illumination using a focused beam,(b) Cross-sectional profiles of electric field intensity (xy -plane) at the InP surface and at 100nm beneath InP surface

Figure 3(b) shows the cross-sectional profiles of electric field intensity in the xy-plane calculated at two points; at InP surface and at depth of 100nm beneath the InP surface. By calculating the full widths at half maximum (FWHM) of each profile, the illumination area at a depth of 100nm below surface (FWHM=190nm) is significantly smaller than the illumination area just below the aperture (FWHM=280nm). This is attributed to the large refractive index of InP which provides a higher spatial resolution than the diameter of aperture.

4.2 Experiment

An amorphous aperture was formed by irradiating the GeSbTe mask with 10 femtosecond laser pulses (λ=800 nm) at a fluence of 27 mJ/cm^2 through a microscope objective (NA=0.8). The diameter of the amorphous region was approximately 300 nm based on confocal laser scanning microscopy measurement. A laser diode (λ=662 nm) was used to recrystallize the amorphous aperture by thermal annealing. During PL measurement, a He–Ne laser was focused onto the amorphous aperture through a microscope objective (NA=0.65) at excitation wavelength 632.8nm. PL spectrum was measured using spectromenter and a liquid-nitrogen-cooled InGaAs photodiode array.

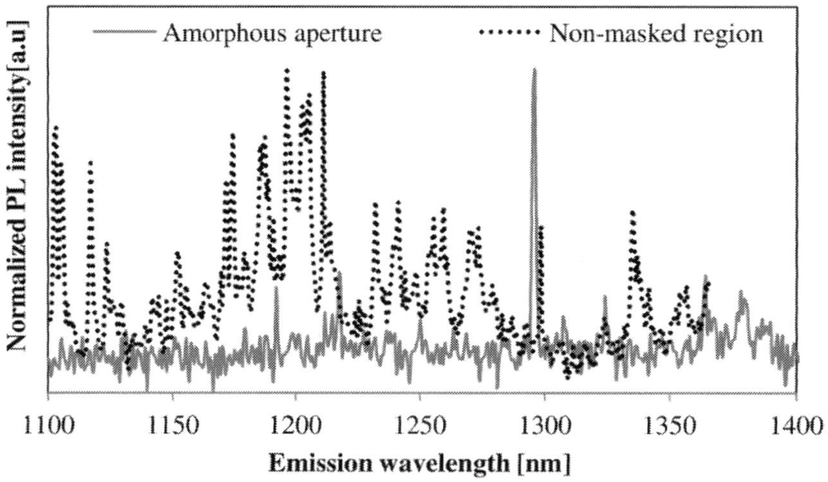

Figure 4. PL spectra of InAs QDs obtained through the amorphous aperture (solid line) and non-masked region (dotted line).

Figure 4 compares the PL spectra of InAs QDs through the amorphous aperture (solid line) and non-masked region (dotted line). In the region without mask layer, one can observe a spectrum with many peaks, which corresponds to an ensemble of QDs. Based on the spatial resolution of measurement and the number of peaks, the QD density of used sample was estimated to be $3.3\times 10^9/cm^2$. On the contrary, the spectrum obtained through the amorphous aperture shows only one peak as a result of large reduction in area of observation. The number of peaks obtained agrees well with the estimated QD density and the size of amorphous aperture. This result demonstrates that the amorphous mask can serve as an aperture which allows high spatial resolution without degrading the collection efficiency of PL signal.

4. Local control of emission energy of QDs using volume expansion of a phase-change material

4.1 Experiment

Nanoscale amorphization was performed by irradiation of a subnanosecond Q-switched solid state laser. Here, a nanosecond Q-switched laser was focused through a microscope objective lense (N.A=0.65) at excitation wavelength 532nm onto the sample surface. During the PL measurement, QDs were excited using a CW laser at excitation wavelength 662nm through the amorphous aperture, and PL signal was detected by a spectrometer and a liquid-nitrogen cooled InGaAs diode array. For local crystallization, a semiconductor nanosecond pulsed laser (λ=662 nm) was used. Crystallization and amorphization processes were repeated and changes in PL peak energy shift were observed. An amorphous region with a diameter of approximately 1 μm was formed under the conditions of the laser wavelength (532 nm) and microscope objective (NA=0.65).

Figure 5 shows the PL spectra of QDs obtained after crystallization and re-amorphization of GST film. The GST film is in the amorphous state at point A and point C, and crystalline state at point B. Redshift was observed after amorphization, and its recovery to its original position was observed after recrystallization. This indicates that this method allows reversible energy tuning of the emission from individual QDs. The maximum energy shift obtained during experiment was approximately 2.3meV. A wider energy shift range is expected by manipulating the thickness of GST film.

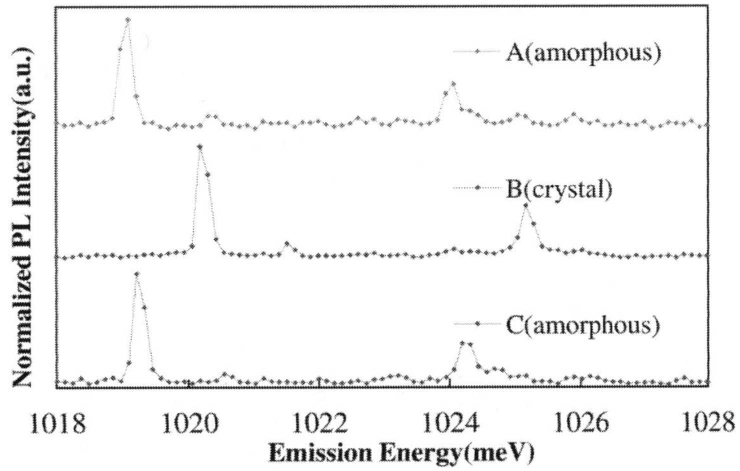

Figure 5. PL spectra of QDs obtained as a result of local amorphization and crystallization of GST film

4.2 Simulation

We developed a two-dimensional finite element (FE) model to investigate the stress distribution and energy shift distribution in the sample. Since the proposed method is considered as a modified approach to nanoindentation with a probe tip, we conducted an indentation test using ANSYS 14.0 simulation software. The geometry of the model is illustrated in Figure 6.

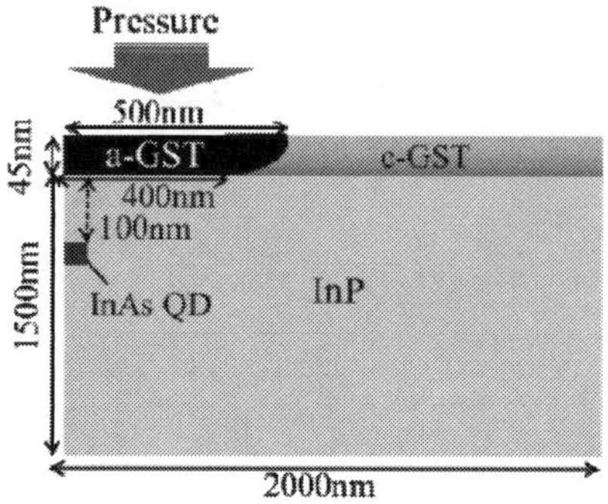

Figure 6. Geometry of model used in calculation

Indentation was simulated assuming the apex of the indenter applies compressive stress on the top of InP surface. The size of the indenter is selected based on the amorphous mark size obtained in experiment. A quantum dot (InAs QD) with width of 20nm and height of 1nm was placed 100nm below the surface of the sample along *y*-axis. These values are taken based on the microscopic observation of the QD sample used in experiment. Young's modulus and Poisson's ratio of InP, InAs and $Ge_{10}Sb_2Te_{13}$ are listed in Table 1. According to our previous experimental results, amorphization of phase change material caused approximately 2meV redshift of PL peak energy. The indentation force was chosen to generate a 2meV redshift at 100nm beneath the top of InP surface.

Table 1. List of material parameters used in calculation

material	InP	InAs	$Ge_{10}Sb_2Te_{13}$ (amorphous)	$Ge_{10}Sb_2Te_{13}$ (crystal)
Young's modulus (Pa)	6.10×10^{10}	5.14×10^{10}	2.02×10^{10}	3.00×10^{10}
Poisson's ratio	0.30	0.36	0.33	0.30

The energy shift was calculated by using the following equation [16], which provides the shift of QD emission energy induced by [001] compression.

$$\Delta E_{e-hh} = \frac{a}{Y}(1-2\nu)3\sigma^h + \frac{2b}{Y}(1+\nu)\frac{3}{2}\sigma^u_{zz} \tag{1}$$

where Y and ν are Young's modulus and Poisson's ratio respectively. $\sigma^h = \frac{1}{3}(\sigma_{xx} + \sigma_{yy} + \sigma_{zz})$ and $\sigma^u_{zz} = \sigma_{zz} - \sigma^h$ correspond to mean stress components and deviatoric stress components respectively. The value of a and b is obtained from Reference [17]. The first term in Equation (1) shows the energy shift due to hydrostatic stress, while the second term in Equation (1) corresponds to the energy shift due to uniaxial stress.

Figure 7 shows the stress distribution in the model obtained through finite element modelling (FEM) simulation. In this model, the indentation force is approximately 100MPa and the stress applied on the QD is at least 50MPa. The energy shift distribution in our model obtained through calculation is illustrated in Figure 8. The energy shift is negative (redshift) at the shallow area beneath the flat part of indenter. On the contrary, the energy shift is positive (blueshift) beneath the edge region of indenter. In previous experiments, we obtained both redshift and blueshift after a local amorphization of GST film. The ratio of redshift and blueshift obtained was 11:2. This simulation result agrees well with our previous experimental results which show both redshift and blueshift as a result of compressive stress applied by volume expansion of phase change material.

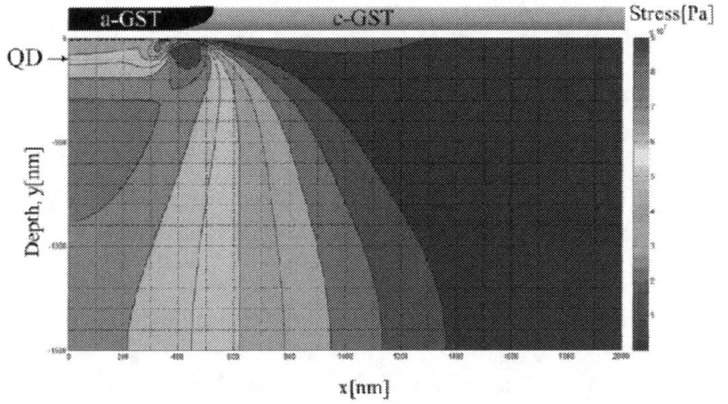

Figure 7. Stress distribution in model.

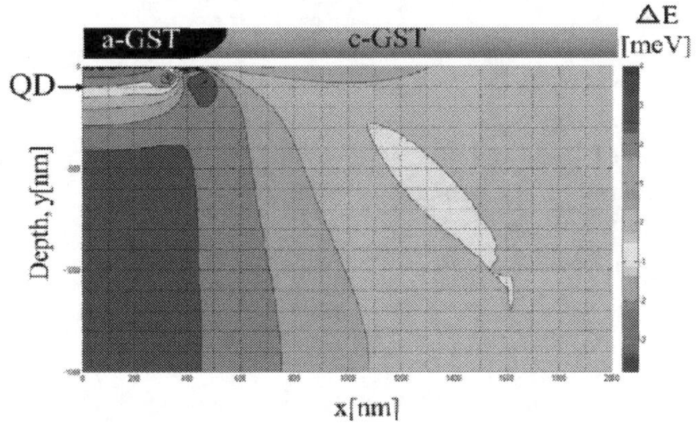

Figure 8. Energy shift distribution in model.

5. Conclusion

We have demonstrated the feasibility of near-field imaging spectroscopy with a phase-change mask layer by performing numerical simulation and PL measurements of QDs. We report that the amorphous mark can act as an aperture; reducing the observation area without degrading the collection efficiency of PL signal. We have also experimentally demonstrated the ability of our method to tune the energy of emission from QDs. Significant shifts in emission energy of QDs indicate that effective emission energy control can be realized through this method. We also showed that this technique allows reversible energy tuning. These results can be used for imaging spectroscopy and energy tuning purposes which is crucial for realizing quantum communication applications.

Acknowledgements

We acknowledge the Ministry of Education, Culture, Science and Technology Japan for financial support; Grant-in-Aid for Scientific Research(B). We would like to thank Keiichiro Yusu for depositing GeSbTe film on the QDs.

References

[1] Takemoto K, Nambu Y, Miyazawa T, Wakui K, Hirose S, Usuki T, Takatsu M, Yokoyama N, Yoshino K, Tomita A, Yorozu S, Sakuma Y and Arakawa Y 2010 *Appl. Phys. Express* **3** 092802

[2] Miyazawa T, Takemoto K, Sakuma Y, Hirose S, Usuki T, Yokoyama N, Takatsu M and Arakawa Y 2006 *Jpn. J. Appl. Phys.* **44** L620

[3] Bimberg D, Stock E, Lochmann A, Schliwa A, Tofflinger JA, Unrau W, Munnix M, Rodt S, Haisler VA, Toropov AI, Bakarov A and Kalagin AK 2009 *IEEE Photonics J.* **1** 58

[4] Saiki T, Nishi K and Ohtsu M 1998 *Jpn. J. Appl. Phys., Part 1* **37** 1638

[5] Matsuda K, Saiki T, Nomura S, Mihara M and Aoyagi Y 2002 *Appl. Phys. Lett.* **81** 2291

[6] Sugimoto Y, Tsumori N, Saiki T and Nomura S 2009 *Opt. Rev.* **16** 269

[7] Tsumori N, Takahashi M, Sakuma Y and Saiki T 2011 *Appl. Opt.* **50** 5710

[8] Stevenson RM, Young RJ, See P, Gevaux DG, Cooper K, Atkinson P, Farrer I, Ritchie DA and Shields AJ 2006 *Phys. Rev. B* **73** 033306

[9] Kowalik K, Krebs O, Lemaitre A, Laurent S, Senellart P and Voison P 2005 *Appl. Phys. Lett.* **86** 041907

[10] Seidl S, Kroner M, Hogele A, Karrai K, Warburton RJ, Badolato A and Petroff PM 2006 *Appl. Phys. Lett.* **88** 203113

[11] Robinson HD, Muller MG, Goldberg BB and Merz JL 1998 *Appl. Phys. Lett.* **72** 2081

[12] Bhallamudi VP, Berger AJ, Labanowski DE, Stroud D and Hammel PC 2012 *J. Appl. Phys.* **111** 013902

[13] Mei E and Higgins DA 1998 *J. Phys. Chem. A* **102** 7758-7563

[14] Tominaga J, Suji H, Sato A, Nakano T, Fukaya T and Atoda N 1998 *Jpn. J. Appl. Phys.* **37** L1323-L1325

[15] Strand D, Tsu DV, Miller R, Hennessey M and Jablonski D 2006, contributed paper presented at European Phase Change and Ovonics

[16] Mintairov AM, Sun K, Merz JL, Li C, Vlasov AS, Vinokurov DA, Kovalenkov OV, Tokranov V and Oktyabrsky S 2004 *Phys. Rev B* **69** 155306

[17] Priester C, Allan G and Lannoo M 1988 *Phys. Rev. B* **37** 8519

Journal of Physics: Conference Series **471** (2013) 012008 doi:10.1088/1742-6596/471/1/012008

Defect Formation and Strain Relaxation in graded GaPAs/GaAs, GaNAs/GaAs and GaInNAs/Ge Buffer Systems for high-efficiency Solar Cells

J Schöne[1,2], E Spiecker[1], F Dimroth[2], A W Bett[2], W Jäger[1]

[1] Institute for Materials Science, Christian-Albrechts-University, Kaiserstrasse 2, 24143 Kiel, Germany

[2] Fraunhofer Institute for Solar Energy Systems ISE, Heidenhofstrasse 2, 79110 Freiburg, Germany

E-mail: wolfgang.jaeger@tf.uni-kiel.de

Abstract. Transmission electron microscopy of cross-section specimens and high-resolution X-ray diffraction analyses have been applied to investigate the formation of defects and the relaxation of layer strain in step-graded GaP_xAs_{1-x} and GaN_yAs_{1-y} buffer layer systems grown by metal-organic vapour phase epitaxy on GaAs (001) substrates with 6° miscut towards (111)A. The investigations have been complemented by characterization of the layer surfaces employing optical microscopy. The comparison of the different buffer concepts reveals characteristic differences in the formation of defects and in the relaxation of tensile layer strain. For GaPAs layers dislocations and microtwins form, releasing the major part of the tensile misfit strain. In contrast, for GaNAs dislocations and microtwins are largely absent, at least in the upper part of the buffer structure, and microcracks are generated. Consequently, during subsequent growth of layers with tensile strain, strain relaxation and defect formation can be effectively hindered by introducing intermediate GaN_yAs_{1-y} layers with concentrations $y > 2\ \%$ into a $GaAs_{1-x}P_x$ buffer structure [1]. A similar concept can be used for layer systems with compressive strain, however, modified by using layers of differing alloy composition. The use of dilute nitride layers appears to offer a new concept for engineering defect distributions and layer strain in lattice-mismatched compound semiconductor layer structures. Such concepts are of particular interest not only but especially also for applications in high-efficiency III-V solar cells.

1. Introduction

For many technological components of the semiconductor industry, heteroepitaxial buffer layer systems with well engineered defect distributions and distributions of layer strain are essential for determining and optimizing device properties. Some examples are high-efficiency lattice-mismatched multi-junction solar cells based on III-V compound semiconductors [2], optoelectronic integrated circuits on Si [3], and vertical cavity surface emitting lasers [4].

The key issue in heteroepitaxial layer growth for device components is the control of the layer strains and of the type and distribution of potentially formed defects which may have detrimental effects on the electronic properties and on the long-term stability. For instance, the alloying of III-V compound semiconductors with defined concentrations of N allows engineering the band gap and lattice constant. The use of such dilute nitride compounds, like GaNAs and GaNP, for the epitaxial growth of buffer layers is particularly challenging under conditions of lattice-mismatch since these alloys show plastic properties which are drastically different from those of conventional ternary alloys, like GaPAs or GaInAs [5]. Already small concentrations of nitrogen may lead to considerable alloy hardening thus aggravating the formation and glide of dislocations [5-8]. High-quality strain-relaxed layers are difficult to obtain [7,8,10,11]. For these epitaxial layers, the strain relaxation is consequently not well described any more by equilibrium models, like the classical model by Matthews and Blakeslee [9].

While the effect of hardening of dilute nitride semiconductor compounds has mainly been recognized as a problem we show in this paper that the phenomenon can actually be used as a concept for controlling defect formation and strain relaxation in lattice-mismatched hetero-epitaxy. From comparing defect formation and strain relaxation in step-graded GaP_xAs_{1-x} and GaN_yAs_{1-y} buffer layer systems composed of similar layer sequences we observe that characteristic differences in the formation of defects and in the strain relaxation of layers with tensile strain occur. The use of layers of ternary compounds GaN_yAs_{1-y} with N-concentrations that exceed a threshold value y of about 2% are effective in suppressing the formation of dislocations. Such layers are thus beneficial for layer epitaxy aiming at a control of layer strain and of defect distributions.

2. Experimental

Two step-graded buffer structures, a GaP_xAs_{1-x} buffer layer system (buffer A) and a GaN_yAs_{1-y} buffer layer system (buffer B), were grown by metal-organic vapour phase epitaxy (MOVPE) in an Aixtron AIX2600-G3 reactor on vicinal GaAs (001) substrates with 6° miscut towards (111)A [12]. Before buffer growth, a GaAs nucleation layer with a thickness of around 350 nm was deposited. Precursors for the phosphor layer growth were trimethylgallium (TMGa), arsine (AsH_3) and phosphine (PH_3). For the nitride buffer layer growth trimethylgallium (TMGa), tertiarybutylarsine (TBAs) and dimethylhydrazine (DMHy) were used. Figure 1 schematically gives the nominal data about the two systems. Both, the GaP_xAs_{1-x}-system and the GaN_yAs_{1-y}-system consist of eight layers for which the P- or N-content, respectively, is increased in fixed concentration steps for subsequent layers. In the following we call these systems step-graded buffer layer systems. In order to improve strain relaxation, the layer system is capped with 2 layers, an individual layer whose P- or N-concentration, respectively, deviates from the grading sequence towards a higher concentration, and a layer repeating the P- or N-concentration, respectively, of the second last layer grown.

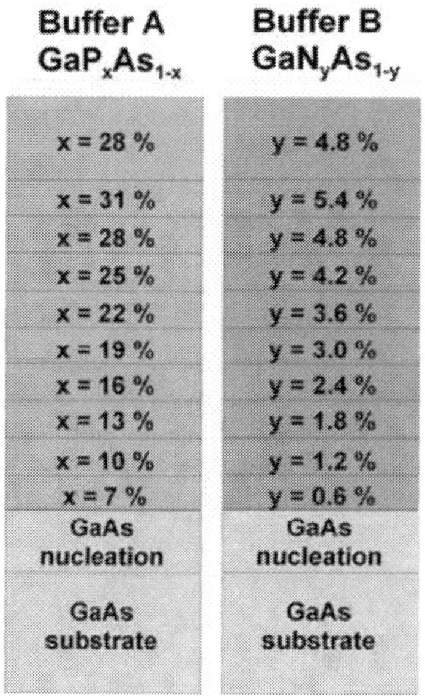

Figure 1: Nominal buffer layer compositions x and y for the GaP_xAs_{1-x}/GaAs buffer system A (left) and the GaN_yAs_{1-y}/GaAs buffer system B (right).

The respective growth temperatures, the reactor pressures and the V/III flux ratios were 580 °C, 70 mbar and ~ 17 for the GaP_xAs_{1-x} layers, and 573 °C, 70 mbar and ~ 3 for the GaN_yAs_{1-y} buffer layers. The gradings of the layer concentrations were chosen such that the two structures possess similar relative misfit profiles. The P- and N- concentrations in the topmost buffer layers correspond to a (negative) lattice misfit of ~ -1 % with respect to the substrate (estimated assuming Vegard's law). The individual layers of the graded P- and N-containing structures possess nominal thicknesses of 170 nm and 200 nm, respectively. Only the topmost layers for buffers A and B are thicker, namely 420 nm and 500 nm.

A third set of compressive strained step-graded buffer structures was grown by MOVPE on a Ge-substrate with 6° miscut towards (111). The buffer contained $GaIn_zAs_{1-z}$ layers with and without integrated GaInNAs intermediate layer. The individual layers had a nominal thickness of 200 nm. The GaInNAs intermediate layer was grown lattice matched onto the buffer layer. Trimethylindium (TMIn) was used as indium precursor.

The strain states of the individual layers of the buffer systems were determined from reciprocal space maps (RSM) measured by a Philips X'Pert PRO high-resolution X-ray diffractometry (XRD) system, i.e. by measuring the diffracted intensity distributions around the (004) reflections and around (2-24) and (224) reflections and to deduce the strain of individual layers from the corresponding lattice distances. The intensity distributions around (004) reflections are sensitive to layer strain whereas the intensity distributions around (2-24) and (224) reflections are sensitive to layer tilt. The (2-24) and (224) RSM measurements, respectively, were taken from the lattice planes parallel and perpendicular to the edges of the crystallographic steps induced by the substrate miscuts. In the following we use the terms "parallel to" or "perpendicular to step-edges". A crystallographic layer tilt correction has been performed for the RS maps measured perpendicular to the "step edges" by applying the data from the symmetric (004) RSM.

The microstructure, the defects and interfaces were investigated in [110] and [1-10] lattice directions by methods of transmission electron microscopy (TEM) on cross-section samples. The TEM experiments were performed using a Philips CM30 electron microscope at 300 kV and, for high-resolution lattice imaging, a FEI TECNAI F30 at 300kV. Complementing information on the surface morphologies of the buffer layers were obtained from optical microscopy.

3. Results

3.1 Step-graded GaP_xAs_{1-x} buffer layer system (buffer A)

The residual strain of the individual buffer layers in the two in-plane <110>-directions is determined from symmetric (004) and asymmetric (2-24) and (224) RSM measurements, respectively. As example, Figure 2 displays the two asymmetric RS maps. From these RSM measurements we obtain quantitative values for the elastic strain of the individual layers. From these strain analyses it can be deduced that for the upper buffer layer, the strain levels in the direction parallel and perpendicular to the "step-edges" amount to about 30 % and 32 % of the total nominal misfit, respectively. From these values of the residual layer strain it can be concluded that buffer A is highly strain-relaxed.

TEM investigations of sample cross-sections indicate that misfit dislocations have been formed throughout the buffer layer system (Figure 3). The density of dislocations decreases towards the surface. Parallel to the "step-edges" (not shown here) occasionally cracks and microtwins are observed to form. It has been observed also before that the formation of microtwins, stacking faults and cracks is a common phenomenon in layer structures grown under conditions of negative misfit [13]. The role which these defects play for layer strain and for the surface morphology in our case will be discussed in Ch. 4.

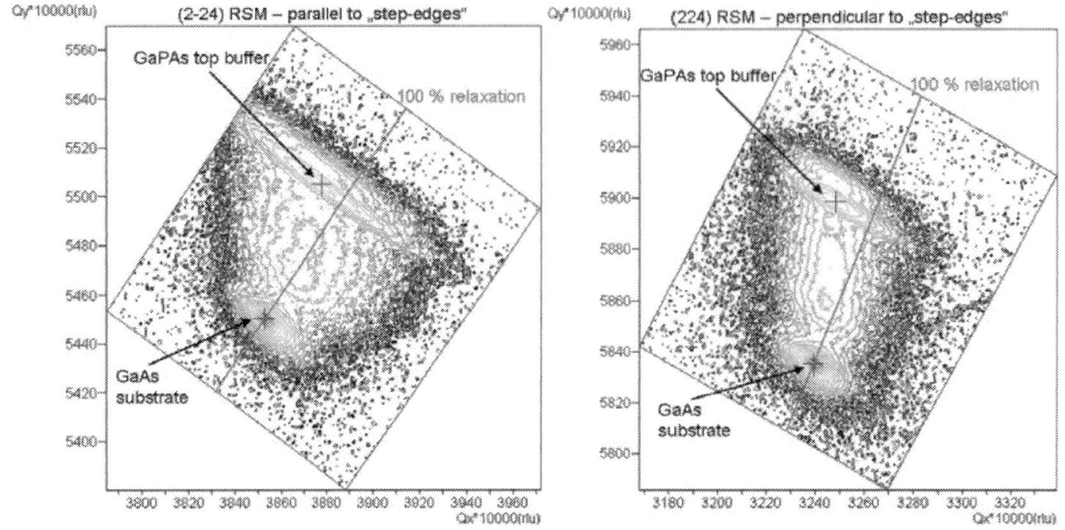

Figure 2: Asymmetric (2-24) and (224) RSM of buffer A. The maps show the distribution of diffracted intensities in projections along [1-10] and [110]-direction. Qx and Qy are the reciprocal lattice units in directions parallel and perpendicular to the sample surface, respectively. The colour code depicts increasing intensities as spectral colours (from blue for low to red for high intensities). The crosses indicate the measured positions of the intensity maxima in reciprocal space for the different regions. The line drawn through the substrate peak depicts the expected distribution of intensity maxima from the layers within the buffer system for the case that the buffer layers show 100 % strain relaxation (analogous to a thick relaxed layer). Note that the buffer system reveals a high degree of strain relaxation in both in-plane <110>-directions as the map shows intensity concentrations close to the line of 100 % relaxation (see text).

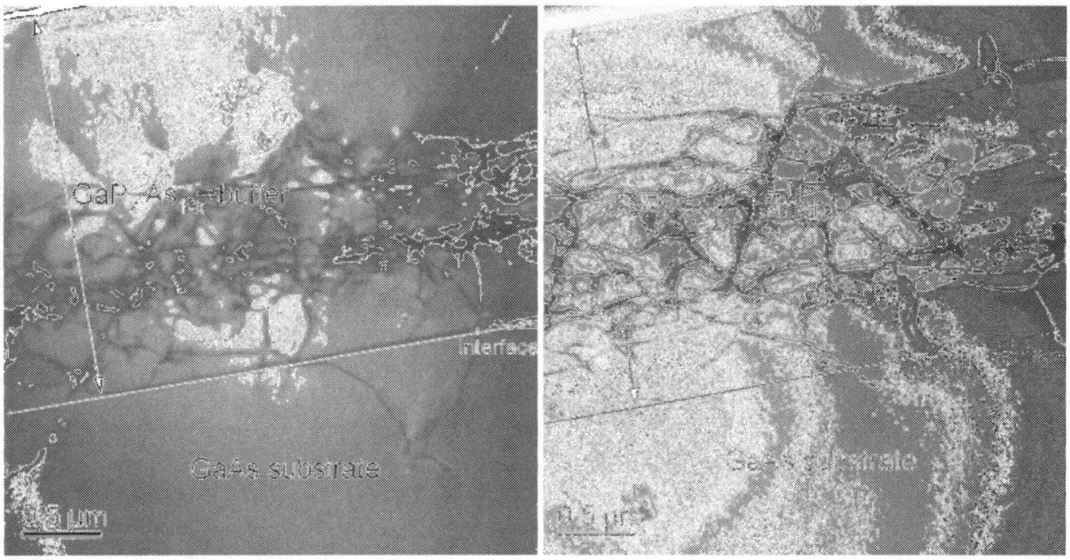

Figure 3: TEM bright-field images of buffer A: overview of layer system imaged in the [1-10]-direction parallel to the "step-edges" (left) and in the [110]-direction perpendicular to the "step-edges" (right). The position of the interface is marked. Note the high density of dislocations. Only few cracks and a number of microtwins can be detected in the direction parallel to the "step-edges" (not visible in the picture shown above).

Analyses by light-optical microscopy allow for the inspection on a larger dimension scale. Figure 4 shows the presence of microcracks. The cracks are extending parallel to the "step-edges" in <1-10> substrate direction. The distances between the cracks range from about 500 μm to about 1000 μm.

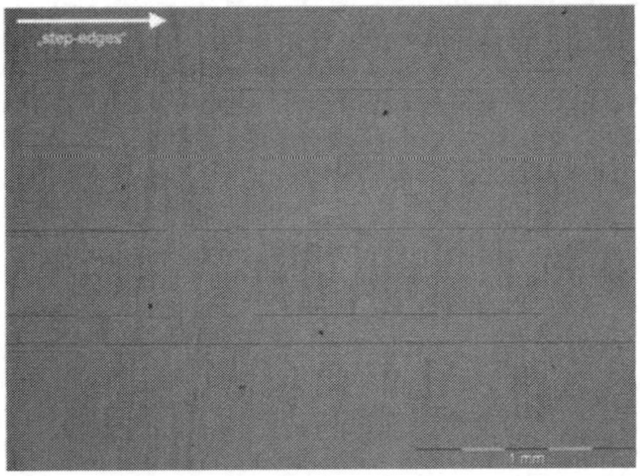

Figure 4: Surface morphology of buffer A illustrated by means of a large-area overview in light-optical microscopy. The cross-hatch pattern parallel to the "step-edges" in [1-10]-direction is induced by dislocations and microtwins. Furthermore, cracks extending along <110> directions are visible.

3.2 Step-graded GaN_yAs_{1-y} buffer layer system (buffer B)

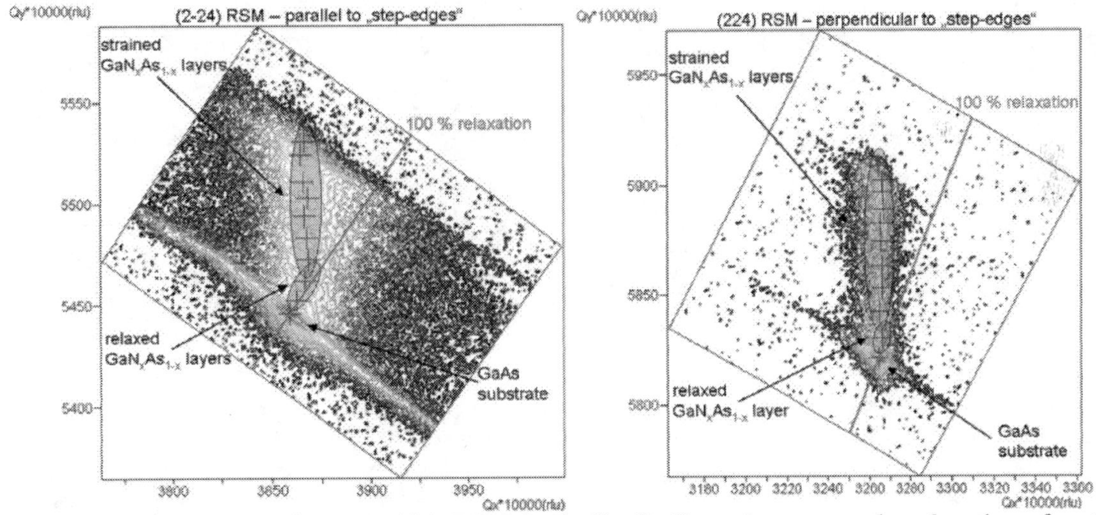

Figure 5: Asymmetric (2-24) and (224) RSM for buffer B. From these maps the relaxation of each individual buffer layer in both in-plane <110>-directions can be determined. In the direction parallel to the "step-edges" the upper buffer layer reveals a degree of strain relaxation of about 14 % and in the direction perpendicular to the "step-edges" of about 23 %, respectively. Note that the RSMs of the buffer system B shows a strong deviation from the intensity distributions expected to occur for complete relaxation of layer strains (see text).

As for the buffer A, the residual strain of the layers of buffer B in the two perpendicular in-plane <110>-directions can be determined from symmetric (004) and asymmetric (2-24) and (224) RSM measurements. Figure 5 shows as example two asymmetric (2-24) and (224) RSM measurements from buffer B. By means of these X-ray measurements the degree of strain relaxation of each individual buffer layer can be estimated as each layer contributes to the distribution of diffracted intensity[#11].

Again, a tilt correction is necessary for the measurement in the direction perpendicular to the "step-edges". By applying RSM measurements in the directions parallel and perpendicular to the substrate off-cut, we detect a slight asymmetry in the strain relaxation of buffer B. Next to the influence of the substrate off-cut on the symmetry, the asymmetry may emerge from the different behaviour of α and β dislocations in diluted nitrides. The broadening of the substrate intensity peak can be attributed to a lattice bending which results from the coating with an epitaxial buffer layer system with layer mismatch. From these RSM measurements we gain an overview about the strain state of the buffer layer system and can obtain quantitative values for the elastic strain of the individual layers of the buffer B. Qualitatively, the measurements reveal that the strain in the lower three GaN_yAs_{1-y}-layers is almost completely relaxed whereas almost no strain relaxation has occurred in the upper layers. The measurements reveal a strain relaxation of more than 70 % for the first two buffer layers in both in-plane <110>-directions. For the subsequent buffer layers, the degree of strain relaxation decreases rapidly. The uppermost buffer layer finally shows a degree of strain relaxation of around 14 % in the direction parallel to the "step-edges" and of around 23 % in the direction perpendicular to the "step-edges", respectively. These observations indicate that significant differences exist in the strain relaxation behaviour for the lower and the upper buffer layers. This behaviour is clearly different from the experimental results obtained for buffer A.

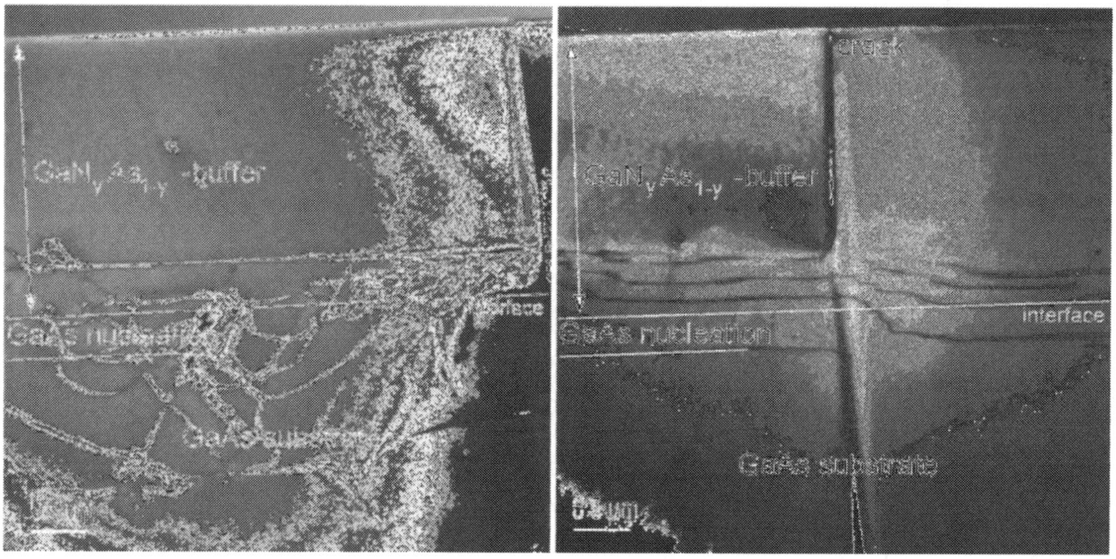

Figure 6: Cross-section TEM dark-field images of buffer B in [1-10]-direction parallel to the "step-edges" (left) and in [110]-direction perpendicular to the "step-edges" (right): overview of layer system with regions showing dislocations and microcracks. The presence of dislocations is observed for layers with a "threshold" concentration of N of around 2 %. Beyond this value, no dislocations are observed. Instead, microcracks are observed whose formation can be attributed to cumulative tensile layer strain. In [1-10]-direction, dislocations are observed which also extend into the GaAs substrate.

The TEM investigations of cross-section samples of buffer B show that essentially dislocations and microcracks have formed parallel to the "step-edges" as well as perpendicular to the "step-edges". Figure 6 shows as example that the dislocations are present only in the layers of the lower part of the buffer, with nominal N concentrations $y < 2$ % (cf. Figure 1). When imaging parallel to the "step-edges", a considerable number of dislocations can be observed in the GaAs substrate. In contrast, no dislocations are formed in the upper part of the buffer system where the layers have a nominal N concentration $y > 2$ %. Instead, in the two mutually perpendicular <110>-directions, microcracks and occasionally also microtwins (not shown here) are formed.

Figure 7: Optical micrograph of surface area of buffer B showing microcracks (dark contrast lines) extending along both <110>-directions.

Observations by light-optical microscopy show that microcracks have formed and propagated along both <110> substrate directions (Figure 7). The distances between the cracks are in the range of 50 µm to 500 µm. Some of the cracks are blocking each other.

3.3 Step-graded $Ga_{1-z}In_zAs$ buffers with integrated nitride blocking layer

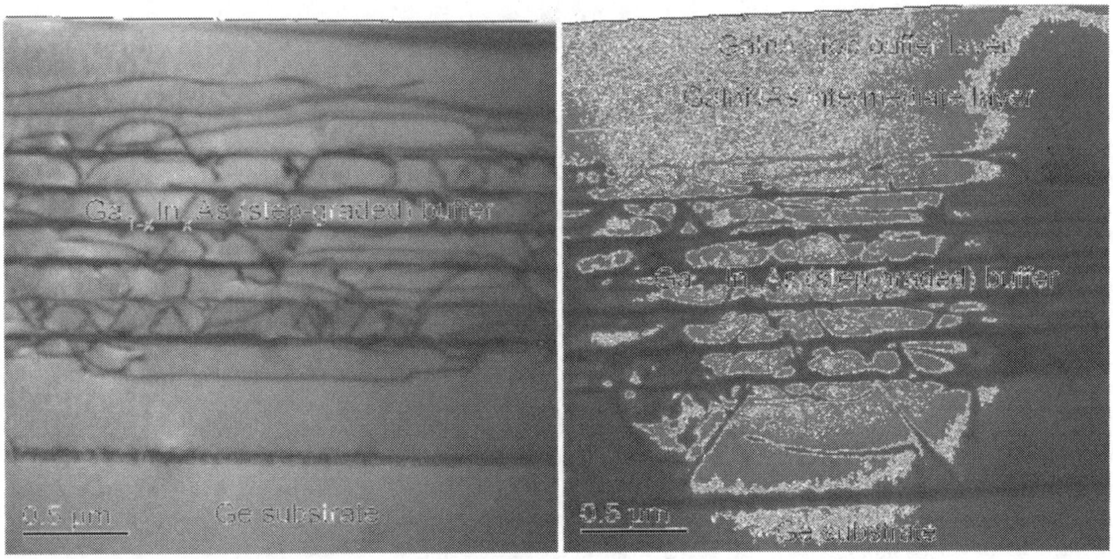

Figure 8: TEM bright-field micrographs of $Ga_{1-x}In_xAs$/Ge step-graded buffer structures without (left) and with (right) GaInNAs intermediate layer deposited onto a step-graded buffer system and capped by a GaInAs top buffer layer. Formation of dislocations in the top buffer layer is not observed (right).

Figure 8 shows a section of the dislocation network of a Ga_{1-z} In_zAs buffer system without (left) and with an integrated GaInNAs intermediate layer (right). In the first case, dislocations are present also in the upper part of the structure. In contrast, dislocations are absent and obviously hindered from gliding into the topmost buffer layer during layer growth when a GaInNAs intermediate layer is introduced.

Figure 9 shows the angular dependence of the diffracted X-ray intensity for the GaInAs buffer region, the Ge layer, and the top buffer region as revealed by high resolution X-ray rocking curves measured for the two cases. The comparison reveals that the full width at half maximum (FWHM) value of the intensity peak generated by the top buffer layer is reduced by about 15 % for the case of the buffer containing a GaInNAs layer. This decrease of the FWHM value can be interpreted as signature of a low density of dislocations within the upper buffer layer since the presence of significant numbers of threading dislocations would cause measurable broadening of the intensity peaks in rocking curve measurements.

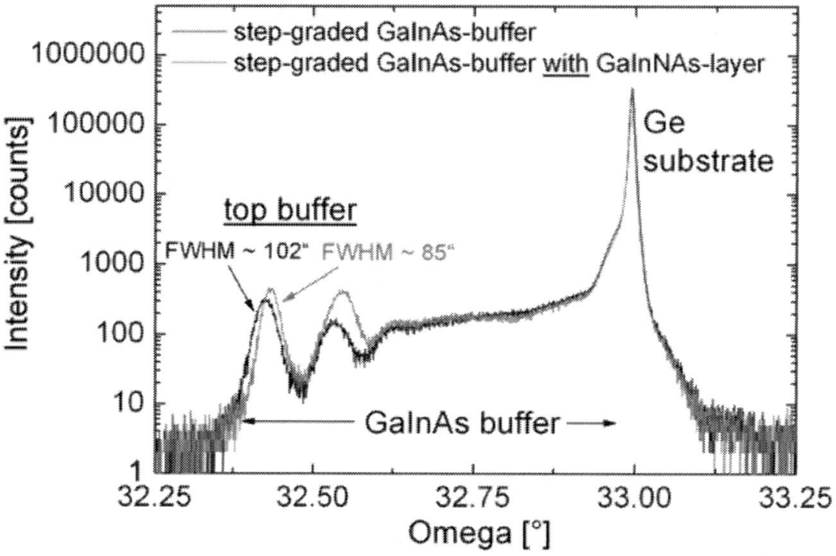

Figure 9: High resolution X-ray diffraction measurement of the two $Ga_{1-x}In_xAs$/Ge step-graded buffer structures (cf. Figure 8). The decrease in the full width at half maximum (FWHM) of the intensity peak originating from the diffraction of the top buffer layers is consistent with the TEM observation of a strongly reduced dislocation density originating from the application of a nitride layer.

4. Discussion

Characteristic for the step-graded GaPAs buffer system is the strain relaxation via formation of misfit dislocations and microtwins (Figure 3).The residual strain stays low enough to avoid crack propagation. In contrast, the formation of misfit dislocations during the growth of the first layers and the subsequent formation of microcracks during further buffer growth is characteristic for the GaNAs buffer system (Figure 6). Obviously, the suppression of dislocation formation in the upper part of the structure leads to an increase of the residual strain (due to the lattice grading) so that the system reaches strain values for which crack formation and propagation becomes possible.

4.1 Development of the buffer microstructures during layer growth

The microstructure observations (Ch.3) reveal significant differences in the strain relaxation behaviour of the investigated buffer systems. The two buffer layer systems which were designed for this investigation represent cases for which the strains of the individual layers are tensile in nature. This situation is achieved in a controlled way by adding P or N to crystalline GaAs and reduces the lattice constants of the resulting alloy to a lesser (by adding P) or a higher degree (by adding N).

For the first case of a step-graded GaP_xAs_{1-x} system (buffer A, Ch. 3.1), we observe that the microstructure develops predominantly by forming misfit dislocations and microtwins. Crystalline defects like dislocations, stacking faults, and twins are known to be effective in releasing misfit strain during the epitaxial growth of layers [14]. In our case, the observation of such defects is regarded as being the direct result of the relaxation of layer strain. From the RSM measurements (Figure 2) it is deduced as well that the buffer A is highly strain-relaxed, with a remaining residual strain of about 30 % of the original strain in the upper layers in the both mutually perpendicular <110>-directions. Only few microcracks are detected in such buffer systems proceeding in the direction parallel to the "step-edges". This clearly indicates that the major part of the stress for this buffer is relieved effectively by forming dislocations and microtwins.

In contrast, for the step-graded GaN_yAs_{1-y} buffer layer system (buffer B, Ch. 3.2) misfit dislocations form only within the first three layers of low N content. The composition grading of this layer system corresponds to a relative increase in lattice misfit by up to about 0.3%. The RSM measurements (Figure 5) reveal indeed a large degree of strain relaxation of around 70 % and 85 % of the original values for the first and second buffer layer, respectively. Misfit dislocations are absent for the upper layers of the buffer system containing N contents $y > 2$ %. Instead, microcracks extending along the two mutually perpendicular <110> directions form in high numbers in these buffer regions (Figures 6, 7). The corresponding RSM measurements (Figure 5) show that these layers are still kept under large tensile layer strain whose magnitude increases stepwise with the distance from the substrate, corresponding to the increase of the lattice-misfit by up to 1 %. Compared with a fully strained layer, i.e., assuming a strain value based on a calculated lattice misfit related to the layer composition, the residual strain in the upper layers is still of the order of 80%. These results indicate that the observed microcracks do not significantly contribute to the relaxation of strain, and that only part of the misfit strain is relaxed by forming cracks. Dislocation formation and movement of dislocations are largely suppressed in this part of the buffer system.

Adding small amounts of nitrogen to GaAs has effects on the electronic properties and on the mechanical properties. On the one hand, the band-gap energy of GaN_yAs_{1-y} decreases, and on the other hand the material becomes harder which may result in a reduced dislocation mobility [5]. The TEM results (Figure 6) and the X-ray measurements (Figure 5) convincingly demonstrate that diluted GaN_yAs_{1-y} layers with $y > 2$ % are effective in hindering misfit dislocations from threading through the bulk material of mismatched semiconductor heterostructures. Hence, strain relaxation is aggravated for the subsequently grown layers. This effect can be attributed to the effect of the substitutional nitrogen atoms which, due to their small covalent radius, lead to local lattice perturbations that decrease the dislocation mobility [5]. Furthermore, alloy hardening may induce a pinning of dislocations which will lead to further contributions to strain hardening.

As can be concluded from the TEM results (Figure 6, Ch. 3.2) and from the RSM measurements (Figure 5, Ch. 3.2), this characteristic behaviour is not observed when the nitrogen concentration is below a threshold value of around 2 %. By applying RSM measurements in the directions parallel and perpendicular to the substrate off-cut, we have detected a slight asymmetry in the strain relaxation of buffer B. Next to the influence of the substrate off-cut on the symmetry, the asymmetry may emerge from the different behaviour of α and β dislocations in diluted nitrides.

Comparing the two buffers A and B, we arrive at the conclusion that for the GaP_xAs_{1-x} buffer the behaviour is nearly reversed. As the material becomes softer by adding phosphorous the dislocations feel a weaker Peierls potential in the upper GaP_xAs_{1-x} buffer layers. This favours dislocation glide and propagation and thus leads to the generation of a very high dislocation density by means of dislocation multiplication.

4.2 Comparison with critical layer thickness for defect formation in single layers

We discuss the observations for buffer systems A and B in regard with simple estimates for a critical layer thickness for defect formation under conditions of negative misfit.

Journal of Physics: Conference Series **471** (2013) 012008 doi:10.1088/1742-6596/471/1/012008

For layer systems with low mismatch, theoretical considerations which are based on the equilibrium theory of Matthews and Blakeslee [9] allow to estimate a critical layer thickness for defect formation and a critical thickness for the fracture of strained brittle materials [15]. The thickness beyond which a single pseudomorphic layer grown epitaxially on a substrate starts to release strain by the formation of misfit dislocations is given by

$$h_{c,dislocation} = \frac{b}{8\pi f \cdot (1+\nu)} \cdot \left(\ln\frac{h}{b} + 1 \right) \tag{1}$$

with b the length of the Burgers vector (assuming the formation of perfect 60° misfit dislocations), h the layer thickness, f the lattice-misfit between layer and substrate and ν the Poisson ratio.

For systems with higher lattice mismatch the critical thickness for the formation of cracks is given by

$$h_{c,crack} = \frac{a \cdot (1-\nu)^2}{5\pi f^2} \tag{2}$$

with a the lattice constant of the respective materials, f the lattice misfit, and ν the Poisson ratio. The critical misfit for spontaneous fracture is described by

$$f_{c,fracture} = \frac{(1-\nu)}{10}. \tag{3}$$

For the cases of a GaPAs layer and a GaNAs layer on a GaAs substrate, respectively, Figure 10 (A) shows the result of the calculated critical thickness for different values of misfit f according to equation (1), assuming a value $\nu = 0.32$ for GaPAs and GaNAs. Figure 10 (B) shows the calculated values (equ. 2) for the critical thickness for cracking for a GaPAs layer and for a GaNAs layer on a GaAs substrate as function of the misfit f, assuming the same values for ν. Figure 10 (C) represents the situation for the critical strain for spontaneous fracture calculated to occur at a misfit value of 7 % according to equ. 3. Such values for the misfit of layers on GaAs substrates can indeed be theoretically achieved for GaNAs layers with sufficiently high N contents but cannot be achieved for GaPAs layers for which the maximum lattice misfit that can be realized will be about 3.6 %. Therefore, spontaneous fracture of GaPAs layers is not expected to occur.

Figure 10 also qualitatively illustrates the different strain relaxation behaviour of the two buffer systems A and B (arrows) observed in our experiments. For buffer A, the average lattice misfit is $f_{GaPAs,av} \sim -0.65$ %. Therefore, the theoretical values for the critical thickness for dislocation and crack formation in GaPAs (with $f = -0.65$ % lattice-misfit to the substrate) are about 5 nm and 400 nm, respectively. Continuous strain relaxation by formation of misfit dislocations and microtwins as observed for this buffer system (Figure 3) therefore leads to the situation that the residual strain stays low enough during layer growth to avoid crack propagation. Starting with a completely strained state (at a layer misfit of around 0.1 %), this development is schematically illustrated by the blue arrow.

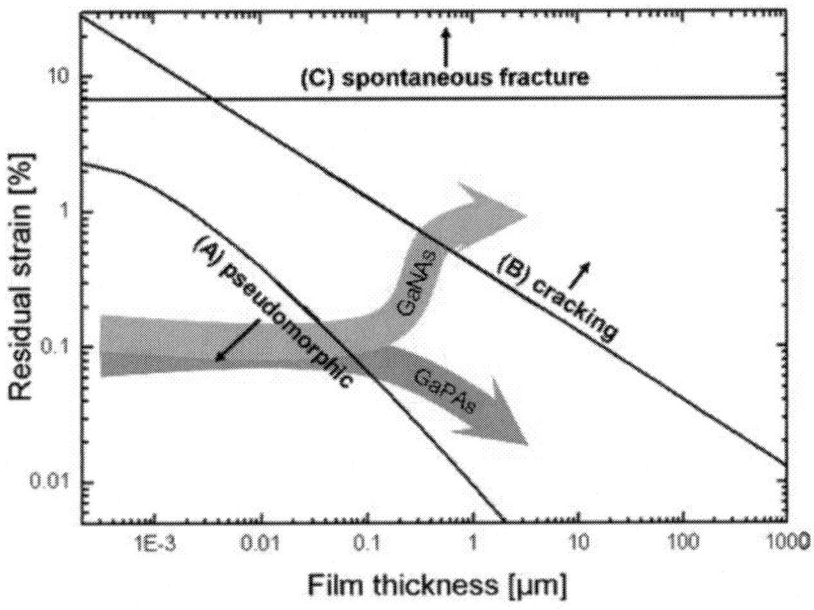

Figure 10: Residual strain, expressed by the misfit f of an individual layer, versus film thickness. Theoretical critical layer thickness for dislocation formation (line A), crack formation (line B) and spontaneous film fracture (line C). Illustrated by arrows are the regions of different microstructure development during layer growth based on the predictions for single layer growth. The observed strain relaxation paths during layer growth for the buffer layer systems A and B are schematically illustrated by the coloured arrows (see text for details).

For buffer B, the estimated value for the average lattice misfit amounts to about -0.5 %. The theoretical estimates for the critical thickness for dislocation and crack formation result in 6 nm and 650 nm, respectively. During growth of the step-graded GaNAs buffer system, the misfit strain starts to relax already during the growth of the first layer by the formation of misfit dislocations (Fig. 6). We could observe that this mode of strain relaxation stops after the deposition of the third layer. Above a layer thickness of around 700 nm, microcracks form that contribute only little to releasing the misfit strain. This suppression of dislocation formation in the upper part of the structure leads to a situation where during further layer growth the residual strain due to the lattice grading increases until finally cracks form and propagate. This development is schematically illustrated by the red arrow. While the overall strain relaxation behaviour can be interpreted qualitatively in this way, a more quantitative analysis would need to consider the grading of layer composition.

4.3 Role of alloying additions: P-based versus N-based buffer systems

The observed differences in microstructure development are likely to originate from the differences in microhardness of the ternary alloy layer materials and the resulting consequences for dislocation formation and dislocation mobility. For the dislocation movement, hard materials possessing high interatomic binding energies have a high Peierls potential. Microhardness values of binary compounds published in the literature yields ranges from 5 GPa to 7 GPa for GaAs [16], from 4 GPa to 6 GPa for GaP [17], and from 10 GPa to 20 GPa for GaN [16,18,19]. Based on these values it can be concluded that, by adding more and more P to the GaAs compound, the resulting ternary GaPAs compound semiconductor will tend to decrease the hardness and become softer. In contrast, by adding more and more N to GaAs, the ternary GaNAs compound semiconductor should increase its hardness.

Based on these considerations we conclude that dislocation gliding, propagation and multiplication is suppressed in the upper harder buffer layers of higher N content, or, in other words, dislocations are blocked from propagating into the upper layers for the GaNAs buffer layer system. As confirmed by our TEM investigations (Figure 6), these phenomena can still occur in the lower buffer layers which are 'softer' due to their low N content. The observation that dislocations are formed ('reflected') even in the GaAs substrate (Figure 6) confirms this point of view. The suppression of dislocation formation and propagation in the upper buffer layers is accompanied by the generation of high tensile strain leading to the observed crack formation. On the other hand, the continuous formation of dislocation arrangements and the gliding of dislocations during layer growth is very effective for the step-graded GaPAs buffer system, resulting in the gradient of dislocation density observed by TEM (Figure 3).

4.4 Strain engineering for high-efficiency solar cells: buffer systems of tensile and compressive strain

Both the GaPAs and the GaNAs buffer systems are multilayer systems under tensile strain. As we have discussed above for the GaNAs buffer layer system, dislocation gliding, propagation and multiplication is suppressed in the upper harder buffer layers of higher N content, and dislocations and microtwins are largely absent. Instead, microcracks form during the growth of the upper layers of the step-graded buffer system. Consequently, we can conclude from these studies that, during subsequent growth of layers with tensile strain, the strain relaxation by defect formation can be effectively hindered when introducing intermediate GaN_yAs_{1-y} layers with adequately tailored N-concentrations $y > 2$ % into a buffer structure. This concept has also been tested and has proven to be successful in a related study of GaP_xAs_{1-x} buffer structures with an incorporated GaN_yAs_{1-y} layer (for values of y ranging from ~ 2.5 % to ~ 4 %) [1]. We conclude that the use of dilute nitride layers offers the possibility for new concepts for engineering defect distributions and layer strain in lattice-mismatched compound semiconductor layer structures. They are therefore of particular interest especially also for applications in high-efficiency III-V solar cells.

It could be shown by our study that a similar growth concept, modified by adding In to GaNAs, can be used as dislocation blocking layer for $Ga_{1-x}In_xAs$/Ge buffer layer systems under compressive strain. This concept is similarly effective in controlling the defect formation and desired defect reduction in compressively strained heterostructures, making it a promising tool for growth of metamorphic high efficiency III-V multi-junction solar cells on GaAs, Ge or Si substrates.

5. Conclusions

Metamorphic step-graded GaP_xAs_{1-x} and GaN_yAs_{1-y} buffer layer systems were grown by metal-organic vapour phase epitaxy (MOVPE) on GaAs (001) substrates with a 6° miscut towards (111)A. Defect formation and strain relaxation have been investigated by transmission electron microscopy and high-resolution X-ray diffraction and were complemented by surface characterization employing optical microscopy.

The comparison of these different buffer systems under tensile strain revealed characteristic differences in the formation of defects and in the strain relaxation behaviour:

- The step-graded GaP_xAs_{1-x} buffer layer system (buffer A) shows effective strain relaxation induced by the formation of dislocations and microtwins. HRXRD analyses reveal relatively small amounts of residual strain.

- The step-graded GaN_yAs_{1-y} buffer layer system (buffer B) in contrast shows only low dislocation generation in the lower part but crack generation induced by cumulative tensile strain in the upper part of the buffer. Thus dislocation formation is suppressed by means of N-induced material hardening for an N-concentration higher than around 2 %.

- Comparative investigations of a step-graded $Ga_{1-z}In_zAs$ buffer layer with integrated nitride blocking layer reveal a lower dislocation density in the upper buffer layer. Such nitride blocking layers therefore appear promising for defect engineering of metamorphic multi-junction solar cells.

Our observations indicate that for the GaPAs-system the relaxation of tensile strain is achieved by forming a high number of dislocations and microtwins. In contrast, for the GaNAs-system the tensile strain induces microcrack formation and only few dislocations and microtwins are generated. The differences in strain relaxation behaviour appear to reflect the differences in microhardness (or the Peierls potential) of the ternary alloy materials.

The usage of dilute nitride layers in buffer systems offers a new concept for engineering defect distributions and layer strain in lattice-mismatched compound semiconductor layer structures and is of particular interest especially also for applications in high-efficiency solar cells. Obviously, strain relaxation and formation of defects, such as misfit dislocations, can be effectively hindered during subsequent growth of layers when introducing intermediate GaN_yAs_{1-y} layers with concentrations $y > 2$ %. For layers under tensile strain, the concept has been tested and has proven to be successful also in a related study of GaP_xAs_{1-x} buffer structures with incorporated GaN_yAs_{1-y} layers (values of y ranging from ~ 2.5 % to ~ 4 %) [1]. On the other hand, the investigations presented in this paper show that successful defect engineering becomes possible as well for layer systems under compressive strain when using layers of differing alloy composition, for instance alloys containing In.

Acknowledgement

This work was supported by the European Commission through the project FULLSPECTRUM (SES6-CT-2003-502620). One of the authors (Jan Schöne) acknowledges financial support by Deutsche Bundesstiftung Umwelt (DBU).

References

[1] Schöne J, Spiecker E, Dimroth F, Bett AW and Jäger W 2008 *Appl. Phys. Lett.* **92** 081905.
[2] Jackrel DB, Bank SR, Yuen HB, Wistey MA, Harris JS, Ptak AJ, Johnston SW, Friedman DJ and Kurtz SR 2007 *J. Appl. Phys.* **101**, 114916.
[3] Kunert B, Volz K and Stolz W 2007 *phys. stat. sol. (b)* **244**, 2730.
[4] Laurand N, Calvez S, Dawson MD, Jouhti T, Konttinen J and Pessa M 2005 *phys. stat. sol. (c)* **2**, 3895.
[5] Momose K, Yonezu H, Fujimoto Y, Ojima K, Furukawa Y, Utsumi A and Aiki K 2002 *Jpn. J. Appl. Phys.* **41**, 7301.
[6] Li NY, Wong WS, Tomich DH, Kavanagh KL and Tu CW 1996 *J. Vac. Sci. Technol. B* **14**(4), 2952.
[7] Fujimoto Y, Yonezu H, Momose K, Utsumi A and Furukawa Y 2001 *J. Cryst. Growth* **227-228**, 491.
[8] Adamcyk M, Schmid JH, Tiedje T, Koveshnikov A, Chahboun A, Fink V and Kavanagh KL 2002 *Appl. Phys. Lett* **80**, 4357.
[9] Matthews JW and Blakeslee AE 1974 *J. Cryst. Growth* **27**, 118.
[10] Young EC, Koveshnikov AN, Tixier S, Kavanagh KL and Tiedje T 2005 *phys. stat. sol. (a)* **15**, 2849.
[11] Pan Z, Wang YT, Li LH, Wang H, Wei Z, Zhou ZQ and Lin YW 1999 *J. Appl. Phys.* **86**, 5302.
[12] Bett AW, Baur C, Dimroth F and Schöne J 2005 *MRS Symp. Proc.* **836**, 223.
[13] Maree PM, Barbour JC, van der Veen JF, Kavanagh KL, Bulle-Lieuwma CWT and Viegers MPA 1987 *J. Appl. Phys.* **62**, 4413.
[14] Li Y, Weatherly GC and Niewczas M 2005 *Phil. Mag.* **85**, 3073.
[15] Matthews JW and Klockholm E 1972 *Mat. Res. Bull.* **7**, 213.

[16] Yonenaga I, Hoshi T and Usui A 2000 *J. Phys.: Condens. Mat.* **12**, 10319.
[17] Grillo SE, Ducarroir M, Nadal M, Tournié E and Faurie JP 2003 *J. Phys. D: Appl. Phys.* **36**, L5.
[18] Nowak R, Pessa M, Suganuma M, Leszczynski M, Grzegory I, Porowski S and Yoshida F 1999 *Appl. Phys. Lett.* **75**, 2070.
[19] Kucheyev SO, Bradby JE, Williams JS, Jagadish C, Toth M, Phillips MR and Swain MV 2000 *Appl. Phys. Lett.* **77**, 3373.

Journal of Physics: Conference Series **471** (2013) 012009 doi:10.1088/1742-6596/471/1/012009

Measuring composition in InGaN from HAADF-STEM images and studying the temperature dependence of Z-contrast

T Mehrtens[1,4], M Schowalter[1], D Tytko[2], P Choi[2], D Raabe[2], L Hoffmann[3], H Jönen[3], U Rossow[3], A Hangleiter[3] and A Rosenauer[1]

[1]Institute of Solid State Physics, University of Bremen, 28359 Bremen, Germany
[2]Max-Planck-Institut für Eisenforschung GmbH, 40237 Düsseldorf, Germany
[3]Institute of Applied Physics, TU Braunschweig, 38106 Braunschweig

E-mail: mehrtens@ifp.uni-bremen.de

Abstract. In this contribution, the indium concentration profile of an $In_xGa_{1-x}N$/GaN five-fold multi quantum well structure is measured from high-angle annular dark field scanning transmission electron microscopy (HAADF-STEM) images. The results are compared with an atom probe tomography study. Indium concentrations in the range of 26 at.% to 33 at.% are measured in the centre of the quantum wells. An additional indium layer of 14 at.% has been found on top of the quantum wells. In the second part, the temperature dependence of measured intensities in GaN is investigated. Here, multislice calculations in the frozen lattice approximation are carried out in dependence of specimen thickness and compared to experimental data. An increase of intensity with specimen temperature is found.

1. Introduction

Semiconducting materials are used in a variety of technological applications due to their electronic properties. For example, $In_xGa_{1-x}N$ with its bandgap ranging from 0.7eV [1] to 3.5eV [2] is particularly suitable for optoelectronic devices such as light emitting diodes or laser diodes that operate in the visible spectrum of the light. As a typical LED contains several layers of different semiconducting materials and layer thicknesses of only a few nanometres, adequate techniques for composition analysis at sub-nanometre scale are required.

Nowadays, in scanning transmissions electron microscopes (STEM) a high spatial resolution well below one nanometre can be achieved and in combination with a high-angle annular dark field (HAADF) detector the image intensity also strongly depends on the atomic number (Z) of the scattering atoms. Thus, HAADF-STEM is an appropriate technique for composition analysis.

Rosenauer et al. [3] demonstrated on an AlGaN/GaN heterostructure that specimen thickness and material composition can be deduced by a comparison of measured intensities with simulated reference data. The comparison is based on a normalisation of image intensities with respect to the scanning probe [4]. Grillo et al. [5] pointed out that static atomics displacements (SAD) due to different covalent radii of different atomic species have to be considered in multislice calculations for an accurate simulation of HAADF intensities.

[4]Corresponding author.

2. Experimental Setup

The investigated specimen was a five-fold InGaN/GaN multi-quantum well (MQW) structure grown by metal-organic vapour phase epitaxy (MOVPE) in a horizontal reactor (Aictron AIX 200RF). The InGaN quantum wells were separated by 20 nm GaN barriers (more details on specimen growth can be found in the publications of Rossow et al. [6] and Hoffmann et al. [7]). For the STEM analysis, a thin lamella was prepared by focused ion beam lift out. Due to the preparation process, the specimen surface is usually covered with an amorphous layer of a few nanometres. Additional etching with argon ions of low energy (400eV) has been applied to remove this layer [8]. STEM investigations were performed with an FEI TITAN 80/300 equipped with an HAADF-detector (Fischione model 3000). A spot size of 6 and a gun lens setting of 9 was used. The integration angle of the HAADF-detector ranged from 36 to 220 mrad.

Atom probe tomography (APT) was conducted on an Imago LEAP™ 3000X HR system. Here, the analysed specimen was also prepared by focused ion beam lift-out. To gain a III:V ratio close to 50:50 the specimen was cooled to 40K and the energy of the triggering laser pulses was reduced to 0.05 nJ.

Figure 1. HAADF-STEM image and APT reconstruction of InGaN/GaN MQW structure.

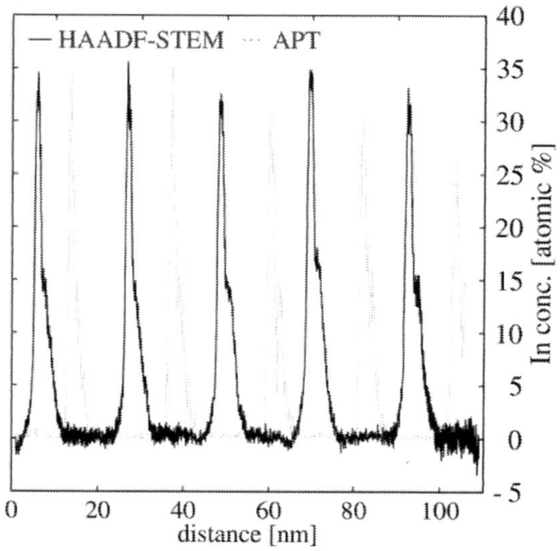

Figure 2. Concentration profiles deduced by HAADF-STEM and APT. The profiles are shifted with respect to each other for better visibility.

3. Composition analysis

For the quantification, reference intensities were simulated for $In_xGa_{1-x}N$ with indium concentrations ranging from $x = 0$ to $x = 0.55$ and specimen thicknesses up to 150 nm. These multislice calculations in the frozen lattice approach were performed using the STEMsim program [9]. The simulations included SADs [5] and the non-uniform sensitivity of the HAADF-detector was also taken into account. A parameterisation of the reference data and a more detailed description of the simulation can be found in [10].

Figure 1 displays an HAADF-STEM image of the five quantum wells. The indium concentration was extracted by normalizing the measured intensities with respect to the intensity of the scanning electron probe [4]. Afterwards, the intensities of the GaN barriers were interpolated over the quantum wells and used to determine the specimen thickness by comparing them with the simulated data for GaN. Then, the intensities measured within the quantum wells were compared with the reference data taking the interpolated thickness into account. This yielded the concentration profile shown in figure 2. Concentrations between 30 and 33 at.% indium were measured within the quantum wells and 14 at.%

indium were detected within an additional layer on top of the wells that formed during specimen growth.

For comparison, the specimen was also analysed by APT (see lower part of figure 1). The deduced concentration profile is also shown in figure 2. Here, concentrations in the range of 26 to 31 at.% were measured within the quantum wells and of 13 at.% in the additional layers.

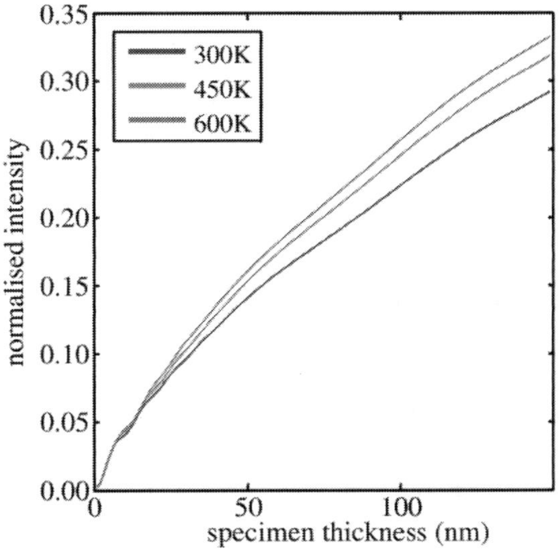

Figure 3. Simulated HAADF intensity of GaN in dependence of temperature.

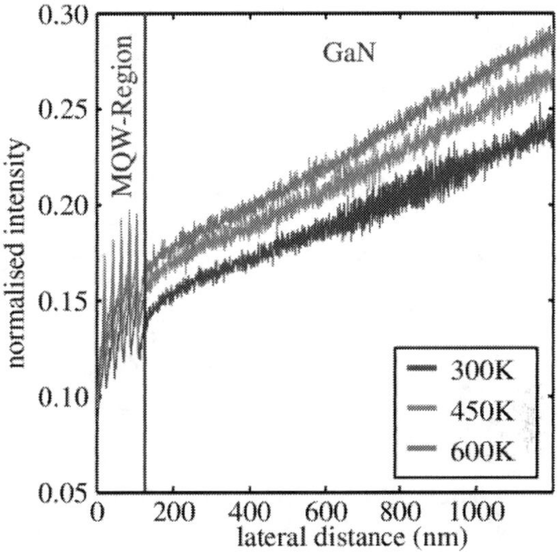

Figure 4. Intensity profiles for different temperatures. The MQW-region corresponds to figure 1.

4. Temperature dependence of HAADF-intensity

A main contribution of the intensity scattered into high angles originates from thermal diffuse scattering due to thermal vibrations of the specimen atoms. In order to investigate the influence of the specimen temperature on the measured intensities frozen lattice simulations were performed for GaN for varying specimen temperature. The specimen temperature was included into the simulations by temperature dependent Debye-Waller factors [11].

In figure 3 the simulation results are shown for specimen temperatures of 300 K, 450 K an 600 K in dependence of the specimen thickness. As can be seen, the intensity increases with increasing specimen temperature due to the higher thermal displacements of the specimen atoms.

This has also been observed experimentally. The specimen was heated in a Gatan double tilt heating holder (Model 652) and HAADF-STEM images were acquired at exactly the same position. Linescans were extracted that are shown in figure 4. For the study of the temperature dependence of the GaN intensities, the GaN region below the MQW-region has been used. The general tendency of all three profiles to increase from the left to right is due to increasing specimen thickness. Nevertheless, it can be also seen that the measured intensities increase with temperature. As expected from the simulations, the difference between the three profiles is increasing with increasing specimen thickness and the difference between the 300 K and 450 K profiles is larger than the corresponding difference between the 450 K and 600 K curves.

5. Conclusions

We have demonstrated the composition measurement by quantitative HAADF-STEM on an InGaN/GaN heterostructure. An average concentration (QW1-QW4) of (31.9±1.0) at.% indium has been measured that is in well agreement with concentrations measured by an atom probe tomography study (30.5±0.9) at.%.

In the second part, the temperature dependence of HAADF-STEM has been studied for GaN. An increase of the measured intensities has been found, which was also predicted by multislice calculations. The increase of the intensities is caused by an increase of thermal diffuse scattering that rises the amount of electrons scattered into higher angles.

Acknowledgements

This work was supported by the Deutsche Forschungsgemeinschaft under Contract No. RO 2057/8-1 and the Bundesministerium für Bildung und Forschung (BMBF) in the frame of the "ERA-SPOT True Green (13N9634)" project.

References

[1] Matsuoka T, Okamoto H, Nakao M, Harima H and Kurimoto E 2002 *Applied Physics Letters* **81** 1246-1248

[2] Vurgaftman I, Meyer J R and Ram-Mohan L R 2001 *Journal of Applied Physics* **89** 5815-5875

[3] Rosenauer A, Gries K, Müller K, Pretorius A, Schowalter M, Avramescu A, Engl K and Lutgen S 2009 *Ultramicroscopy* **109** 1171-1182

[4] LeBeau J M and Stemmer S 2008 *Ultramicroscopy* **108** 1653-1658

[5] Grillo V, Carlino E and Glas F 2008 *Physical Review B* **77** 054103

[6] Rossow U *et al.* 2013 *Journal of Crystal Growth* **370** 105-108

[7] Hoffmann L, Bremers H, Jönen H, Rossow U, Schowalter M, Mehrtens T, Rosenauer A and Hangleiter A 2013 *Applied Physics Letters* **102** 102110

[8] Mehrtens T, Bley S, Satyam P V and Rosenauer A 2012 *Micron* **43** 902-909

[9] Rosenauer A and Schowalter M 2008 STEMsim-a New Software Tool for Simulation of STEM HAADF Z-Contrast Imaging ed A G Cullis and P A Midgley ([S.l.]: Springer Netherlands)

[10] Mehrtens T, Schowalter M, Tytko D, Choi P, Raabe D, Hoffmann L, Jönen H, Rossow U, Hangleiter A and Rosenauer A 2013 *Applied Physics Letters* **102** 132112

[11] Schowalter M, Rosenauer A, Titantah J T and Lamoen D 2009 *Acta Crystallographica Section A* **65** 227-231

Journal of Physics: Conference Series **471** (2013) 012010 doi:10.1088/1742-6596/471/1/012010

Composition variation of $In_{1-x}Ga_xAs$ epitaxially grown in narrow trenches on Si

P Favia, O Richard, J Geypen, N Waldron, C Merckling, W Guo, M Caymax, H Bender

Imec, Kapeldreef 75, B-3001 Leuven, Belgium

E-mail: paola.favia@imec.be

Abstract. In this work we investigate the indium content in $In_{1-x}Ga_xAs$ narrow trenches on Si by transmission electron microscopy (TEM), energy dispersive spectroscopy (EDS) and nano beam diffraction (NBD). We find a higher indium content in wider trenches and by scanning a trench from bottom to top we observe an increase of indium up to a maximum value close to the level of the surface of the shallow trench isolation oxide.

1. Introduction

The use of III-V compound semiconductors as n-type channel replacement materials has attracted great attention because of their excellent bulk electron mobilities [1]. In particular, indium gallium arsenide compounds, with their intrinsically superior electron mobility and high saturation velocity, are considered as candidates for nMOS devices beyond the 14 nm node technology [2].

The material investigated in this work is $In_{1-x}Ga_xAs$ epitaxially grown in narrow trenches on (001) silicon wafers by metal organic chemical vapour deposition (CVD). Different trench widths and nominal In concentrations are considered. The In content is a critical parameter in InGaAs devices because it determines the bandgap of the channel and, as a consequence, the two-dimensional charge carrier density [3]. It is therefore important having an insight on its distribution in the trenches to better understand the device performances. We investigate, by TEM based techniques, how the indium content varies with the width of the trenches and in the lateral and vertical dimension inside the same trench width.

2. Experimental

2.1. Samples and characterisation techniques

TEM specimens were prepared by focused ion beam (FIB Strata, FEI) perpendicular (and parallel) to the trench directions by using the in-situ lift-out technique. In order to avoid damage during FIB ion milling the specimens were protected with a layer of CVD oxide and ion beam deposited platinum. The InGaAs trenches are grown on InP layers deposited over Ge seeds on Si substrate. Specimens with different trench widths as listed in Table 1 were prepared in a single experiment in order to investigate the dependence of the In content on the trench dimensions.

TEM, EDS and NBD analysis were performed on an F30 Tecnai TEM operating at 300kV. The In content was measured using the automatic quantification procedure included in the TIA version 4.5 software (FEI). NBD results were processed by using the TrueCrystal software (FEI), the mismatch is calculated with respect to the Si substrate.

Table 1. Specimens investigated in this study.

Specimen label	Nominal trench width (nm)	Measured trench width at top (nm)	Cross section	Nominal In concentration (at %)
1	20	10	perpendicular	50
2	40	27	perpendicular	50
3	60	40	perpendicular	50
4	80	58	perpendicular	50
5	100	76	perpendicular	50
6	500	500	perpendicular	50
7	blanket on InP		wedge	53

2.2. Wedge sample: absorption effects and thickness dependence

The blanket InGaAs layer is defect free and shows perfect epitaxy on the InP substrate which implies that it has a $In_{53}Ga_{47}As$ composition. This is also confirmed by XRD analysis. A specimen with sloped thickness ranging from 100 nm to 300 nm was prepared to understand absorption effects and thickness dependence in the EDS quantification process.

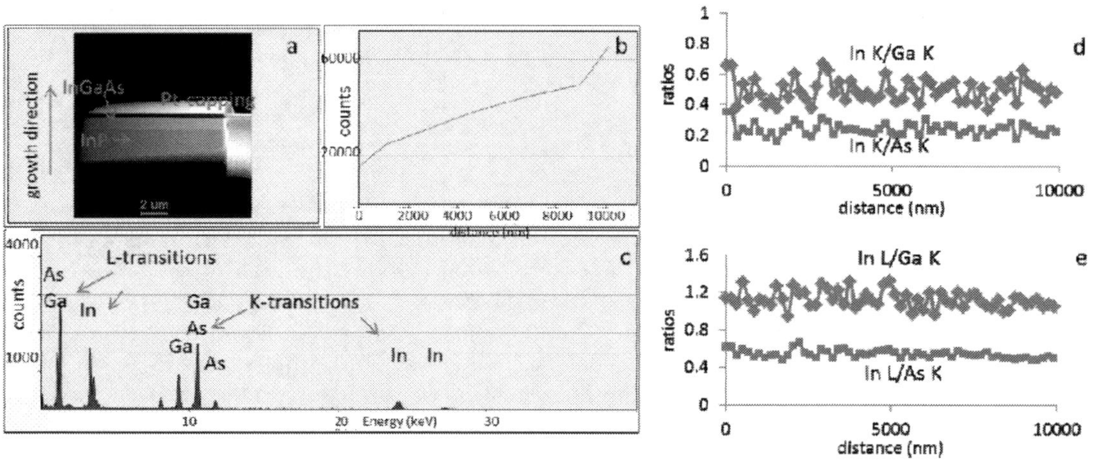

Figure 1. a) HAADF-STEM image of the FIB prepared wedge sample, b) HAADF-detector intensity profile along the red line in a), c) EDS spectra, d) In/Ga, In/As for K transitions and e) In-L/Ga-K and In-L/As-K ratios.

An EDS profile was acquired along the full length of the slope as shown in the upper left corner of Figure 1, with 50s acquisition time per point so that the S/N on the In-K transition is high enough to be well processed by the software. Care was taken to position the TEM specimen in the TEM holder in such way to minimize the Cu stray signal due to the Cu grid and that the sample could be oriented at the optimal angle for best EDS signal detection. The detector take-off angle was 20°. The curves at the bottom of figure 1 represent the ratios between the K In/Ga, In/As peaks (Figure 1d) and the ratios between the In L-peak and the Ga and As K-peaks (Figure 1e). The curves are noisy but on average constant indicating no or little dependence on thickness and absorption for the K peaks, some absorption may be revealed for L-lines in agreement with literature [4].

EDS quantification with the K peaks for this reference sample with nominal 53% In concentration gives ~56%. Therefore a systematic overestimate of the EDS quantification values is to be taken into account but for relative comparison between measurements this is not a limitation.

3. Results and discussion

The specimens investigated in this study are shown in Figure 2. Most 20nm and 40nm trenches are not fully filled while the 60, 80 and 100nm are all well filled and end up with a large facetted cap above the shallow trench isolation (STI) surface. Differently, the 500nm line is facetted but does not develop a cap on top. All trenches are narrower at the STI surface than at the bottom. Also, the wider trenches are more deeply etched in the Si. Misfit dislocations are revealed at the Si/Ge and Ge/InP interfaces and a few nanotwins and stacking faults are also observed in the trench and in the cap.

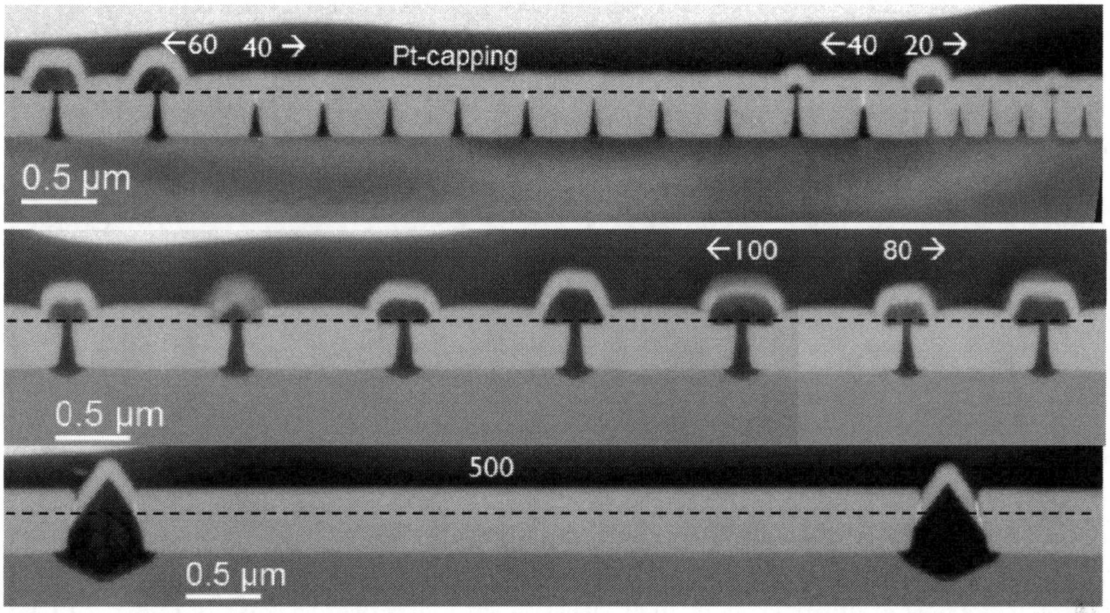

Figure 2. TEM cross-sectional images of the different trench sizes. The dashed lines indicate the STI surface. The material stack is labelled in Figure 4.

EDS analysis shows that, in the trenches, the In content increases from bottom to top with a maximum value close to the surface of the STI layer and then decreases again as shown in Figure 3a. The average In content depends on the size of the trench; it increases exponentially with the trench width as shown in Figure 3b. The lateral variations of the In content in the trenches are small and vary from trench to trench in a non-systematic way.

NBD measurements are performed to evaluate the variation in lattice parameter due to the In content inhomogeneity. In order to compare NBD and EDS results the horizontal and vertical mismatches measured by NBD are converted to In concentrations by using the Vegard's law and then averaged assuming a fully relaxed trench as shown in Figure 4b. The NBD measurements confirm the EDS trend but the values do not fully coincide. This is probably due to the presence of a certain degree of strain in the trench as well as to the effect of the lattice defects on the accuracy of the NBD measurements. In fact, for a fully relaxed InGaAs layer we would expect the same values for the horizontal and vertical mismatch. This is not the case and may therefore be the cause for the discrepancy between EDS and NBD results.

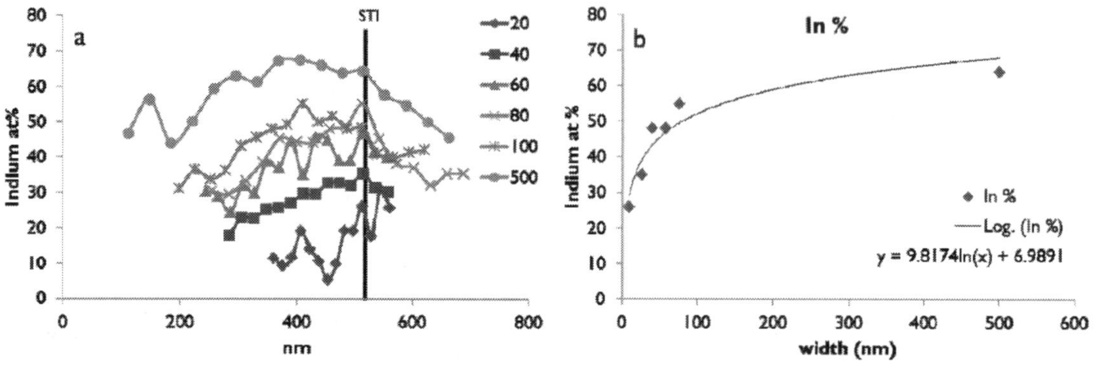

Figure 3. a) The In content increases from bottom to top in all size trenches. Curves are aligned to the top of the STI, b) average In content increases with trench width.

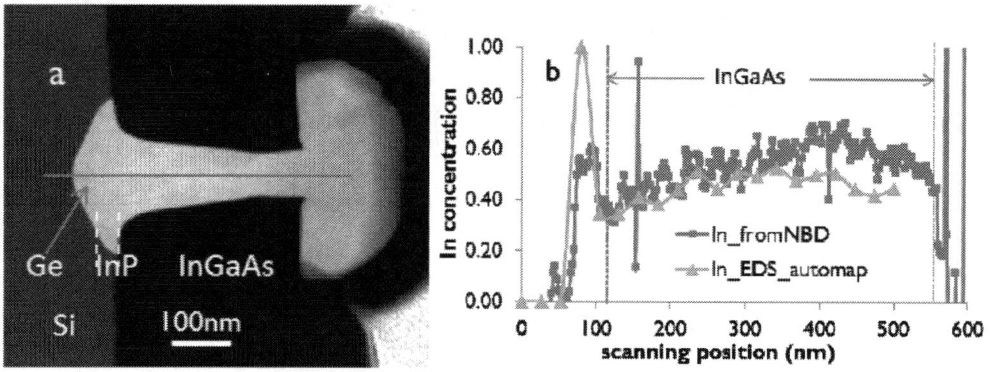

Figure 4. a) HAADF-STEM image, the red line indicates the NBD and EDS line profile. b) Indium concentration obtained by averaging the horizontal and vertical mismatch measured by NBD from bottom to top along the red line compared with the Indium content measured by EDS. The peak near 80 nm corresponds to the InP layer.

4. Conclusions

In this study we investigated the In content along the trench height for trenches with different widths. The In content increases from the bottom to the top of the trench up to the STI surfaces and increases with the InGaAs trench width. Awareness of this inhomogeneity can help to better understand the electrical behaviour of InGaAs devices and optimization of the growth process.

Acknowledgement

We thank Patricia Van Marcke for TEM specimen preparation.

References

[1] *International Technology Roadmap for Semiconductors* 2011
[2] Heyns M et al., 2011 Proc. *IEEE International Electron Devices Meeting (Washington)* p 299
[3] Wagner J W 1970 *Journal Electrochemical Society* **117**(9) p 1193-1196
[4] Walther T 2010 *Journal of Physics: Conference Series* **209** 012029

Journal of Physics: Conference Series **471** (2013) 012011 doi:10.1088/1742-6596/471/1/012011

Quantitative Composition Evaluation from HAADF-STEM in GeSi/Si Heterostructures

M Tewes[1], F F Krause[1,3], K Müller[1], P Potapov[2], M Schowalter[1], T Mehrtens[1] and A Rosenauer[1]

[1] Institut für Festkörperphysik, Universität Bremen, Otto-Hahn-Allee 1, 28359 Bremen, Germany
[2] GLOBALFOUNDRIES, Wilschdorfer Landstr. 101, 01109 Dresden, Germany

f.krause@ifp.uni-bremen.de

Abstract. High-angle annular dark field scanning transmission electron microscopy has been successfully used for composition evaluation in various material systems. In this work, the quantitative applicability of this method to GeSi/Si heterostructures was studied. Reference images were simulated by frozen lattice multislice simulations for different Ge concentrations accounting for static atomic displacements and biaxial strain due to pseudomorphic growth. Specimen thickness and composition are obtained by comparison of simulated and normalised experimental intensities. The measured thickness of a pure Si wedge specimen is compared to thickness determined from Pendellösung fringes in dark field micrographs. The deviation is below 10 nm coinciding with the accuracy of prior works. The composition of a GeSi-layer structure was measured in a calibration sample of known concentration and good agreement is found. Two-dimensional concentration maps of a GeSi/Si transistor structure were created. Measured concentrations agree with nominal values. However, strain fields in the Si lead to a variation of the image intensity causing an artificial fluctuation of the measured concentrations of ±4%.

1. Introduction

Germanium-Silicon (GeSi) heterostructures are of technological importance for modern semiconductor devices, e.g. metal-oxide semiconductor field effect transistors (MOSFETs) [1] that are widely used in modern electronics. Hence, the measurement of the Ge concentration is of high interest in those systems. The evaluation of the chemically sensitive high-angle annular dark field (HAADF) signal of scanning transmission electron microscopy (STEM) has been successfully used for accurate chemical composition measurement at Ångström resolution within many other material systems such as InGaN, SiGaAs, AlGaN or GaNAs [2-5].

In this work the applicability of quantitative HAADF-STEM for composition evaluation was studied in GeSi/Si heterostructures [6] by investigation of several specimens of this material system.

2. Simulation of reference images

To get reference data for evaluation of the measured HAADF-intensities, STEM-images were simulated by frozen lattice multislice simulations conducted with the STEMsim Software [7].

[3] corresponding author

A supercell laterally consisting of 7×7 orthorhombic unit cells in [110] zone axis was used with a numerical grid of 2100×1500 pixels.

The mean square thermal displacements used in these simulations to account for thermal diffuse scattering (TDS) were derived from density functional theory calculations as described in reference [8]. Static atomic displacements caused by the different covalent radii of Ge and Si, which have influence on the diffuse signal [9], were computed with an empirical potential suggested by Tersoff [10]. With this potential, the supercells for the simulations were relaxed by energy minimization with the LAMMPS software [11]. Furthermore, to account for the biaxial strain due to pseudomorphic growth of GeSi layers on bulk Si, the supercells were strained according to the results of elasticity theory calculations. Thus, various effects were taken into account whereas prior works assumed a Rutherford scattering model [6].

These simulations were executed for various specimen thicknesses and Ge concentrations, where for each concentration twenty individual supercell configurations with statistically distributed atoms and thermal displacements were used. The non-uniform sensitivity of the HAADF-detector, which was used in the experiments, was also considered. Thus, a dataset of HAADF-intensities was obtained as a function of specimen thickness and Ge concentration and was used as a reference in the evaluation of the experimental data.

3. Experimental procedure

The specimens used in this work were, except for the purchased calibration sample in section 4.2, prepared as lamellae with a focused ion beam (FIB) and treated with low energy Ar ion milling to remove amorphous surface layers [12]. The experimental investigation was performed with a FEI Titan 80/300 (S)TEM. To ensure comparability to the reference simulations, the measured HAADF-intensities are normalised with respect to the incident electron beam intensity, that were measured by scanning the focused beam directly over the detector as described in reference [13].

The specimen thickness is then determined from regions of known composition, in most cases of pure Si, by comparison with the reference. For regions with unknown concentration the thickness is polynomial inter- or extrapolated, respectively. Using the resulting thickness map and the measured intensities local compositions are determined for the entire STEM image [2].

4. Results

4.1. Thickness determination in a pure Si wedge sample

To validate the thickness measurement by HAADF-STEM with the simulated reference, a wedge specimen of pure Si was FIB-prepared in [110] zone axis and micrographs were acquired by HAADF-STEM and by dark-field conventional transmission electron microscopy (DF-CTEM) with the (004)-beam under imaging conditions that allowed the detection of Pendellösung thickness contrast fringes. From these fringes the specimen thickness was determined by comparison with Bloch-wave simulations conducted for the same sample orientation as used for the acquisition.

Figure 1 shows a comparison of the measured specimen thickness from both methods. The values from HAADF-STEM show a good agreement with those from DF-CTEM; the deviation is below 10 nm. This coincides with the accuracy of prior works on quantitative STEM comparing simulations and experiment, e.g. reference [5].

4.2. Concentration evaluation of a calibration sample

Furthermore, a purchased commercial calibration sample, MAG*I*CAL® from Norrox Scientific Ltd., containing multiple GeSi-layers with a known nominal Ge concentration of 19% embedded in Si and conventionally prepared in [110] zone axis was examined. The determined Ge concentration was (18±4)%, which is in good agreement with the nominal value. The evaluated composition of single layers from high resolution HAADF-STEM images coincided with the results of strain state analysis of high resolution CTEM micrographs which were evaluated for comparison.

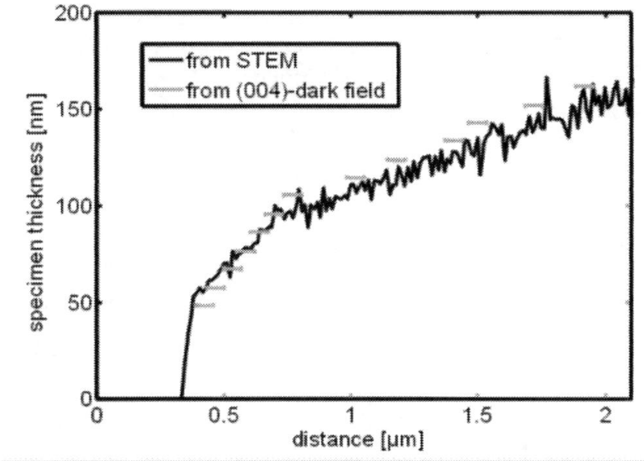

Figure 1. Comparison of specimen thickness of the pure Si wedge sample determined from HAADF-intensity and from Pendellösung thickness fringes. A good agreement is found; the deviation is below 10 nm.

4.3. Investigation of a GeSi/Si-MOSFET structure

In addition we investigated a Si-based p-MOSFET structure in which source and drain contain Ge to improve charge carrier mobility by straining the Si in the gate region [1]. The resulting two-dimensional composition map from the HAADF signal is depicted in figure 2. The mean Ge concentrations in the lower and upper regions are (22±3)% and (28±3)%, respectively. These are slightly higher than the nominal values of 19% and 27%, but fall within statistical error.

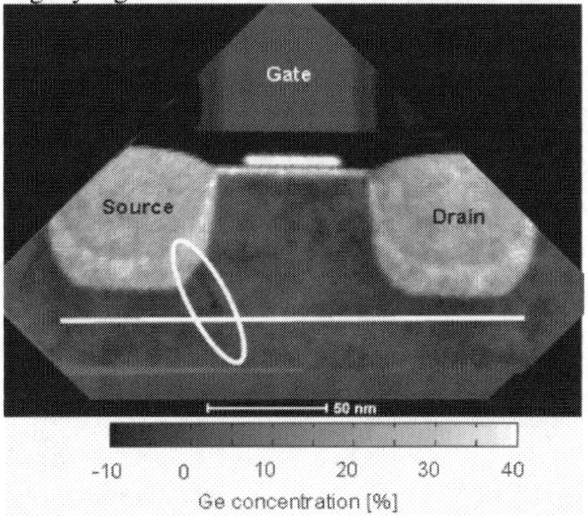

Figure 2. Concentration map of a GeSi/Si-MOSFET structure with two regions of different Ge concentrations in source and drain. The line marks the position of the linescan shown in figure 3. In the encircled area an artificial negative Ge concentration caused by a stacking fault is observed.

4.4. Influence of strain on measured concentrations and thicknesses

In the concentration map regions of artificial negative Ge concentrations also appear especially near GeSi/Si interfaces. This is partially caused by stacking faults as the one marked in figure 2, which occur in this highly strained structure. A linescan below the transistor in the pure Si region, as shown in figure 3, however, reveals artificial Ge concentration fluctuations of ±4% that cannot totally be caused by dislocations.

Another possible reason are effects of strain in the Si induced by the GeSi-regions: For the reference simulations described in section 2 pseudomorphic growth of the GeSi on bulk Si was assumed, which is legitimate for layer samples, but in this highly strained structure, the GeSi in source and drain could actually cause strain in the Si. This mismatch between simulations and experiment would cause a systematic error. To investigate its influence, simulations were conducted for strained Si supercells. The results showed a significant intensity difference to unstrained Si, that can lead to an underestimation of the specimen thickness of up to 10 nm for 100 nm thick specimens or, if the thickness was estimated correctly, negative Ge concentrations in regions of pure Si of up to -2%. This systematic error is unavoidable as long as the strain state of both the investigated and the reference area is unknown. It partly explains the observed fluctuations. The remaining artifacts are most likely caused by surface strain effects as described in reference [2].

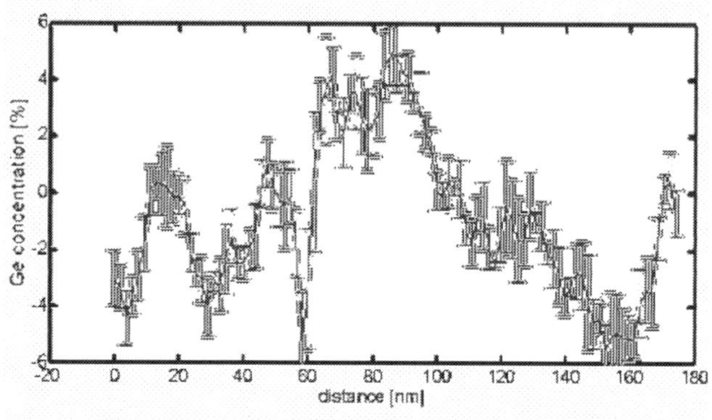

Figure 3. Linescan of 10 nm width from the concentration map of the MOSFET shown in figure 2 taken in the pure silicon area below the transistor (red line in figure 2). A HAADF-intensity corresponding to a Ge concentration in the range of ±4% is measured. Below source and drain artificial negative concentrations are obtained. This is caused by strain effects and, as in the dip at 60 nm, by dislocations.

5. Summary

The applicability of chemical composition evaluation with normalised HAADF-STEM intensity by comparison to multislice simulations in the GeSi/Si material system was investigated. The measurement of thickness in a pure Si wedge specimen and the concentration determination in a layer structure yielded results which are in good agreement with both reference measurements and nominal values.

For a GeSi/Si MOSFET structure two-dimensional composition maps were determined. The concentrations were also in good agreement with nominal values. Within these highly strained structures, however, artificial Ge concentration fluctuations were observed in pure silicon regions, which could partly be explained by strain effects. To avoid these, the simultaneous measurement of strain as suggested by reference [14] would be necessary.

References

[1] Auth M et al. 2008 *Intel Technology Journal* **12** 77-85
[2] Rosenauer A et al. 2011 *Ultramicroscopy* **111** 1316-1327
[3] Carlino E and Grillo V 2005 *Phys. Rev B* **71** 235303
[4] Rosenauer A, Gries K, Müller K, Pretorius A, Schowalter M, Avramescu A, Engl K and Lutgen S 2009 *Ultramicroscopy* **109** 1171-1182
[5] Grieb T, Müller K, Fritz R, Schowalter M, Neugebohrn N, Knaubb N, Volz K and Rosenauer A 2012 *Ultramicroscopy* **117** 15-23
[6] Walther T and Humphreys C J 1999 *Journal of Crystal Growth* **197** 113-128
[7] Rosenauer A and Schowalter M 2007 *Microscopy of Semiconducting Materials 2007* Cullis A G, and Midgley P A (Springer Netherlands) pp 170-172
[8] Schowalter M, Rosenauer A, Titantah J T and Lamoen D 2009 *Acta. Cryst. A* **65** 5-17
[9] Grillo V, Carlino E and Glas F 2008 *Phys. Rev. B* **77** 054103
[10] Tersoff J 1988 *Phys .Rev. B* **37** 6991-7000
[11] Plimton S 1995 *J. Comput. Phys.* **117** 1-19
[12] Mehrtens T, Bley S, Satyam P V and Rosenauer A 2012 *Micron* **43** 902-909
[13] Le Beau J M and Stemmer S 2008 *Ultramicroscopy* **108** 1653-16581
[14] Müller K, Rosenauer A, Schowalter M, Zweck J, Fritz R and Volz K 2012 *Microsc. Microanal.* **18**, 995-1009

Journal of Physics: Conference Series **471** (2013) 012012 doi:10.1088/1742-6596/471/1/012012

Compositional analysis of InAs-GaAs-GaSb heterostructures by Low-Loss Electron Energy Loss Spectroscopy

A M Beltrán[1,5,6], T Ben[1], A M Sánchez[2], M H Gass[3], A G Taboada[4], J M Ripalda[4] and S I Molina[1]

[1] Departamento de Ciencia de los Materiales e I.M. y Q.I., Facultad de Ciencias, Universidad de Cádiz, Campus Río San Pedro, s/n, 11510 Puerto Real, Cádiz, Spain

[2] Physics Department, University of Warwick, Coventry CV4 7AL, United Kingdom

[3] UK SuperSTEM, Daresbury Laboratory, Daresbury WA4 4AD, United Kingdom

[4] Instituto de Microelectrónica de Madrid (CNM, CSIC), Isaac Newton 8, 28760 Tres Cantos, Madrid, Spain

E-mail: ana.m.c.beltran@ww.uni-erlangen.de

Abstract. As an alternative to Core-Loss Electron Energy Loss Spectroscopy, Low-Loss EELS is suitable for compositional analysis of complex heterostructures, such as the InAs-GaAs-GaSb system, since in this energy range the edges corresponding to these elements are better defined than in Core-Loss. Furthermore, the analysis of the bulk plasmon peak, which is present in this energy range, also provides information about the composition. In this work, compositional information in an InAs-GaAs-GaSb heterostructure has been obtained from Low-Loss EEL spectra.

1. Introduction

Self-assembled InAs(Sb)GaAs quantum dots (QDs) heterostructures have received much attention since they are able to emit at 1.3-1.55 μm even at room temperature [1,2]. This emission range is useful for telecommunication lasers and opto-electronic applications. Since opto-electronic properties depend on the structural quality and compositional distribution, their full characterization is essential to establish a relationship among structure, real composition and properties.

Compositional distribution can differ from the nominal one due to the concurrence of different effects during the epitaxial growth such as segregation and surface diffusion [3,4]. In fact, previous Scanning-Transmission Electron Microscopy (STEM) studies of InAs QDs grown by Molecular Beam Epitaxy (MBE) over a GaAs (001) substrate suggest the formation of the $In_xGa_{1-x}As_ySb_{1-y}$ alloy when a GaSb layer is directly grown on the QDs [5]. These compositional changes are very important for the application and production of QDs. This emphasizes the importance of very high resolution analytical techniques for the study of these heterostructures in order to understand the origin of the inherent decompositions and contribute to their improvement.

Classical analytical transmission and/or scanning transmission electron microscope techniques, such as Energy Dispersive X-Ray (EDX) and Electron Energy Loss Spectroscopy (EELS), are not

[5]Present address: Center for Nanoanalysis and Electron Microscopy (CENEM). Department Werkstoffwissenschaften, Universität Erlangen-Nürnberg, Cauerstrasse 6, 91058, Erlangen, Germany

[6]To whom any correspondence should be addressed

useful for analysing In-Ga-As-Sb heterostructures due to overlapping signals associated with several elements. As an alternative, Low-Loss EELS is a suitable method for the analysis of these systems since in this energy range (0-60 eV), the edges corresponding to these elements are better defined. Besides, the bulk plasmon peak is also present in this range, providing additional information about compositional distribution and strain of the heterostructures [6]. In this work, compositional distribution of an InAs-GaAs-GaSb QDs heterostructure is estimated by the analysis of Low-Loss EEL spectra.

2. Experimental details

The studied heterostructure consists of 2.2 ML of InAs QDs grown by MBE at 510ºC on a GaAs (001) substrate. QDs have been capped by 6 ML of GaAs intermediate layer at 480°C, followed by 3 ML of GaSb and 100 nm of GaAs grown at 590ºC. Electron-transparent TEM samples were prepared following classical methods of mechanical grinding, dimpling and Ar milling at low voltage to minimize the damage. Only the thinnest areas of the sample (30-35 nm) were analysed to avoid the broadening and the shift of the plasmon peak due to surface excitation.

Spectrum Images (SI) have been acquired in SuperSTEM-1 (Daresbury Lab, U.K.), a VG HB501 UX FEG STEM equipped with a Nion spherical aberration corrector operating at 100 kV and a Gatan ENFINA™ parallel electron energy loss spectrometer system. The nominal probe size is 0.13 nm. The EELS data were acquired, from 0 to 60 eV, using a collection aperture with a semi-angle of 5 mrad and a convergence semi-angle of 24 mrad. A typical EELS map contained 700 spectra. The acquisition time for each spectrum was 0.35 s. In the Low-Loss region of the spectrum, using an energy dispersion of 0.05 eV per channel, the reached energy resolution was 0.35 eV. Although the use of a microscope with very small probe size allows a precise estimation of the composition, a 10% of error is expected. Apart from SI, high angle annular dark field (HAADF) images before and after the spectrum acquisition have been recorded to check the drift and/or damage caused in the sample by the beam, which have been negligible.

To analyse a SI, Zero-Loss peak and plural scattering must be removed using a standard Fourier-log deconvolution routine to obtain the Single Scattering Distribution which is related to the energy-loss function and contains the plasmon peak and the edges corresponding to the M and N signals. Then, the Kramers-Kronig (K-K) transformation is applied over the SSD to obtain the energetic dependence of the real, ε_1, and imaginary, ε_2, parts of the dielectric function. We will focus on the imaginary part, because it contains information of the 3d and 4d transitions. These main signals are 3d transitions for As ($M_{4,5}$ ~41 eV) and Ga ($M_{4,5}$ ~20-22 eV) and 4d for In ($N_{4,5}$ ~18 eV) and Sb ($N_{4,5}$ ~31 eV) [6]. After the background removal, signals have been integrated with a range of around 0.5 eV each side, to minimize the contribution of other near-by signals

The bulk plasmon peak is one of the most important signals in the Low-Loss region and it is the collective oscillation of the loosely bound electrons. Its shape and energy are characteristic of each solid. It can be simply described by the Drude-Lorentz model [7] which includes the single damped harmonic oscillator energy and the dielectric function of the core electrons. The bulk plasmon peak is modified by compositional and/or strain variations, even by the thickness of the TEM sample, among other factors. It has been observed that the change in the electron density via lattice parameter is the dominating factor determining the shift of the plasmon peak, where changes due to the band structure are a second-order effect. A full description of the methodology, including the relation among the parameters, as the lattice and the core dielectric constants can be found in reference [8].

3. Results and discussion

EEL spectra were acquired from the green rectangle marked in the HAADF of a QD (Figure 1a). After checking that the drift is reduced, steps described in the previous section have been followed to extract the imaginary part of the dielectric function. Then, the background is removed and signals are integrated over a range of ±2 eV. Extracted compositional maps are shown in Figures 1b)-1d).

Considering the complementary relationships between composition of groups III and V, from the compositional map of Ga and As, the distribution of In and Sb could be estimated as C_{In}=100-C_{Ga} and C_{Sb}=100-C_{As} where C_{In}, C_{Ga}, C_{Sb} and C_{As} represent In, Ga, Sb and As concentrations, respectively. Ga and As elemental maps are normalized with respect to Ga and As signals from GaAs substrate. The maximum of Ga is located on the substrate and surrounding the QD (Figure 1b). Inside the QD, Ga content is around 40% due to the well-known segregation phenomenon in InGaAs QDs systems [15]. Following a similar estimation procedure, As concentration inside the QD is estimated over 75-80%. Therefore, As is partially replaced by Sb. The Sb signal has been also extracted but, due to the high segregation coefficient of Sb and considering the inherent error of this analytical technique (around 10%) its content is expected to be below the error limit, which has been already confirmed by the analysis of this and similar heterostructures by other advanced techniques [9].

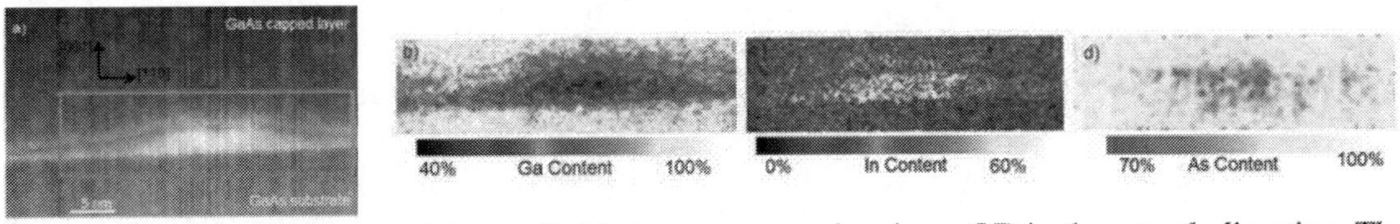

Figure 1. a) HAADF image of the studied heterostructure, showing a QD in the growth direction. The observed vertical stripes are due to the scan of the microscope. b), c) and d) show the Ga, In and As maps, respectively, corresponding to the green colour marked area on Figure 1a).

Furthermore, Ga and In elemental profiles have been acquired along the grown direction in a bi-dimensional area between QDs and integrated across the interface, in order to increase the signal/noise ratio (Figure 2). Maxima of Ga at the beginning and at the end of the profile are observed, corresponding to the GaAs substrate and final GaAs layer, respectively. After the first Ga maximum, there is a minimum on the Ga profile which corresponds to a maximum In content (InAs wetting layer). There should also be another Ga-maximum (from the GaAs intermediate layer) but instead, there is a relative maximum Ga peak, indicating a modification of the composition from the nominal one. It is followed by a relative minimum before the final maximum value. In every case, a Ga concentration decrease corresponds with an In increase, which confirms the complementarity of both elements, as we can observe even at the end of the analysed area; in this area it is more difficult to appreciate this behaviour due to the intensity decrease of the EEL signal. Although the existence of an In segregation process in these heterostructures is well-known, this profile does not correspond to a typical one in which, after a maximum, the In concentration decreases along the GaAs intermediate layer [10]. Instead, two In peaks are observed due to the presence of segregated Sb from the GaSb layer, which strongly competes with In and effectively blocks its incorporation into the lattice [9].

Concerning the bulk plasmon signal, bi-dimensional areas between QDs have been analysed. Figure 3 shows the variation of the plasmon energy $E_{p'}$, determined by a Gaussian fit, along the growth direction and integrated across the interface. For a GaAs-based heterostructure, the bulk plasmon is located around 16 eV; the observed shift is due to the relationship between plasmon energy and electron density, which is represented through the variation of the lattice volume. Changes of the plasmon energy are mainly originated by differences of the unit-cell size [11]. $E_{p'}$ moves to low energetic values with respect to GaAs, following a tendency similar to the experimental Ga profile taken from a Low-Loss EELS (Figures 2 and 3). Therefore, even when the slight variations of $E_{p'}$ observed are within the energy resolution of the experiments, they can be associated to the modifications of the lattice parameters. Further studies will provide more details about the relationship between the plasmon energy and the variation of the composition and strain of the heterostructure.

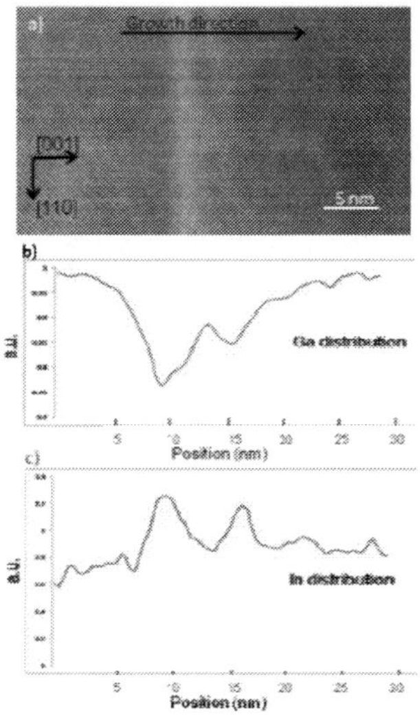

Figure 2. a) HAADF image of a bi-dimensional area. b) and c) represent Ga and In elemental distribution profiles, respectively.

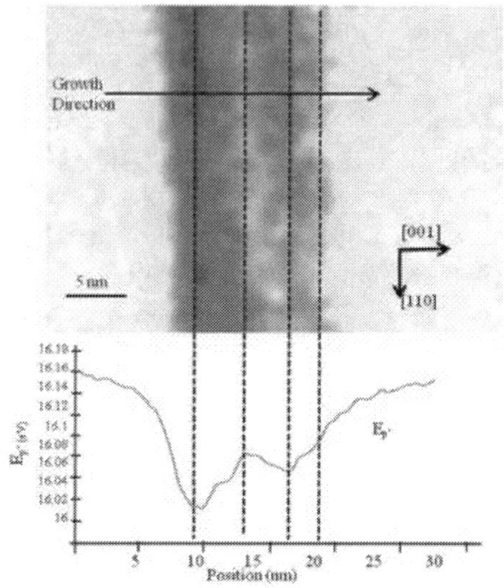

Figure 3. Image of the plasmon peak map and $E_{p'}$ profile integrated signal from a Low-Loss spectra.

4. Conclusions

Low-Loss EELS is an alternative method for compositional analysis of quaternary semiconductor heterostructures for alloys with convoluted core-loss peaks. Analyses show the existence of In-Ga atomic intermixing in the studied In-Ga-As-Sb heterostructure. Specifically, the formation of $In_xGa_{1-x}As$ and $In_xGa_{1-x}Sb_yAs_{1-y}$ alloys has been detected because of In segregation in spite of the presence of 6 ML GaAs intermediate layer.

Acknowledgements

This work was supported by the Spanish MINECO (project TEC2011-29120-C05-03 and CSD2009-00013) and the Junta de Andalucía (TEP-120 and INNANOMAT TEP-946 PAI research groups). Co-financing from UE-FEDER is also acknowledged.

References

[1] Akahane K, Yamamoto N and Ohtani N 2004 *Phys. E* **21** 295
[2] Ripalda J M, Granados D, González Y, Sánchez A M, Molina S I and García J M 2005 *Appl. Phys. Lett.* **87** 202108
[3] Moison J M, Houzay F, Barthe F, Gerard J M, Jusserand B, Massies J and Turco-Sandroff F S 1991 *J. Cryst. Growth* **111** 141
[4] Taboada A G et al 2010 *Phys. Rev. B* **82** 235316
[5] Molina S I et al 2007 *Appl. Phys. Lett.* **91** 263105
[6] Egerton R F 1996 *EELS in the electron microscope* (New York Plenum Press)
[7] Ahn C C 2004 *Transmission electron energy loss spectrometry in materials science and the EELS atlas* (California Wiley VCH)
[8] Sánchez A M, Beanland R, Papworth A J, Goodhew P J and Gass M H 2006 *Appl. Phys. Lett.* **88** 051917
[9] Sánchez A M , Beltrán A M, Beanland R, Ben T, Gass M H, De la Pegna F, Walls M, Taboada A G, Ripalda J M and Molina S I 2010 *Nanotechnology* **21** 145606
[10] Muraki K, Fukatsu S and Shiraki Y 1992 *Appl. Phys. Lett.* **61** 557
[11] Sánchez A M, Beanland R, Gass M H, Papworth A J, Goodhew P J and Hopkinson M 2005 *Phys. Rev. B* **72** 075339

Journal of Physics: Conference Series **471** (2013) 012013 doi:10.1088/1742-6596/471/1/012013

Evidence of perfect dislocation glide in nanoindented 4H-SiC

M Texier[1], A De Luca[1], B Pichaud[1], M Jublot[2], C Tromas[3], J-L Demenet[3] and J Rabier[3]

[1] Aix Marseille Université, CNRS, IM2NP UMR7334, Faculté des Sciences et Techniques, av. Escadrille Normandie Niémen, F-13397 Marseille Cedex, France

[2] CP2M, Aix-Marseille Univ, Faculté des Sciences et Techniques, Campus de Saint-Jérôme, Av. Escadrille Normandie Niémen, F-13397 Marseille Cedex, France

[3] Institut Pprime, UPR3346 CNRS - Université de Poitiers, bât. SP2MI, bd. Marie et Pierre Curie, Téléport 2, F-86962 Futuroscope - Chasseneuil du Poitou Cedex, France

E-mail: michael.texier@univ-amu.fr

Abstract. Plastic deformation of a 4H-SiC wafer has been produced by nanoindentation at room temperature. The superficial layer of the specimen in the indented area has been lifted out thanks to a specific focussed ion beam micromachining method and the deformation microstructure close to the imprints has been investigated by means of both conventional and high-resolution transmission electron microscopy (TEM). The analysis of the images revealed the presence of various types of extended defects. Perfect dislocations and single stacking faults bounded by isolated Shockley partial dislocations have been observed on the basal plane. Perfect dislocations have also been evidenced out of the basal plane. These results highlight the competition between various activated systems involved during the plastic deformation of 4H-SiC in the brittle regime.

1. Introduction

Plastic deformation in covalent semiconducting materials is known to proceed by the motion of dissociated dislocations in the ductile regime, but very little is known about the nature of gliding dislocations in the brittle domain, particularly at low temperatures. In silicon, both calculations and experiments indicate that mobile dislocations below the brittle-to-ductile transition temperature are likely perfect dislocations gliding in the shuffle set. In 6H and 4H-SiC, deformation experiments performed above the brittle-to-ductile transition and microstructure analyses suggest that plastic deformation involves weakly-dissociated dislocations [1-2] whereas Shockley partial dislocations were evidenced in 4H-SiC deformed in the brittle regime [3-7,9]. However, some studies have also reported the presence of perfect or weakly-dissociated dislocations in prismatic and basal planes in 4H-SiC single crystals deformed in the brittle regime by microindentation [8] or under confining pressure [9-12].

In this work, a (11-20)4H-SiC wafer was deformed by nanoindenting the surface at room temperature. TEM thin foils were then prepared by dual-beam focused ion beam (DBFIB) micromachining parallel to the indented surface in order to perform conventional and high-resolution transmission electron microscopy imaging of the deformation microstructure viewed in plan view.
The TEM observations revealed various types of defects in the investigated areas.

2. Experimental details

Plastic deformation of 4H-SiC wafer was achieved by nanoindenting the $(11\bar{2}0)$ surface. A periodic array of nano-imprints was performed in force-controlled mode using the CSM Instruments model NHT device equipped with a Berkovich indentor. The applied force was fixed to 15 mN leading to an average penetration depth (elastic plus plastic) of about 150 nm. After unloading, the dimensions of the residual nano-imprints were estimated to 60 nm in depth and 420 nm in lateral size. In order to facilitate the localisation of the nano-imprints, micro-indents were performed at the vicinity of the nano-indented area.

Samples for TEM imaging in plan view were prepared by focussed ion beam (FIB) micromachining using an original method specifically developed for this study. The principle of this sample preparation method rests on the use of a Si shield fixed onto the sample surface during the micromachining process. This precaution permits to protect the superficial layer of the specimen during the ion beam milling and to obtain a thin lamellae suitable for TEM analyses in plan view, *i.e.* suitable for studying the microstructure just below the surface. Details of the method will be shortly described [13]. Final back-side thinning and cleaning of the FIB lamellae was carried out by Ar^+ ion milling at very low voltage (down to 100 V) using the Fischione model 1040 Nanomill system, allowing to investigate the deformation microstructure around the nano-imprints by TEM imaging.

C-TEM and HRTEM observations were respectively performed using a FEI Tecnai G2 microscope operating at 200 kV and a FEI Titan 80-300 Cs-corrected microscope operating at 300 kV.

3. Results and discussion

Plan view C-TEM observations performed in the thickest areas of the FIB lamellae allowed us to get an overview of the deformation microstructure around the nano-indents. Characterization of the observed defects by the classical method based on their stereographic analysis and the use of extinction conditions has not been carried out on account of the very strong bending of the lamellae in the vicinity of the nano-indents. However, different types of extended defects can be distinguished in dark-field images, as shown in figure 1. On the one hand, numerous straight lines are observed perpendicular to one of the three facets of the imprint. The direction parallel to these defects is the $[1\bar{1}00]$ direction, indicating that these defects extend in the (0001) planes, which are seen edge-on for this observation orientation. On the other hand, dislocation half loops out of the basal plane are also visible (see figure 2). It must be underlined that the observed dislocation loops do not drag planar defects indicating that they consist of perfect non-dissociated (or weakly-dissociated) dislocations, in agreement with previous observations reported in [4] where perfect dislocations have been identified in prismatic planes in 4H-SiC micro-indented below the brittle-to-ductile transition temperature. In addition, the observed dislocation loops are very facetted; one of the segments of the largest loop shown in figure 2, parallel to the $[0001]$ direction, exhibits a very large kink suggesting that the growth of this loop preferentially proceeded by the nucleation and propagation of kinks along this direction.

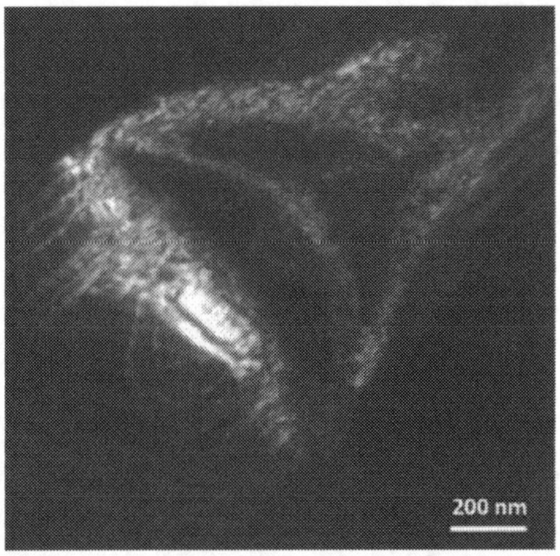

Figure 1. Dark-field TEM image of a nano-imprint seen in front view. Numerous extended defects parallel to the $[1\bar{1}00]$ direction are visible.

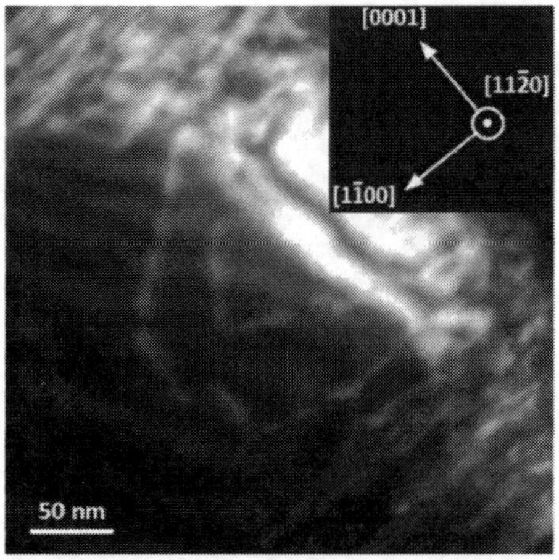

Figure 2. Detail of the dark field image shown in figure 1. Two perfect dislocation loops extend out of the basal plane which is seen edge on for this crystal orientation.

HRTEM analyses have also been performed in very thin areas of the sample, close to the nano-indents. The HRTEM images and the associated strain mapping obtained by the geometrical phase analysis (GPA) method applied to the experimental images, revealed the presence of many isolated stacking faults (SFs) extended in the basal plane, as shown in figure 3. A careful examination of the observed stacking disorder introduced in the 4H-SiC structure (see for instance the disrupted sequence of the C-centred tetrahedra superimposed to the image in figure 3) indicates that these defects systematically correspond to single SFs. It must be underlined that multiple SFs have not been identified in this study, contrary to previous studies performed in 4H-SiC deformed in the brittle regime [6-9].

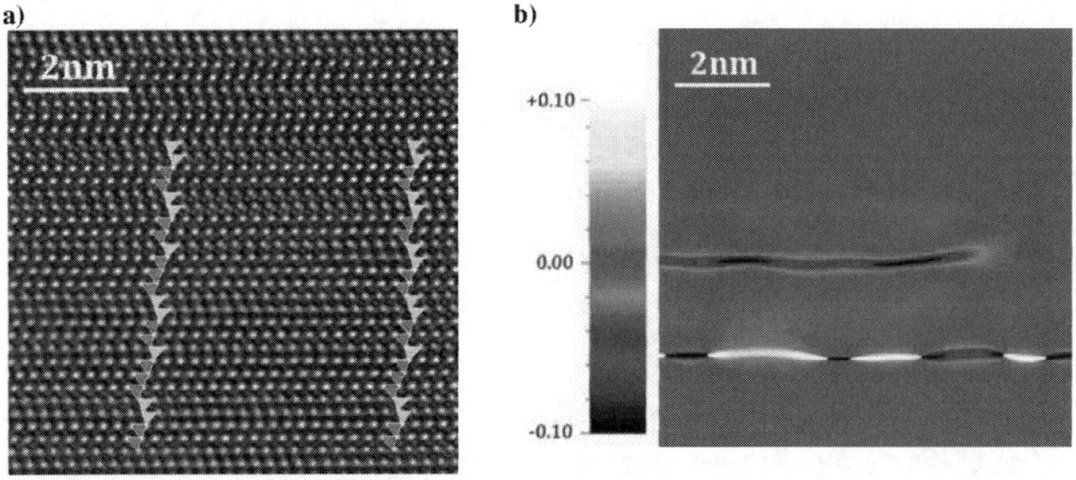

Figure 3. a) HRTEM image of the area close to a nano-indent. Two single SFs are visible in this image. One can note that the comparison between the stacking sequences at the left and right sides of the image evidenced the presence of a Shockley partial dislocation in the field of view of the image. b) Shear map (ε_{xy}) measured using the GPA method ($e_x = [1\bar{1}00], e_y = [0004]$) in the same area.

Journal of Physics: Conference Series **471** (2013) 012013 doi:10.1088/1742-6596/471/1/012013

In addition, the presence of non-dissociated perfect dislocation in the basal plane was also revealed by HRTEM imaging. The structure shown in figure 4 contains a dislocation parallel to the observation direction which can clearly be evidenced thanks to the GPA strain mapping. In order to unambiguously identify the character of this dislocation, multislice simulations based on elastic calculations (isotropic approximation) for various dislocation types have been compared with the experimental image. From these simulations, this dislocation has been identified as being a 60° perfect dislocation.

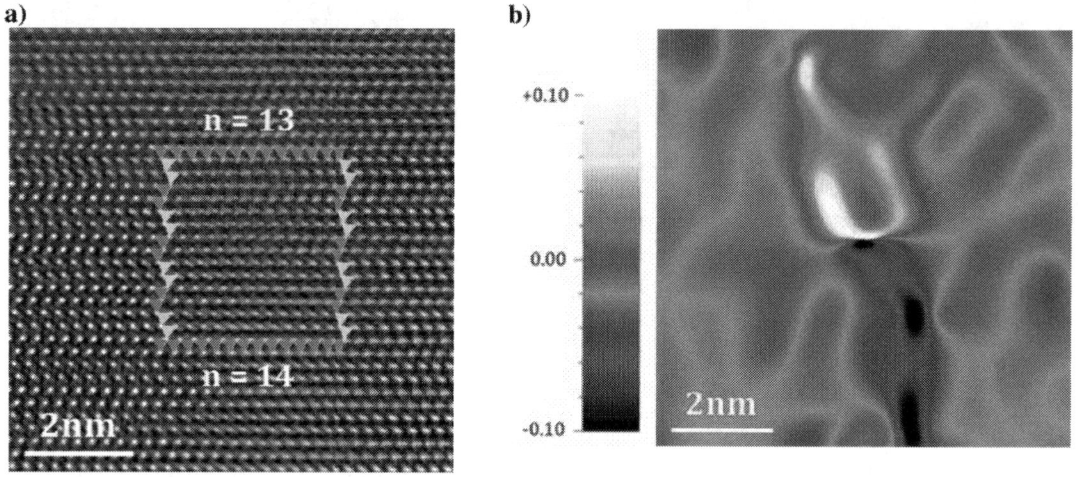

Figure 4. a) HRTEM image of a perfect 60° dislocation. b) Corresponding strain map ($\varepsilon_{1\bar{1}00}$) measured using the GPA method.

4. Conclusion

Defects produced by nanoindentation of 4H-SiC at room temperature have been studied by TEM imaging. This work evidenced the concomitant nucleation and propagation of various types of defects for the low-temperature plastic deformation of this material, in agreement with [9]. Isolated Shockley partial dislocations and resulting single SFs have been frequently observed in the basal plane. Perfect dislocations have been identified in and out of the basal plane. These observations highlight the possible motion of perfect dislocations in the shuffle set in the brittle regime.

References

[1] Samant A V, Zhou W L and Pirouz P 1998, *Phys. Stat. Sol.* A **166**, 155
[2] Samant A V, Hong M H and Pirouz P 2000, *Phys. Stat. Sol.* B **222**, 75
[3] Ning X J, Huvey N and Pirouz P 1997, *J. Am. Ceram. Soc.* **80** 1645
[4] Mussi A, Rabier J, Thilly L and Demenet J-L 2007, *Phys. Stat. Sol.* C **4** 2929
[5] Regula G, Lancin M, Idrissi H, Pichaud B and Douin J 2005, *Phil. Mag. Lett.* **85** 259
[6] Texier M, Regula G, Lancin M and Pichaud B 2006, *Phil. Mag. Lett.* **85** 529
[7] Lancin M, Texier M, Regula G and Pichaud B 2009, *Phil. Mag.* **89** 1251
[8] Maeda K, Suzuki K, Fujita S, Ichihara M and Hyodo S 1988 *Phil. Mag.* A **57** 573
[9] Demenet J-L, Milhet X and Rabier J 2005, *Phys. Stat. Sol.* C **2** 1987
[10] Demenet J-L, Rabier J, Milhet X, Hong M H, Pirouz P, Stretton I and Cordier P 2002, *J. Phys.: Condens. Matter.* **14** 12961
[11] Rabier J, Demenet J-L, Denanot M-F and Milhet X 2005, *Mat. Sci. & Eng.* A **400-401** 97
[12] Mussi A, Demenet J-L and Rabier J 2006, *Phil. Mag. Lett.* **86** 561
[13] Jublot M and Texier M, to be published

The nature of electrostatic potential fluctuations in Cu_2ZnSnS_4 and their role on photovoltaic device performance

BG Mendis[1*], MD Shannon[2], MCJ Goodman[1], JD Major[3], AA Taylor[1], DP Halliday[1], K Durose[3]

[1] Dept of Physics, Durham University, South Road, Durham, DH1 3LE, UK
[2] SuperSTEM, Daresbury Laboratories, Keckwick Lane, Daresbury, WA4 4AD, UK
[3] Stephenson Institute for Renewable Energy, University of Liverpool, L69 7ZF, UK

E-mail: b.g.mendis@durham.ac.uk

Abstract. Aberration corrected STEM EELS is used to investigate point defects in Cu_2ZnSnS_4 (CZTS). Nano-scale clusters of Zn_{Cu} anti-site donors are observed with the donor concentration being sufficiently high to degenerately dope the semiconductor. Uncompensated donors and acceptors result in electrostatic potential fluctuations within the material. The effect of these potential fluctuations on the photovoltaic device properties is discussed.

1. Introduction

Cu_2ZnSnS_4 (CZTS) shows considerable promise as a thin-film photovoltaic absorber layer for Tera-Watt energy production since it has none of the materials issues, such as abundance, toxicity and cost, that are expected to be bottlenecks for CdTe and $Cu(In,Ga)Se_2$ (CIGS) technologies. Despite limited thermodynamic stability of the CZTS phase causing secondary phase precipitation the record cell efficiency for CZTS is at present 11.1% [1]. Defect free CZTS forms in one of two tetragonal crystal structures, named kesterite and stannite after the parent mineralogical phases [2]. Kesterite and stannite can be derived from the chalcopyrite CIGS phase by replacing two group IIIb elements (In/Ga) with a group IIb (Zn) and group IVb element (Sn), thereby satisfying the octet rule; ordering of the Cu and Zn cations within the (004) lattice planes gives rise to the two CZTS polymorphs (figure 1a). Since Cu and Zn are neighbouring elements in the periodic table, the changes in bonding are subtle, with the result that the theoretical energy difference between the two structures is only 3 meV/atom [3]; this allows for Cu, Zn cation intermixing as confirmed by neutron diffraction [4]. Uncompensated point defects give rise to electrostatic potential fluctuations within the material as well as energy states within the band-gap which are Shockley-Read-Hall (SRH) recombination sites. Indeed Romero *et al.* [5] observed a blue shift of the main CZTS luminescence peak during carrier injection, either by increased temperature or illumination intensity, and attributed this to a transition from band tail recombination to band-band recombination at the high carrier densities where the amplitudes of the potential fluctuations are diminished. Density functional theory studies for CZTS [6] shows Cu_{Zn},V_{Cu} to be acceptors with low formation energy while Zn_{Cu} is a low formation energy donor. Point defects associated with the multi-valent Sn ion on the other hand tend to have energy levels deep within the band-gap [6,7].

* To whom any correspondence should be addressed.

In this contribution aberration corrected, analytical scanning transmission electron microscopy (STEM) is used to identify point defects in CZTS. Structural information is extracted at the nano-scale that complements results from neutron diffraction [4], a technique that requires much larger sample volumes. The experimental results have been reported previously [8] and will therefore be only summarised here for completeness. This precedes a detailed discussion on the role of potential fluctuations on photovoltaic device properties. The discussion is a synthesis of extensive literature results for grain boundary potential fluctuations [9,10] as well as potential/band-gap fluctuations due to composition inhomogeneities in CIGS [11,12].

2. Sample and methods

Bulk CZTS samples were synthesised by annealing elemental powders at 800 °C for 24 hours, the starting composition of the weighed material being $Cu_{1.5}ZnSnS_4$ (i.e. Cu-poor). Sulphur was placed separately to the other elements in an evacuated quartz ampoule during annealing. The crushed powder was dispersed onto a holey carbon grid and analysed using an aberration corrected Nion UltraSTEM operating at 100 kV. The composition was measured using electron energy loss spectroscopy (EELS); probe convergence and EELS semi-collection angles were 31 mrad.

3. Results and Discussion

3.1 Aberration corrected STEM EELS

Figure 1(a) is a schematic of the kesterite and stannite crystal structures. As mentioned previously, the two phases differ in the ordering of Cu and Zn cations within the (004) planes. Viewing the tetragonal crystal along the *a*-axis we find the pure Sn atom columns to be separated by three atom columns of either pure Cu or pure Zn in a direction parallel to the *c*-axis. Sn appears brighter in a STEM high angle annular dark field (HAADF) image due to its larger atomic number and therefore provides unique reference points of the atoms within the unit cell. Chemical analysis of the three Cu/Zn atomic columns sandwiched between the Sn is sufficient to distinguish between pure kesterite and stannite phases. For the former the composition of the atom columns varies as Cu, Cu, Zn (or its reverse), while in the latter the variation is as Cu, Zn, Cu. An atom column containing both Cu and Zn shows evidence for Cu, Zn disorder within the unit cell (i.e. either Zn_{Cu} or Cu_{Zn} anti-site atoms).

Figure 1(b) is a HAADF image of CZTS in the *a*-axis orientation. Cu and Zn L-edge EELS spectra for the probe positioned on atom columns labelled atom1, atom2 and atom3 were extracted from a line spectrum image acquired along the direction marked by the arrow (these atom columns are sandwiched between the Sn columns labelled Sn1 and Sn2). The individual EELS spectra are shown superimposed in figure 1(c). The shape of the EELS spectra is similar for all three atom columns and reveals the Zn L-edge clearly but little or no intensity for the Cu L-edge at ~930 eV. Since for both kesterite and stannite two atom columns must be pure Cu this implies the presence of a large concentration of Zn_{Cu} donor anti-site atoms. There is evidence for the measurement being influenced by electron beam broadening within the sample, since the EELS signal extracted from a Sn column (e.g. Sn1) shows a similar spectrum shape to that in figure 1(c), despite no (or very little) Zn being expected within the Sn column. The geometric spreading of the electron probe is of the order of a few nanometres, suggesting that the composition inhomogeneity observed is a particularly large point defect cluster. If it is assumed that all Cu atoms in the unit cell are replaced by Zn (i.e. no Cu vacancies) the Zn_{Cu} donor concentration is estimated to be $\sim 10^{22}$ cm^{-3}. This is considerably larger than typical values for the effective density of states in the conduction band so that the material is degenerately doped locally. It should be mentioned however that most STEM EELS measurements revealed atom columns with mixed Cu, Zn compositions, although beam spreading is always present as in figure 1. Further details of the STEM EELS results can be found in [8].

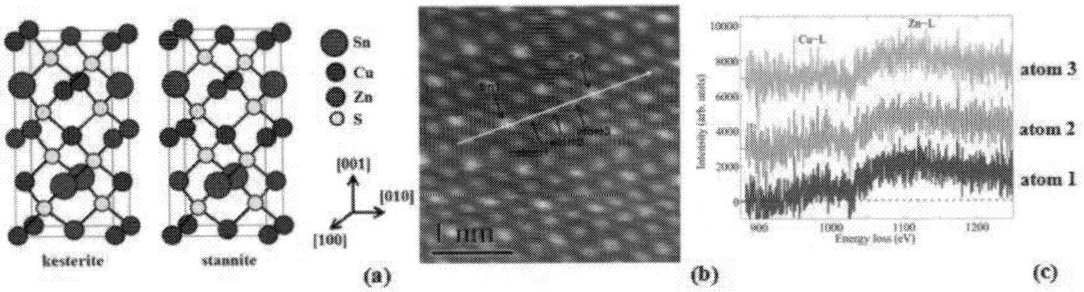

Figure 1: (a) crystal structure of kesterite and stannite, (b) STEM HAADF image of CZTS along the *a*-axis and (c) EELS spectra for the probe positioned on the indicated atom columns.

3.2 Band structure of a point defect cluster

A schematic band structure of the experimentally determined point defect cluster is shown in figure 2a. The Zn_{Cu} anti-site atoms are assumed to be ionised, with the result that positive charge is accumulated over the cluster of size *d*. The increase in potential leads to downward bending of the conduction and valence bands. The point defect cluster therefore acts as a sink for electrons and barrier for holes; consequently it is less *p*-doped than the neighbouring bulk (*p*-type) CZTS. Minority electrons within CZTS screen the positive charge of the cluster over a distance governed by the Debye screening length. Zn_{Cu} anti-site atoms are shallow donors [6] but, due to degenerate doping, the discrete energy level is broadened into a continuous band that extends into the conduction band. Note that the band structure for the point defect cluster has features similar to charged grain boundaries [9,10].

In the above analysis any change in lattice parameter due to Zn_{Cu} anti-site atoms has not been taken into account. Strain gives rise to band-gap fluctuations superimposed on to the electrostatic potential band bending. In CIGS, the band-gap depends on the Ga content, so that band-gap fluctuations can also be generated by variations of the Ga/(In+Ga) ratio. Similar effects may occur in CZTS as well. For example, the highest performing CZTS devices have mixed S and Se anions [1,2], with Se-rich compositions giving smaller band gaps compared to S-rich compositions.

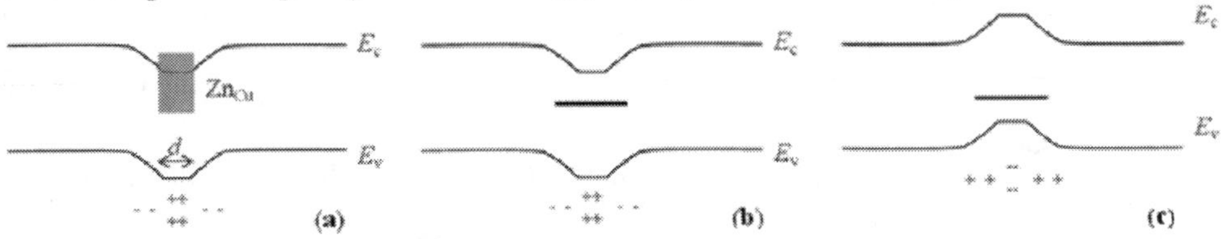

Figure 2: Schematic band structure diagrams for (a) Zn_{Cu} cluster observed experimentally, (b) single energy level uncompensated donors and (c) single energy level uncompensated acceptors.

3.3 Effect of electrostatic potential fluctuations on device performance

The donor/acceptor energy level of an electrostatic potential fluctuation is a SRH recombination site and can therefore affect photovoltaic device parameters. SRH recombination is largest when (i) the electron and hole concentrations at the defect are equal and (ii) the defect energy level lies deep within the band-gap. Figures 2(b) and 2(c) are band structure diagrams for single energy level uncompensated donors and acceptors, respectively. The first criterion for SRH recombination means that in *p*-type CZTS uncompensated donors are stronger recombination sites than uncompensated acceptors (the latter is a sink for majority carrier holes so that the cluster region is more *p*-doped than the neighbouring CZTS). The requirement of deep defect levels means that point defects of Sn are more likely to be deleterious than (say) shallow Zn_{Cu} anti-site atoms. Furthermore, electrical bias also has an important effect on SRH recombination. On biasing, the electron and hole quasi-Fermi levels move towards the conduction and valence band respectively, such that their separation is proportional to the applied voltage. Under high injection conditions the potential change (i.e. band bending) at the cluster will also diminish, in a manner similar to grain boundaries [9], due to changes in the defect level occupancies. The comparable electron and hole populations result in stronger recombination under bias for both uncompensated donors and acceptors.

On a macroscopic level the device *J-V* characteristics can be described as:

$$J(V)=\left[J_{\mathrm{diode}}(V)+J_{\mathrm{SRH}}(V)\right]-J_{SC} \qquad (1)$$

where J_{diode} is the current due to radiative recombination, J_{SRH} the current due to SRH recombination within the space charge region and J_{SC} is the photogenerated short circuit current. From previous studies on grain boundaries [10], point defect clusters are not expected to significantly affect J_{SC}; this is because materials such as CZTS absorb most of the light near the space charge region, so that only carriers generated from long wavelength photons are likely to be lost to recombination in the diode quasi-neutral region. Furthermore, for high efficiency cells, the fraction of defect-free regions is still high enough for generating a large J_{SC}. On the other hand, recombination at point defect clusters in the space charge region will increase J_{SRH} so that the open circuit voltage (V_{OC}) is reduced. The higher diode ideality factor for J_{SRH} also leads to a reduction of the fill factor. The decrease in V_{OC} can be offset via a series resistance [12], so that the voltage across the point defect cluster is lowered, thereby reducing SRH recombination rates. The series resistance can be due to point defect clusters which are obstacles to majority carrier flow, i.e. uncompensated donors within the quasi-neutral diode region (figure 2b), or as a more extreme example, current blocking secondary phases such as ZnSe [13]. Nevertheless, the series resistance also deteriorates the fill factor so that an optimum series resistance exists where the two effects are balanced [12]. Finally, any band-gap fluctuations will affect J_{diode} as this is governed by radiative recombination; a decrease in band-gap tends to equalise electron and hole populations resulting in a larger J_{diode} and hence smaller V_{OC}.

4. Conclusions

Aberration corrected STEM EELS has provided the first direct evidence for electrostatic fluctuations in CZTS in the form of a nano-scale Zn_{Cu} donor anti-site atom cluster that is degenerately doped. Reference to the literature indicates that such point defect clusters are expected to be a site for Shockley-Read-Hall recombination, with the result that the open circuit voltage and fill factor are reduced simultaneously. Uncompensated donors in *p*-type material are stronger recombination centres than uncompensated acceptors. Furthermore an optimised series resistance can be beneficial in minimising the deleterious effects of point defect cluster recombination.

References

[1] Todorov TK, Tang J, Bag S, Gunawan O, Gokmen T, Zhu Y and Mitzi DB 2013, *Adv. Energy Mat.* **3** 34.
[2] Mitzi DB, Gunawan O, Todorov TK, Wang K and Guha S 2011, *Sol. Energy Mat. Sol. Cells* **95** 1421.
[3] Chen S, Gong XG, Walsh A and Wei S.-H 2009, *Appl. Phys. Lett.* **94** 041903.
[4] Schorr S, Hoebler H.-J and Tovar M 2007, *Eur. J. Miner.* **19** 65.
[5] Romero MJ, Du H, Teeter G, Yan Y and Al-Jassim MM 2011, *Phys Rev B* **84** 165324.
[6] Chen S, Yang J.-H, Gong XG, Walsh A and Wei S.-H 2010, *Phys Rev B* **81** 245204.
[7] Biswas K, Lany S and Zunger A 2010, *Appl. Phys. Lett.* **96** 201902.
[8] Mendis BG, Shannon MD, Goodman MCJ, Major JD, Claridge R, Halliday DP and Durose K 2012, *Prog. Photovolt: Res. Appl.* DOI: 10.1002/pip.2279.
[9] Wong J, Huang JL, Varlamov S, Green MA, Evans R, Keevers M and Egan RJ 2011, *Prog. Photovolt: Res. Appl.* **19** 695.
[10] Gloeckler M, Sites JR and Metzger WK 2005, *J. Appl. Phys.* **98** 113704.
[11] Werner JH, Mattheis J and Rau U 2005, *Thin Solid Films* **480** 399.
[12] Grabitz PO, Rau U and Werner JH 2005, *Phys. Stat. Sol. (a)* **15** 2920.
[13] Wätjen JT, Engman J, Edoff M and Platzer-Björkman C 2012, *Appl. Phys. Lett.* **100** 173510.

Journal of Physics: Conference Series **471** (2013) 012015 doi:10.1088/1742-6596/471/1/012015

ZnO layers deposited by Atomic Layer Deposition

B Pécz[1,2], Zs Baji[1], Z Lábadi[1] and A Kovács[3]

[1] Institute for Technical Physics and Matl. Sci., Research Centre for Natural Sciences, Hungarian Academy of Sciences, 1121 Budapest, Konkoly-Thege u. 29-33
[2] Doctoral School of Molecular- and Nanotechnologies, Faculty of Information Technology, University of Pannonia; Egyetem u.10, Veszprém, H-8200 Hungary
[3] Ernst Ruska-Centre for Microscopy and Spectroscopy with Electrons, Peter Grünberg Institute, Forschungszentrum Jülich, D-52425, Germany

E-mail: pecz.bela@ttk.mta.hu

Abstract. The structure of 40 nm thick epitaxial ZnO layers grown on single crystalline sapphire and GaN substrates by atomic layer deposition has been studied using transmission electron microscopy. The growth is carried out between 150°C and 300°C without any buffer layer using di-ethyl zinc and water precursors. The ZnO layer on sapphire is found to be polycrystalline, which is probably due to the large misfit (~15 %) and the relatively low deposition temperature. However, the small misfit (~1.8 %) between the ZnO layer that is deposited on GaN at 300°C resulted in a high quality single crystalline layer.

1. Introduction

ZnO is a wide band gap (~3.37 eV at room temperature) semiconductor and one of the transparent conductive oxides (TCO) that has recently gained interest for solar cells as a replacement of indium tin oxide partly for safety, partly for low cost reasons. For this kind of applications, the polycrystalline ZnO thin film is usually sputter deposited, but other deposition methods such as metalorganic chemical vapour deposition (MOCVD), molecular beam epitaxy (MBE) and atomic layer deposition (ALD) are applied when ZnO is to be used as a semiconducting layer. The principle of the ALD process is that the precursors are introduced into the reaction chamber as time-separated pulses; therefore their reactions can take place only at the gas-solid interface, and never in the gas phase [1]. Consequently the growth can be controlled on an atomic level, and the grown layers are extremely conformal and uniform. The ALD growth, however, is time consuming since several cycles are needed to deposit a complete monolayer. However, due to this growth mechanism a well-controlled epitaxial growth is possible although ALD is used most commonly for the deposition of oxides, nitrides, and other polycrystalline composite layers.

ALD ZnO layers are usually deposited on Si and glass substrates, in which case the grown layers are polycrystalline. There are a few reports published on ALD deposited epitaxial ZnO layers [2-4] on GaN, or Al_2O_3 (sapphire). (In such a case the method is also called atomic layer epitaxy).

2. Experimental

In this work ZnO films are deposited using ALD on sapphire and GaN substrates in a Picosun Sunale R-100 type ALD reactor. Di-ethyl zinc (DEZ) and water were used as precursors for growth of ZnO.

The deposition was carried out in the temperature range of 150 - 300°C. 500 ALD cycles were done, which gave a layer with thickness of about 40 nm. The deposited layers were studied using conventional and aberration-corrected transmission electron microscopy (TEM). High-resolution TEM images were taken in a JEOL 3010 microscope at 300 kV, while the aberration-corrected TEM images were recorded in a FEI Titan microscope operated at 300 kV. The aberration functions were corrected up to fourth order. The spherical aberration function (Cs) was set up to be negative (~ -15 μm) in order to enhance the contrast of the images. The TEM specimens were prepared using conventional methods, namely mechanical polishing and Ar ion milling. The specimen preparation process was finished using low-energy (< 1 keV) Ar ion milling in order to reduce the surface damage.

3. Results

3.1 ZnO on sapphire

Figure 1 shows a cross section of a ZnO layer grown on sapphire at 300°C. The image shows ZnO columns with different orientations. X-ray diffraction results also showed a very small (10$\bar{1}$1) ZnO peak beside the strong (0002) peak [5].

Figure 1. High-resolution cross-sectional TEM image of ZnO/sapphire. One can see crystalline grains in ZnO with different orientations.

The same sample was investigated in plan-view as well. Figures 2 and 3 show the bright field (BF) TEM and the dark field (DF) TEM images, respectively. The morphology, the contrast and the moiré fringes all confirmed a polycrystalline ZnO layer formation. The grain size values are measured in the 10-50 nm range. The selected area diffraction (SAED) pattern is shown in Figure 4 and one can see immediately that some of the grains are oriented with respect to the single crystalline substrate, while other grains give diffraction rings due to their random orientation. Reflections of the Al_2O_3 substrate are marked with dots and indexed with hexagonal indexes. The ZnO diffraction rings are indexed with italic style. One should note that the ZnO is highly textured. The intensity along the 10$\bar{1}$0 ring has 6 maxima close to Al_2O_3 reflections indicating that most of the ZnO is nearly in epitaxial orientation (as (11$\bar{2}$0) Al_2O_3 || (10$\bar{1}$0) ZnO AND (0001) Al_2O_3 || (0001) ZnO). Double diffraction can also be seen forming a hexagon around the Al_2O_3 reflections.

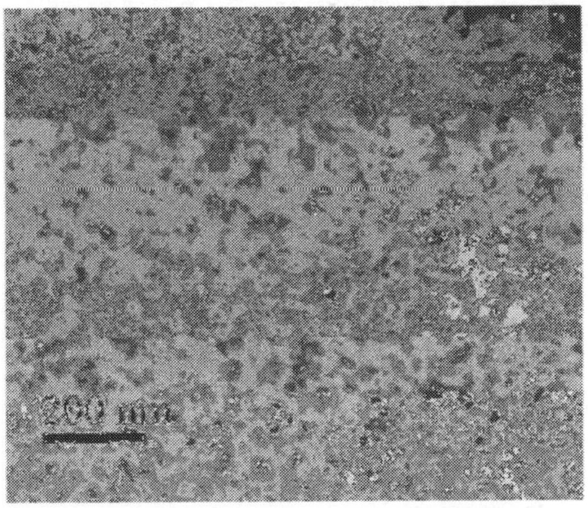

Figure 2. BF TEM image taken on a plan view ZnO/sapphire sample.

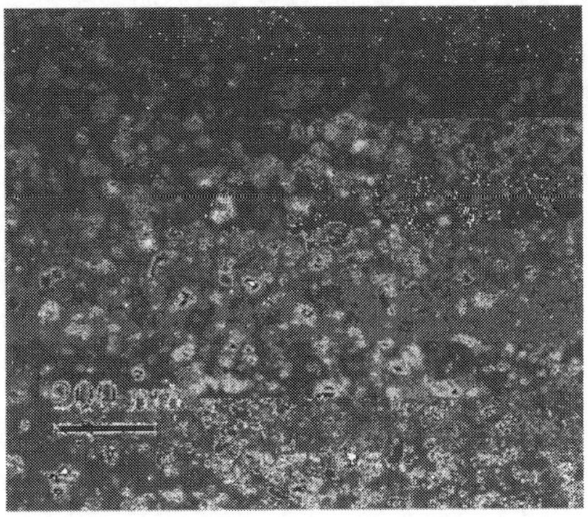

Figure 3. DF TEM image of ZnO grains.

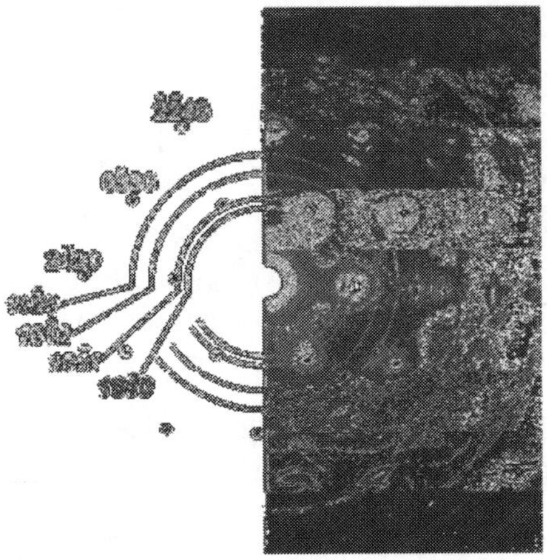

Figure 4. Indexed SAED pattern taken on a plan view ZnO/Al_2O_3 sample grown at 300°C. Al_2O_3 reflections are marked with dots and indexed with hexagonal indexes. The ZnO diffraction rings are indexed with italic style.

3.2 ZnO on GaN

ZnO is deposited onto GaN as well and the structural properties have been investigated using cross-sectional TEM. X-ray diffraction results showed a dominant (0002) peak of ZnO, and a $(10\bar{1}1)$ also appears in the layer that was deposited at low temperature [5]. However, the $(10\bar{1}1)$ peak is not observed in the case of the ZnO layer deposited at 300°C. X-ray results [5] indicate a single crystalline ZnO layer, epitaxial to the GaN substrate. Indeed, the high resolution TEM image in Fig. 5 confirmed the epitaxial growth of ZnO on GaN. The inset is a Fast Fourier transformation (FFT) of the experimental image that shows one set of single crystalline diffraction pattern, as they are not separated due to the very low (1.8%) misfit. Figure 6 shows a plan-view BF TEM image of the same sample showing a network of moiré fringes. The moiré fringe periodicity is ~15.5 nm, which agrees well with the calculated value, when we consider the overlapping of $(10\bar{1}0)$ lattices in ZnO and GaN. However, taking SAED patterns in different regions (of about 5 μm area) sometimes we observed extra spots in them, which we could identify definitely as $(10\bar{1}1)$ reflections. Therefore we concluded that ZnO grains with different orientations can be traced in some regions despite X-ray diffraction measurements that suggested a completely single crystalline nature of ZnO.

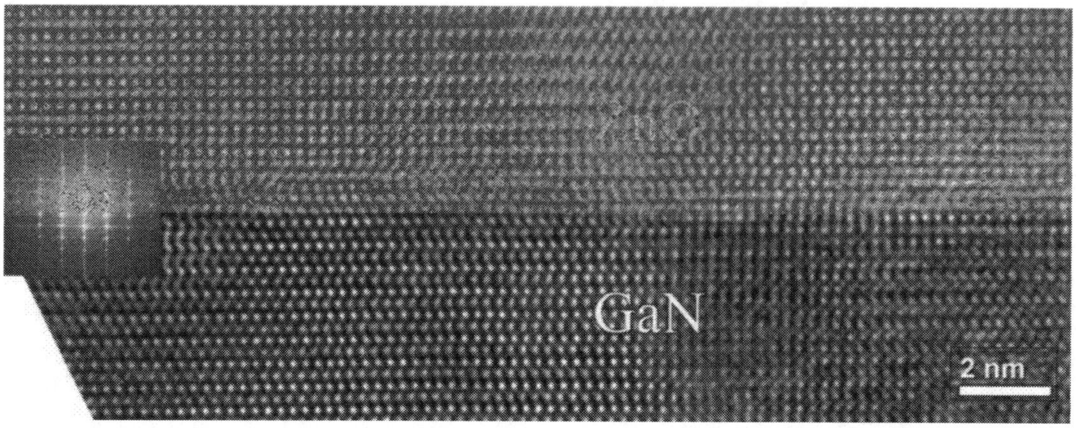

Figure 5. High resolution TEM image showing the ZnO/GaN interface. The inset is an FFT pattern recorded from the interface region.

Figure 6. Plan view TEM image of ZnO/GaN showing a moiré pattern.

4. Conclusions

The ZnO layer on sapphire was found to be polycrystalline, which is explained by the large misfit and by the relatively low deposition temperature. ZnO deposited on GaN is single crystalline due to the low misfit. However, occasionally ZnO domains with different orientations were traced by selected area diffraction images, showing again that SAED is more sensitive for detecting small grains than X-ray analysis.

Acknowledgments

This work was supported by the National Development Agency grant TÁMOP-4.2.2/B-10/1-2010-0025.

References

[1] Leskela M, and Ritala M 2002 *Thin Solid Films* **409** 138
[2] Chen HC, Chen MJ, Liu TC, Yang JR and Shiojiri M 2010 *Thin Solid Films* **519** 536
[3] Wachnicki L, Krajewski T, Luka G, Witkowski B, Kowalski B, Kopalko K, Domagala JZ, Guziewicz M, Godlewski M and Guziewicz E 2010 *Thin Solid Films* **518** 4556
[4] Ku CS, Huang JM, Lin CM and Lee HY 2010 *Thin Solid Films* **518** 1373
[5] Baji Zs, Lábadi Z, Horváth ZsE, Molnár Gy, Volk J, Bársony I and Barna PB 2012 *Cryst. Growth Des.* **12** 5615

SEM/EDS/EBSD study of the behaviour of Ge, Mo and Al impurities in complex-doped crystals of higher manganese silicide

A S Orekhov[1,2] and F Y Solomkin[3]

[1]Shubnikov Institute of Crystallography RAS, Russia, Moscow
[2]National Research Center «Kurchatov Institute», Russia, Moscow
[3]Ioffe Physical Technical Institute of RAS, Russia, St. Petersburg

E-mail: andrey.orekhov@gmail.com

Abstract. The structure of Al, Ge, Mo doped higher manganese silicides (HMS) grown by the Bridgman technique has been studied by SEM/EDS/EBSD methods. It is shown that dopants are partially integrated into the HMS crystal lattice. Some inclusions with sizes of 0.1-100 μm and different shapes (round, irregular, elongated) are formed. The precipitation of tetragonal $MoSi_2$ and Si-Ge solid-solution has been observed. $MoSi_2$ inclusions hundreds of microns in size form a multicomponent texture. The inclusions of Si-Ge solid solution have an irregular shape. The orientation relationship between these inclusions and matrix crystal is determined.

1. Introduction

Higher manganese silicides (HMS) of stoichiometry $MnSi_{1.71\text{-}1.75}$ attract the attention as potential middle-temperature thermoelectric materials which are rather cheap and non-toxic [1-3]. There are four phases with approximate composition Si/Mn=1.7 corresponding to HMS. The HMS has a figure of merit of $ZT=\alpha^2\sigma T/\kappa \sim 0.4$ (α – Seebeck coefficient, σ – electrical conductivity, κ – thermal conductivity, T – absolute temperature).

As a rule, when HMS crystal was grown by the Czochralsky or Bridgman technique the ordered precipitation of MnSi cubic phase (striations) has been observed [4, 5]. The MnSi phase is a good conductor with high thermal conductivity, thus it causes the degradation of HMS figure of merit [6]. The possible way to improve thermoelectric properties is doping. Aoyama et al. [7] showed that Ge doped HMS crystals possessed higher *ZT* value, and the microstructure of doped crystals was changed, MnSi cubic phase was not revealed. Al impurity has been shown to increase the HMS figure of merit [8]. Mo reduces the lattice component of thermal conductivity [9]. The most significant increase in the figure of merit is observed with combined doping by Ge, Mo and Al [2, 8]. However, there is not enough information about the microstructure of doped HMS crystals. The aim of this work is to reveal structural features of complex doped HMS crystals.

2. Materials and methods

A complex doped HMS crystal of nominal composition $(Mn_{0.98}Mo_{0.02})[(Si_{0.98}Ge_{0.02})_{1.75}]_{0.99}Al_{0.01}$ was grown by modified Bridgman method described in detail in [8].

Advanced electron microscopy techniques were applied for characterization of elements distribution in matrix crystal. Scanning Electron Microscopy (SEM), Energy-Dispersive X-ray Spectroscopy (EDS) and Electron Backscattered Electron Diffraction (EBSD) were used for morphology and microstructure investigation as well as for chemical analysis, phase composition study and texture analysis. A dual-beam scanning electron microscope FEI Quanta 200 3D equipped with X-ray EDS and EBSD detectors (EDAX, Ametek) was used.

The samples were prepared by mechanical polishing with diamond lapping films (grain size down to 0.5 μm). For EBSD technique the final step was the mechanical polishing by Al_2O_3 colloidal suspension with grain size of 50 nm. The upper disordered layer was removed by Ar^+ ion milling.

3. Experimental Results

3.1 SEM / EDS data

Analysis of SEM images of the complex-doped HMS crystal does not reveal any ordered precipitates of MnSi phase. However, the matrix crystal contains inclusions with sizes of 0.1-100 μm that have different shapes (round, irregular, elongated) and compositions (Fig. 1).

In accord with EDS microanalysis data obtained at 15 keV the manganese/silicon ratio in the doped matrix crystal is 1.7. This ratio corresponds to HMS composition. The dopants observed at concentrations above detection limits were Al and Mo (0.5 ± 0.3 at%) and Ge (1.5 ± 0.4 at%).

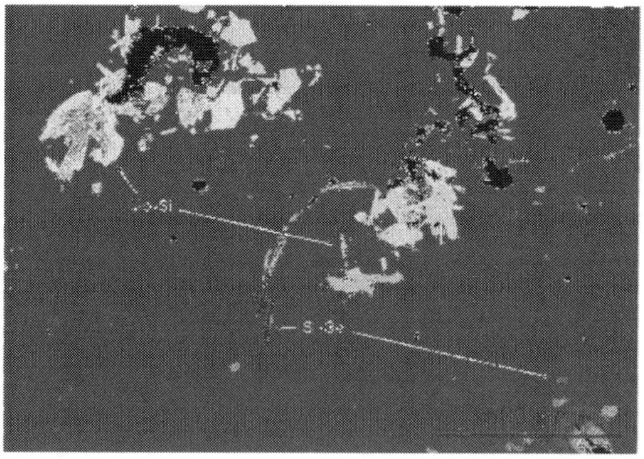

Figure 1. Backscattered SEM image of complex-doped higher manganese silicide crystal

The elemental distribution analysis reveals Si-Ge and Si-Mo inclusions are present in the sample. The EDS map of Al distribution coincides with O map, thus it can be supposed that alumina is formed. The Si-Ge solid-solution inclusions have irregular shape with different ratios of elements. In accordance with microanalysis data, two types of Si-Ge inclusions are present - Ge-rich and precipitates with composition close to SiGe. The first one consist of ~ 90 at% Ge and the second one have a ratio of Si:Ge = 1:1 in the central part and are Ge-rich at the boundary. In addition, separate faceted Mo-containing inclusions hundreds of microns in size are also observed, with a Mo:Si ratio of 1:2. These inclusions were revealed only in a separate area of matrix crystal, while Si-Ge inclusions of both types were homogeneously distributed.

3.2 EBSD data

The EBSD study revealed three types of EBSD patterns from different micro-areas in Fig. 2. They are indexed with low fit factor (it varies from 0.2 to 0.4) and high confidence index (0.7-0.9). The diffraction pattern (DP) from matrix crystal is indexed as tetragonal crystal lattice Mn_4Si_7 with space group $P\bar{4}c2$ and unit cell parameters a=0.553 nm and c=1.752 nm [10] (Fig. 2a). This is one of four known phases of higher manganese silicides. The analysis of 350,000 DPs obtained from an area 2000μm x 3000 μm in size revealed grain structure of matrix crystal. The grains consist of several millimeter large blocks misoriented to each other by ~ 2 degrees.

The DPs from Si-Ge inclusions were indexed as cubic crystal lattice with space group $Fd\bar{3}m$ and unit cell parameter a=0.53 nm [11], with experimental error ~1% (Fig. 2b). The Si-Ge solid-solution inclusions of both types have a similar single crystal structure. These inclusions are oriented relative to the matrix crystal as (121)[10$\bar{1}$]Si-Ge ∥ (010)[100]Mn_4Si_7. Also, a few polycrystalline inclusions of Ge-rich type are observed.

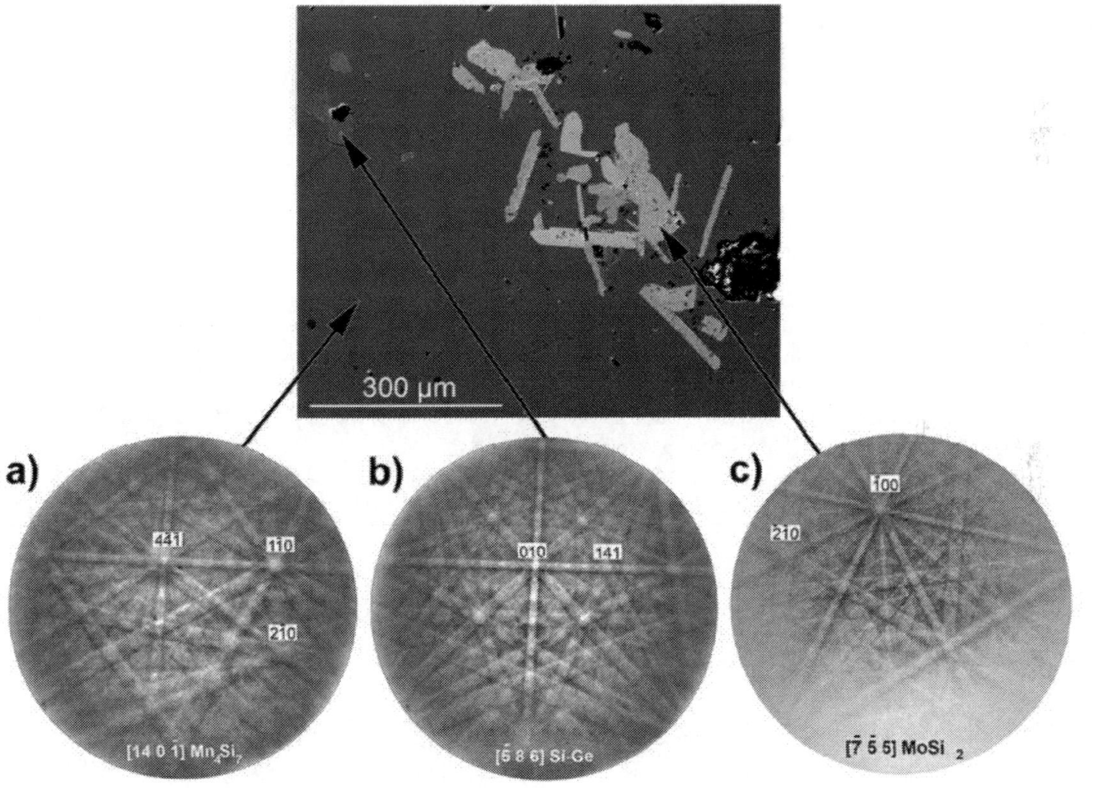

Figure 2. EBSD diffraction patterns from (a) matrix crystal, (b) Si-Ge solid-solution inclusion and (c) Mo-containing precipitates.

The experimental Kikuchi diffraction pattern from inclusions with Mo:Si=1:2 ratio (Fig. 2c) is in a good agreement with simulated DP for tetragonal $MoSi_2$ phase (space group *I4/mmm,* cell parameters a = 0.32 nm and c = 0.78 nm [12]). These inclusions have a large-scale polycrystalline grain structure. The preferred orientation of these inclusions has been examined in the approximation of "one grain - one orientation" to exclude the influence of the size factor. The texture analysis is made along [001] Mn_4Si_7 zone axis of the matrix crystal because it is the main direction relevant for thermoelectric properties (Fig. 3). Inverse pole figure analysis reveals three maxima corresponding to (110), (118) and (507) planes (Fig. 3b).

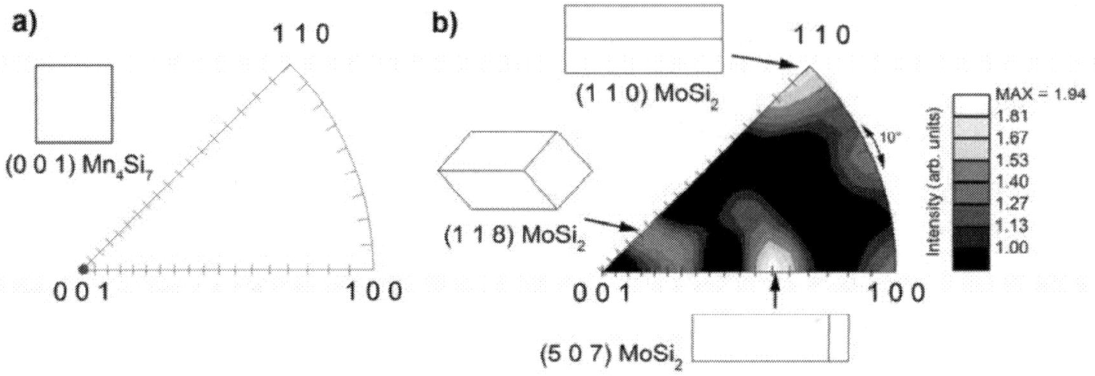

Figure 3. Inverse pole figure (a) of (001) orientation of Mn_4Si_7 matrix crystal and (b) colour coded texture analysis data of $MoSi_2$ inclusions.

The weight of each preferred orientation was estimated ±5° area around maxima. Weight fractions are 0.25 for (118), 0.35 for (110) and 0.40 for (507) plane. Thus, we can conclude that along [001] Mn_4Si_7 direction $MoSi_2$ inclusions form a multicomponent texture.

4. Conclusions

The microstructure of complex doped crystals of higher manganese silicide has been studied by several methods. EDS microanalysis reveals the manganese/silicon ratio in the doped matrix crystal is ~1.7, which corresponds to a HMS composition of $(Mn_{0.98}Mo_{0.02})[(Si_{0.98}Ge_{0.02})_{1.75}]_{0.99}Al_{0.01}$. The Al, Mo, Ge dopants are present in following concentrations: ~0.5±0.3 at% Al and Mo and ~1.5±0.4 at% Ge. Addition of Al, Ge, Mo leads to the formation of Si-Ge and $MoSi_2$ inclusions. MnSi ordered striations typical of undoped HMS crystals are not observed.

The work was partly supported by the RFBR grant №12-02-31444 and grant of Leading Scientific Schools - NSh-2883.2012.5

References

[1] Fedorov M I, Zaitsev V K, Solomkin F Y and Vedernikov M V 1997 *Tech. Phys. Lett.* **23** p 602
[2] Fedorov M I and Zaitsev V K 2006 *Thermoelectrics Handbook* Ed. DM Rowe, London: Taylor and Francis (ch. 31)
[3] Kim S W, Cho M K, Moshima Y and Choi D C 2003 *Intermetallics* **11** p 399
[4] Levinson L M 1973 *J. Sol. State Chem.* **6** p 126
[5] Kojima T, Nishida I and Sakata T 1979 *J. Cryst. Growth* **47** p 589
[6] Itoh I and Yamada M 2009 *J. Electron. Mater.* **38** p 925
[7] Aoyama I, Fedorov M I, Zaitsev V K, Solomkin F Yu, Eremin I S, Samunin A Yu, Mukoujima M, Sano S and Tsuji T 2005 *Jpn. J. Appl. Phys.* **44** p 8562
[8] Aoyama I, Kaibe H, Rauscher L, Kanda T, Mukoujima M, Sano S and Tsuji T 2005 *Jpn. J. Appl. Phys.* Vol. **44** p 4275
[9] Petrova L I, Fedorov M I, Zaitsev V K and Engalychev A E 2013 *Inorganic Materials* **49** p 355
[10] Gottlieb U, Sulpice A, Lambert-Andron B and Laborde O 2003 *J. Alloys Comp.* **361** p 13
[11] Parrish W 1960 *Acta Cryst.* **13** p 838
[12] Norlund Christensen A 1993 *J. Cryst. Growth* **129** p 266

Analytical Electron Microscopy of Semiconductor Nanowire Functional Materials and Devices for Energy Applications

V P Oleshko[1], E H Williams[1,3], A V Davydov[1], S Krylyuk[1,4], A Motayed[1,4], D Ruzmetov[2,4], T Lam[2], H J Lezec[2] and A A Talin[2,5]

[1]Material Measurement Laboratory and [2]Center for Nanoscale Science and Technology, National Institute of Standards and Technology, Gaithersburg, Maryland 20899, USA
[3]Department of Chemistry and Biochemistry and Department of Electrical and Computer Engineering, George Mason University, Fairfax, Virginia 22030, USA
[4]Institute for Research in Electronics and Applied Physics, University of Maryland, College Park, Maryland 20742, USA
[5]Sandia National Laboratories, Livermore, California 94551, USA

E mail: vladimir.oleshko@nist.gov

Abstract. Functionalized individual semiconductor nanowires (SNWs) and 3D SNW arrays attract a continuously growing interest for applications in optoelectronics, sensing, and energy storage. High-resolution field-emission analytical (scanning) transmission electron microscopy ((S)TEM) enables critical insights into the morphology, crystalline and electronic structures and chemical composition of single-crystalline high-aspect-ratio SNWs as prospective building blocks suitable for both a large scale-up synthesis and fabrication. Furthermore, SNW-based lab-on-a-chip devices may allow direct correlation between functional properties tailored for specific performance and the heterostructure morphology and atomic arrangement of the nanoscale structure being analyzed in various (S)TEM modes.

1. Introduction

High-aspect-ratio semiconductor nanowires (SNWs), often referred to as unconstrained length 1D nanostructures, reveal interesting physical properties that are not seen in bulk semiconductors. SNWs have therefore recently attracted much attention for exploring new physics in reduced dimensions and complex geometries. From a practical point of view, there is a growing interest in the fabrication of SNW-based lab-on-a-chip (opto) electronic devices and chemi-resistive chemical and biological sensors, which have the ability to selectively detect traces of certain chemical compounds or bind biomolecules through resistivity-dependent mechanisms. Functionalized individual Si, SiC and GaN

SNWs and 3D SNW arrays have recently shown potential as (a) a new biosensing platform for selective protein immobilization [1], (b) UV-assisted SNW-MO_2(-Pt) (M=Ti, Sn) hybrid chemical sensors for detection of environmental industrial pollutants and traces of explosives [2,3], and (c) scaffolding templates for renewable energy transducers and storage systems, i.e. miniature all solid-state SNW core – multishell rechargeable Li-ion batteries [4].

Here we report on the characterization of functionalized SNWs and energy related devices by high-resolution field-emission analytical (S)TEM, electron diffraction, and tomography to provide insights into morphology, crystalline and electronic structures, and chemical composition of prospective nano-sized building blocks that are suitable both for a large scale-up synthesis and fabrication using existing semiconductor technologies in a variety of emerging nanotechnology applications.

2. Experimental details

Si NWs were grown in a horizontal hot-wall reactor by chemical vapor deposition (CVD) with metal catalyst particles (Au) utilized to facilitate the vapor-liquid-solid (VLS) growth mechanism [1]. For fabrication of Li-ion nanobatteries (NW-LiBs), as-grown Si NWs were coated with 10 nm of Ti, followed by about 30 nm of Pt, 40 nm of Ti, and a 180 nm-thick layer of $LiCoO_2$. All layers have been deposited using physical vapor deposition in the same chamber without exposure to air. The coated NWs were then annealed in oxygen at 700° C for 2 hours. Following annealing, the samples were sputter-coated with a nitrogen-doped Li_3PO_4 solid electrolyte (LiPON) of 110 to 180 nm in thickness, and finally with about 35 nm of a phosphorus-doped amorphous Si anode layer. A FEI Nova Nanolab 600 dual beam FIB/SEM (FEI, Hillsboro, OR, USA) [5] was used to expose the inner Ti/Pt/Ti cathode current collector and electrically isolate the anode from the cathode. Electron beam induced deposition was used to connect the electrodes to the Pt contact pads.

(S)TEM imaging, selected-area electron diffraction (SAED), and energy-dispersive X-ray (EDXS) and electron energy-loss spectroscopy (EELS) of functionalized SNWs and related devices were performed at a 300 kV accelerating voltage using a Schottky field-emission FEI Titan 80-300 analytical (S)TEM (FEI, Hillsboro, OR, USA) equipped with S-TWIN objective lenses. For high spatial resolution nanoanalyses in the STEM mode, the instrument was supplied with a Fischione 3000 model HAADF detector, BF- and ADF-STEM detectors, a 30 mm^2 EDAX Si/Li EDX detector with a 0.13 srad acceptance angle and a Gatan Enfina EEL spectrometer. To ensure optimal counting rates, the specimens were tilted 15 degrees towards the EDX detector.

3. Results and discussion

3.1. Uniform VLS-grown single-crystalline Si NWs

Device applications require reliable approaches for fabricating SNWs with controlled shapes and orientation. Single-crystalline <111> Si NWs were prepared via a two-stage VLS growth procedure for producing uniform NW arrays with a predominantly vertical orientation (Fig. 1a). Narrow diameter distributions (130 nm to 170 nm) of Si NWs were provided by the use of monodispersed Au nanoparticles (Fig. 1b). Typical cross-sections of NWs are hexagonal in shape with {112} facets as shown in right bottom inset of Fig. 1b. A SAED pattern in the [111] zone axis displays {220} point reflections indicating a single-crystalline structure of the diamond cubic type ($Fd\bar{3}m$ space group, left upper inset). HRTEM (left bottom inset, Fig. 1c) displays a defect-free area with two-dimensional 0.20-0.21 nm {220} lattice fringes near the NW's edge. Z-contrast STEM (Fig. 1c) shows a side view of a NW and an enlarged area with 0.32 nm (111) Si lattice spacing and a 4.3 nm-thick oxide layer confirmed by EDXS (right inset, Fig. 1e). A 3D plot of low-loss EEL spectra (Fig. 1d) was obtained from a line scan across an isolated NW. The spectra at the ends near 0 nm and 190 nm are aloof in vacuum, those at 25 nm and at 130 nm are just grazing the NW surface, and those in between

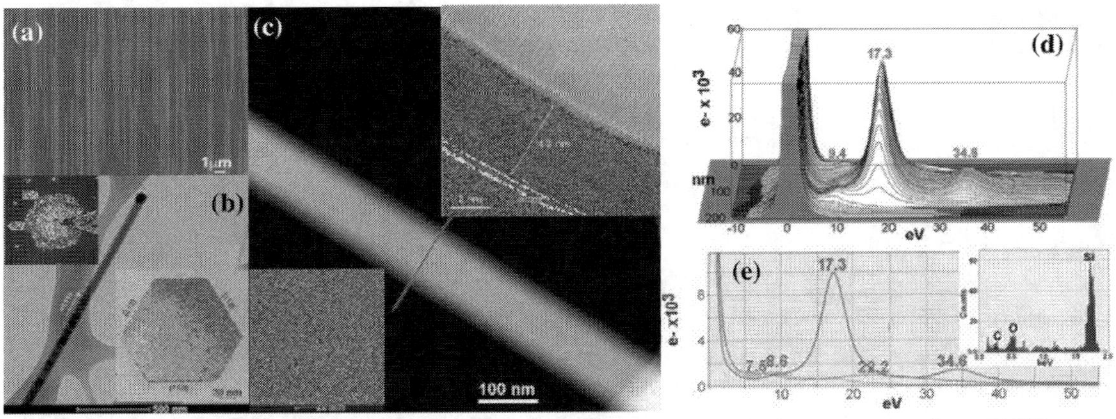

Figure 1. VLS-grown single-crystalline <111> Si NWs: (a) FESEM of a vertically aligned NW array. (b) BF-TEM, SAED (upper inset), and NW's cross-section (bottom inset) prepared by cutting, polishing and ion beam etching of NWs embedded into an epoxy resin. (c) HAADF STEM, 0.32 nm silicon (111) lattice fringes near the edge of the NW with no visible defects (upper inset), HRTEM, NW's cross-section (bottom inset). (d) 3D plot of STEM-EELS line profiling along the green line in (c). (e) Low-loss EEL spectra extracted from (d). X-ray spectrum (right inset in graph e) acquired at the oxide layer showing the Si-K peak at 1.74 keV and the O-K peak at 0.52 keV.

are penetrating the NW's bulk. The low-loss spectra are dominated by bulk plasmons at integer multiples of 17.3 eV (the first plasmon) and of 34.6 eV that correspond to the second plasmon (red line, Fig. 1e). As the beam reaches and passes the edge of the NW, the bulk plasmons disappear very rapidly and are replaced with surface plasmons between 8 eV to 10 eV energy losses (green dot line, Fig.1e). The peak at 7.5 eV is assigned to a Si/SiO_x interface plasmon (blue dashed line, Fig. 1e), which in the aloof mode also may be due in part to retardation effects [6]. The wide peak at about 22.2 eV (blue dashed line) corresponds to a bulk plasmon in SiO_x itself. These results illustrate possibilities for monitoring NW structures, including intentionally induced or native surface oxide layers that can be used for further functionalization or fabrication of semiconductor-oxide interfaces.

3.2. All-solid-state SNW core – multishell rechargeable Li-ion batteries

SNWs are highly attractive as building blocks in novel 3D multifunctional architectures for LiBs that maximize areal energy density. SNWs can be grown directly on the metallic current collector providing improvements in rate capabilities in metal oxide cathodes due to efficient 1D electron transport, reduced defects, and facile strain relaxation. With a focus on a miniature NW-LiB, we employ single-crystalline Si NWs as a scaffold for a complete (S)TEM–transparent all-solid-state nanobattery designed as a testing tool to correlate electrochemical performance with its hetero-structure arrangement and capture material transformations under cycling at the nanoscale *in situ* or *ex situ* [4]. STEM enables visualization of the internal microstructure of deposited active layers due to significant contrast variations even in 1.0 µm to 1.5 µm–thick areas, including textured $LiCoO_2$ crystallites of 160 nm to 200 nm in length and 30 nm to 60 nm in width that are tilted by 70 to 77° degrees towards the core and buried electrode–electrolyte interfaces as well (Fig. 2a). The texture of thermally grown hexagonal $LiCoO_2$ crystals ($R\bar{3}m$ space group) as found by HRTEM appeared with preferential orientations of Li-containing layers towards the core which is favorable for direct Li ion transport from the cathode to the current collector. Using high-tilt angle (±70°) HAADF STEM tomo-

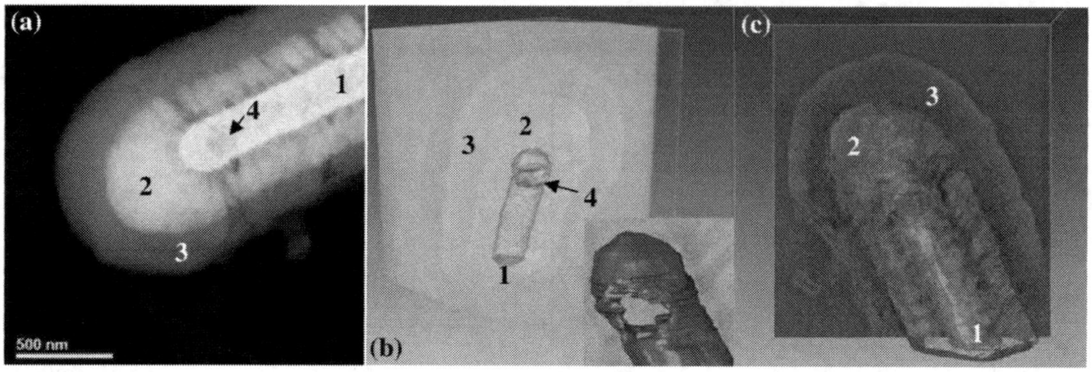

Figure 2. A NW-LiB: (a) HAADF-STEM tomography, a single frame extracted from a ±70 degree tilt-series. (b) The reconstructed 3D view of a NW-LiB's tip showing lateral distribution of $LiCoO_2$ crystallites around the segmented metallized NW core and a void (4) near its tip (inset). (c) 3D overlay of the segmented $LiCoO_2$ shell and the NW core. 1 - Si NW with a Ti/Pt/Ti metal collector layer, 2 - $LiCoO_2$ cathode layer, 3 - LiPON electrolyte layer covered with a n-Si amorphous anode layer.

graphy followed by a 3D reconstruction of the faceted metallized core with the simultaneous iterative reconstruction technique (SIRT), we revealed a void near the NW tip (bottom inset, Fig. 2b) that likely formed during annealing due to the reaction with the Pt-Ti metal collector (a Kirkendall effect), which was barely visible in 2D images. Furthermore, a 3D overlay of the segmented $LiCoO_2$ shell allows visualizing the complex morphology of the $LiCoO_2$ – LiPON interface (shown in green in Fig. 2c), where major electrochemical processes occur during cycling.

4. Summary

Analytical (S)TEM techniques enable critical insights into the 3D morphology, crystalline and electronic structures and chemical compositions of functionalized SNWs as prospective building blocks that are suitable for both a large scale-up synthesis and fabrication using existing technologies for a variety of emerging applications. Furthermore, new designs of SNW-based lab-on-a-chip devices may allow for direct correlation between electrical transport and (electro)chemical properties tailored for specific performance with the unique axial and radial heterostructure morphology and atomic arrangement on the same nanoscale structure being analyzed in various (S)TEM modes.

Acknowledgements

The support by NIST under Grant MML12-1053-N00 is gratefully acknowledged (VPO).

References

[1] Williams E H, Schreifels J A, Rao M V, Davydov A, Oleshko V P, Lin N, Steffens K, Krylyuk S, Bertness K, Manocchi A, Koshka Y 2013 *J Mater Res* **28** 68

[2] Aluri G S, Motayed A, Davydov A V, Oleshko V P, Bertness K A, Sanford N A, Rao, M V 2012 *Nanotechnol* **23** 175501

[3] Bajpai R, Motayed A, Davydov A V, Oleshko V P, Aluri G S, Bertness K A, Rao M V, Zaghloul M E 2012 *Sensors Actuators B: Chem* **171-172** 499

[4] Ruzmetov D, Oleshko V P, Haney P M, Lezec H, Karki K, Baloch K H, Agrawal A K, Davydov A V, Krylyuk S, Liu Y, Huang J Y, Tanase M, Cumings J, Talin A A 2012 *Nano Letters* **12** 505

[5] Commercial equipment and materials are identified in this document to adequately describe experimental procedures. This does not imply endorsement by NIST.

[6] Moreau P, Brun N, Walsh C A, Colliex C, Howie A 1997 *Phys Rev B* **56** 6774

Characterisation of 3D-GaN/InGaN nanostructured Light Emitting Diodes by Transmission Electron Microscopy

I J Griffiths[1], D Cherns[1], X Wang[2], A Waag[2] and H-H Wehmann[2]

[1] School of Physics, H H Wills Physics Laboratory, University of Bristol, Tyndall Avenue, Bristol, BS8 1TL, UK
[2] Institute of Semiconductor Technology, Braunschweig University of Technology, Hans-Sommer-Strasse 66, 38106 Braunschweig, Germany

i.griffiths@bristol.ac.uk

Abstract. Transmission and scanning electron microscopy have been used to characterise GaN/InGaN 3D nanostructures grown on patterned GaN/sapphire substrates by MOVPE. It has been found that the growth of well ordered arrays of such nanostructures, containing multiple quantum wells on non-polar side-facets, can be achieved with a low density of defects. Growth changes and surface morphology play a major role in the nucleation of any defects present. The nanostructure morphology has been investigated and non-uniform growth on adjacent facets studied.

1. Introduction

The growth of 3D-GaN/InGaN nanostructures is of interest for the development of high efficiency Light Emitting Diodes (LEDs) for solid state lighting[1-3]. Utilising a core-shell structure, as illustrated in Fig 1, should allow the active InGaN multiple quantum wells (MQW) to have not only a lower defect density, but also an increased area, compared to conventional (0001)-oriented thin films. Due to the non-polar orientation of the MQW, piezoelectric fields will also be reduced or eliminated, promising a further increase in light output and efficiency. In this paper, we examine patterned GaN/InGaN nanostructures of the type illustrated in Fig 1. To fully understand the structure, and any defects present, the nanostructures have been characterised using scanning and transmission electron microscopy (SEM and TEM).

2. Experimental

The GaN/InGaN nanostructures were grown on c-plane sapphire using continuous flow metal organic vapour phase epitaxy (MOVPE). A layer of n-type GaN was initially grown, followed by the deposition of a SiO_x masking layer by plasma-enhanced chemical vapour deposition (PECVD). The mask was subsequently etched to produce an ordered array of circular holes with diameters from 250nm to 400nm, and pitches of 750nm to 2μm. Growth of the n-type doped GaN was then resumed under growth conditions, producing uniform arrays of high quality vertical-sided nanorods. The growth conditions were then changed to favour lateral growth, producing the MQW and p-type doped GaN growth. More details of the growth conditions can be found in ref. [4].

The nanostructures were initially studied using SEM, prior to the preparation of TEM samples. For TEM characterisation both cross-section and plan-view samples were prepared, allowing a better understanding of the structure and any defects present. To prepare the cross-section samples described here, the nanostructures were removed from the substrate, by sonication, and dispersed in an ethanol suspension. This was deposited onto holey carbon films and dried to create a dispersed film, where the nanorods can be studied free of contamination, Fig 2b, 4b.

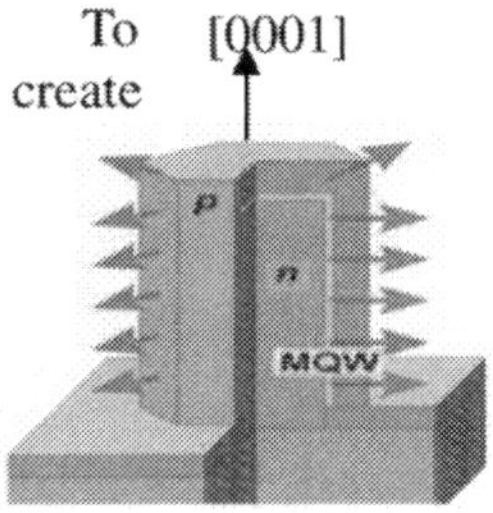

Fig. 1. Schematic of 3D nanostructure. Copyright "GECCO"

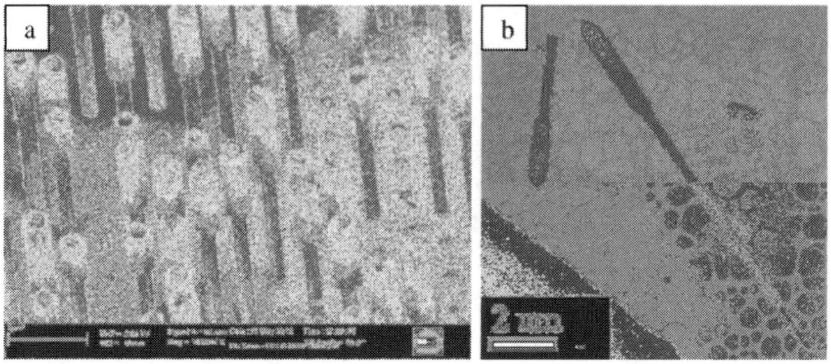

Fig. 2. (a) SEM image showing nanostructures with partial overgrowth. (b) Sample prepared for cross-section analysis free of contamination.

plan-view samples, a novel preparation technique has been developed using dual-beam focused ion beam (FIB) thinning. Samples were prepared by attaching two pieces together with conductive epoxy resin. The FIB was then used to cut a section perpendicular through the nanostructure array, Fig 3. This technique allows the internal structure to be studied as a function of the height in the nanostructures. The prepared section was lifted from the remaining material and deposited on a conventional holey carbon film for imaging, Fig. 3c.

The samples were studied using a Philips EM430 TEM operating at 250kV and a JEOL 2010 operating at 200kV.

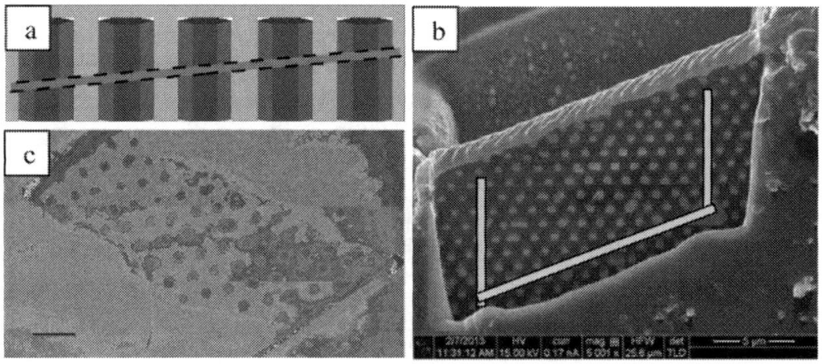

Fig. 3. (a) Schematic illustrating plan-view preparation. (b) SEM image of the sample taken during preparation. (c) The sample viewed in the TEM.

3. Results

Initial studies of the samples using SEM show the masking layer creates a regular array of nanostructures, with similar aspect ratios, which can be controlled during the growth process. It has been found that the switch to growing the MQW plays an important role in the overall crystalline quality of any subsequent growth of p-GaN. Fig 2a shows how the growth of p-GaN has only occurred at the very top of the nanostructure. In this sample the p-GaN layer has been found to contain a large density of both I_1 and I_2 type basal plane stacking faults (BSFs), Fig 5a, which are bounded by partial dislocations. The faults are dislocation half-loops seen both end on and in projection in Fig 5a. Further studies of this layer show that the MQWs have grown with an island like morphology, Fig 5b. The rough island-like MQW can be seen to propagate under the p-GaN overgrowth, suggesting that the

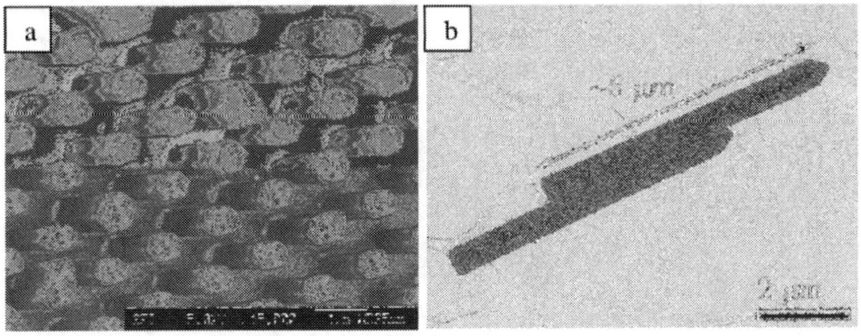

Fig. 4. (a) SEM images showing regular patterned nanostructures. (b) Bright field TEM image showing complete nanostructures.

morphology is not a post-growth artefact. Examination of the n-GaN part of the nanostructure has shown that the initial growth is free of defects, both BSFs and threading dislocations. In contrast, studies of nanostructures where the MQW growth has formed smooth layers show a much reduced defect density. The few defects which have been observed are similar to those seen in the other samples. However it appears that the half-loop, bounding the defect, nucleates below the MQWs, Fig 5c. This suggests that the defects nucleate at the transition between the axial n-GaN growth, and the laterally grown material.

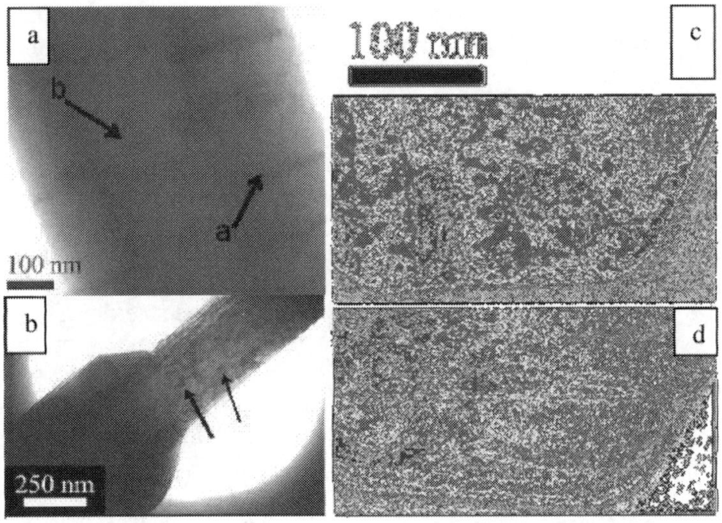

Fig. 5. (a) TEM images showing the presence of BSFs. The BSFs are seen in projection (arrowed a) and end on (arrowed b). (b) The island-like growth of the MQWs can be seen. (c) Plan view image showing a defect nucleating at the n-GaN interface, and tilted (d) to show the half-loop bounding partials.

Plan-view analysis of the samples allows the radial sequence of growth to be examined more closely. From these samples it has been found that the growth rates of different facets are non-uniform, leading to not only non-uniform profiles across the rods, but also to non-uniform spacings within the

MQWs for different facets. This is clearly seen in Fig 6a, where the spacings of the QWs on one facet are roughly double that of the adjacent facet (see inset). Further studies of these samples allowed the quantum well thicknesses to be determined. Figure 6b shows three QWs which are buried within the bulk material. From the line plot we can see that the QWs have a thickness of ~ 2nm and a separation of ~ 4nm. The change in contrast across the MQWs indicates the presence of InGaN.

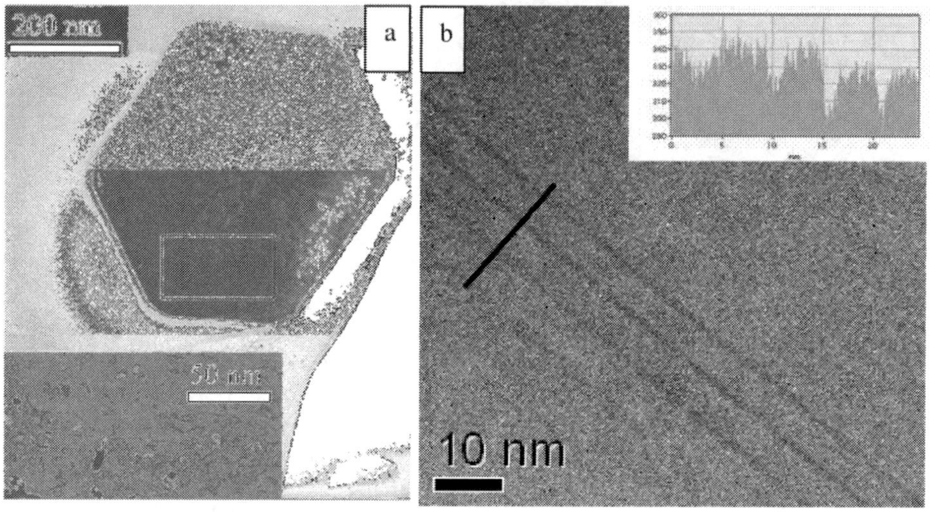

Fig. 6. (a) Plan view images showing the presence of QWs, along with the non-uniform growth of p-GaN on differing facets (inset). Closer examination of other MQWs shows that the InGaN layers have a thickness of ~2nm. A line plot (inset), taken where indicated.

4. Discussion

The growth of highly ordered GaN/InGaN 3D nanostructures on sapphire substrates by MOVPE has been achieved. The nanostructures have been found to consist of a core-shell structure with InGaN MQWs buried between the n-GaN core and the p-GaN shell. It has been found by TEM studies that the initial n-GaN core is free of both threading defects and basal plane stacking faults. The transition to lateral growth in preserving this crystalline quality was found to be important: growth leading to rough MQWs has been associated with the nucleation of BSFs in the subsequent p-GaN layer, whilst growth leading to smooth MQWs correlated with significantly fewer defects. Plan-view studies of the structures have revealed the non-uniform growth of p-GaN on differing facets, along with changes in QW thickness and spacing.

Acknowledgements

This work has been funded by the EU project no. 280694 "GECCO". Thanks to T. Eriksson of Obducat AB, Sweden and EU project 228999 "SMASH".

References

[1] Waag A *et al* 2011 *Phys. Stat. Sol.* C **8** 2296
[2] Bergbauer W *et al* 2010 *Nanotechnology* **21** 305201
[3] Hong Y J *et al* 2011 *Adv. Mater.* **23** 3284
[4] Wang X *et al* 2012 *Cryst. Growth & Des.* **12** 2552

Polarization fields in GaN/AlN nanowire heterostructures studied by off-axis holography

M den Hertog[1], R Songmuang[1] and E Monroy[2]

[1]Institut Néel-CNRS, 25 rue des Martyrs, 38042 Grenoble cedex 9, France
[2]CEA-Grenoble, INAC / SP2M, 17 rue des Martyrs, 38054 Grenoble cedex 9, France

E-mail: martien.den-hertog@grenoble.cnrs.fr

Abstract. In this work, we present an off-axis holography study of GaN/AlN heterostructured nanowires grown by plasma-assisted molecular-beam epitaxy. We discuss the sample preparation of nanowire samples for electron holography and combine potential profiles obtained using holography with theoretical calculations of the projected potential in order to gain understanding of the potential distribution in these nanostructures. The effects of surface states are discussed.

1. Introduction

Semiconductor nanowires (NWs) have become a powerful and diverse family of functional materials for nano-electronics, optoelectronics and biotechnology. III-nitride semiconductors, with their direct band gap, flexibility for band engineering, and capability of operating in extreme environments, are promising materials for single-NW devices, namely high-resolution chemical sensing or high-spectral-contrast photodetectors. A particularity of nitride materials is the very strong polarization-related internal electric field present in heterostructures, which can be assessed through its effect on optical or electrical properties. A direct measurement of the internal electric field profile is more challenging but could in principle be realized by off-axis electron holography. Holography on nitride materials was already reported in [1-4]. In addition to the technical difficulties to measure the field distribution, these direct measurements are necessarily perturbed by the superposition of information on surface depletion due to surface states and by the effect of strain on the internal field distribution. Taking these phenomena into account is critical to understand the physical properties of such nanostructures.

In this paper we use medium resolution off-axis electron holography to measure the projected potential in GaN/AlN NW heterostructures and compare our experimental results with three-dimensional (3D) potential simulations to discriminate the 3D field distribution and the effect of surface charge. Furthermore, we describe the challenges of holography on (nitride) NW heterostructures, which have to be faced to allow quantitative mapping of the potential.

2. Experimental

Off-axis medium resolution electron holography was performed on an FEI Titan 80-300 working at 200 kV. Plasma-assisted molecular beam epitaxy (PAMBE) is a well-established technique to synthesize nanometer-size catalyst-free GaN NWs growing along the *c* crystallographic axis [5,6]. This growth method allows *n*-type and *p*-type doping [7], as well as band engineering by heterostructuring [8-12]. For this study GaN NWs containing a 10-period GaN/AlN (7 nm / 7 nm) superlattice, followed by 30 nm of GaN and 30 nm of AlN, were grown on Si(111) patterned pillars by PAMBE (see Fig. 1). Growth proceeded under N-rich conditions (Ga/N ratio ~ 0.2–0.4) at a substrate temperature ~ 790 °C. An AlN buffer layer is used to improve the vertical alignment of the NWs [13]. The NWs under consideration have a length of 1.2–1.5 μm and a diameter of 30–80 nm. The sample for electron holography was prepared using the cleavage technique [14,15], which allowed studying epitaxial NWs while they were still attached to their substrate and suspended in vacuum, without any modification (electrical or structural) of the NW.

The sample preparation is crucial to be able to obtain good holography data from NW samples. In samples with a high density of NWs, the post geometry has several advantages over conventional dispersed NW samples on a carbon grid. First of all, a very precise orientation of the NW several degrees off the zone axis, to minimize diffraction contrast, is possible as the NWs are grown epitaxially on the Si post. Second, it is very easy to find back the same single NW as the posts also serve as markers. It should be noted that for lower density NW systems, a cleaved sample of NWs on a planar substrate can be sufficient to study free-standing NWs without overlap with other NWs close to the cleaved edge [14,15]. However, as GaN NWs grown by PAMBE are very closely packed, we need to find NWs growing at the edge of the post to prevent superposition of other NWs.

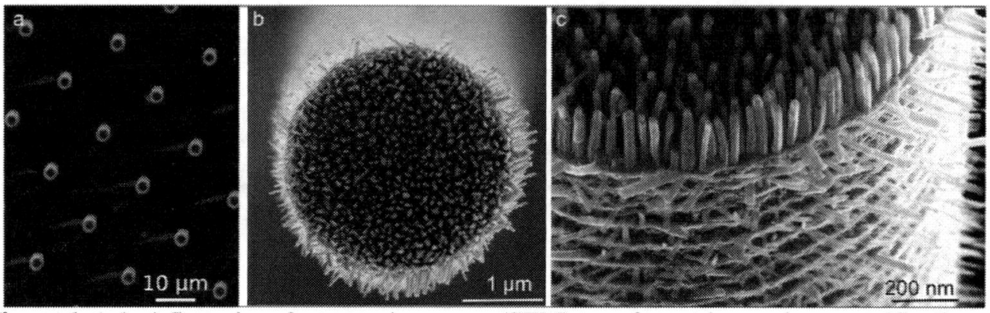

Figure 1. (a,b,c) Scanning electron microscopy (SEM) top view at increasing magnification of GaN/AlN NW heterostructures grown on Si (111) patterned pillars.

Previously we have demonstrated that GaN NWs are hexahedral structures with relatively flat {10-10} *m*-plane sidewall facets, and they exhibit N-polarity, as identified by annular bright field (ABF-) and high angle annular dark field scanning transmission electron microscopy (HAADF-STEM) [16]. As off-axis holography is sensitive to the projected potential we expect to observe the effect of the polarity on the potential in the structure. Figure 2(a) presents a high angle annular dark field (HAADF) STEM image of a GaN/AlN heterostructured NW. The axial superlattice is enveloped by an AlN shell arising from the lateral growth during the deposition of the AlN sections. As a result of the elevated growth temperature, a certain interdiffusion is observed at the interfaces. This phenomenon is particularly important when starting the growth of AlN on GaN, giving rise to a graded interface, which extends along about 1.5 nm. Figure 2b shows the phase image obtained by off-axis electron holography, tilted several degrees off the Si [112] zone axis (ZA). The fringe spacing was 1 nm, giving a spatial resolution in the reconstructed phase image of around 3 nm. The various GaN and AlN regions can be clearly resolved. In the absence of magnetic fields and diffraction contrast, the phase difference $\Delta\varphi$ between electrons that have passed through vacuum and electrons that have traversed the specimen is related to the potential $V(x, y, z)$, by

$$\Delta\varphi = C_E \int_0^t V(x,y,z)dz, \qquad (1)$$

where C_E is an interaction constant which depends on the acceleration voltage (7.29×10^6 rad V^{-1} m^{-1} at 200 kV), and t is the thickness of the sample in the beam direction z. $V(x, y, z)$ is a combination of the mean inner potential (MIP) of the pure components, namely GaN and AlN (MIP = 16.8 V and 14.2 V, respectively [17]), and of the electrostatic potential, related to the electric field, reflecting the combined effects of spontaneous and piezoelectric polarization and surface charges. In Fig 2(c), a projected potential profile is obtained from the phase image in Fig. 2(b) along the white vertical arrow using eq. 1 and the measured NW thickness.

The experimental potential profile is compared to two simulated potential profiles obtained from projected 3D simulations of a NW structure with the layer sequence of the sample in Fig. 2(a) using the nextnano3 software [18] with the material parameters in ref. [19], one including a negative surface charge density of -2×10^{13} cm^{-2}, which is one order of magnitude higher than a value from literature [20], and one without surface charge. These simulations take into account the NW hexagonal cross section with *m* sidewall facets, with the dimensions extracted from our STEM measurements, including the presence of the AlN shell. The strain distribution is calculated to minimize the elastic energy in the system considering a free-standing NW in vacuum. The calculation of the 3D potential profile takes into consideration the local variations of the band gap due to strain, as well as the 3D internal electric field induced by spontaneous and piezoelectric polarization. However, our simulations do not include the variation in MIP between GaN and AlN; thus, the experimentally obtained potential profile was shifted along the y-axis to allow comparison to the simulated profiles. We verified that the step in phase over the AlN top region is consistent with the thickness and the MIP of AlN.

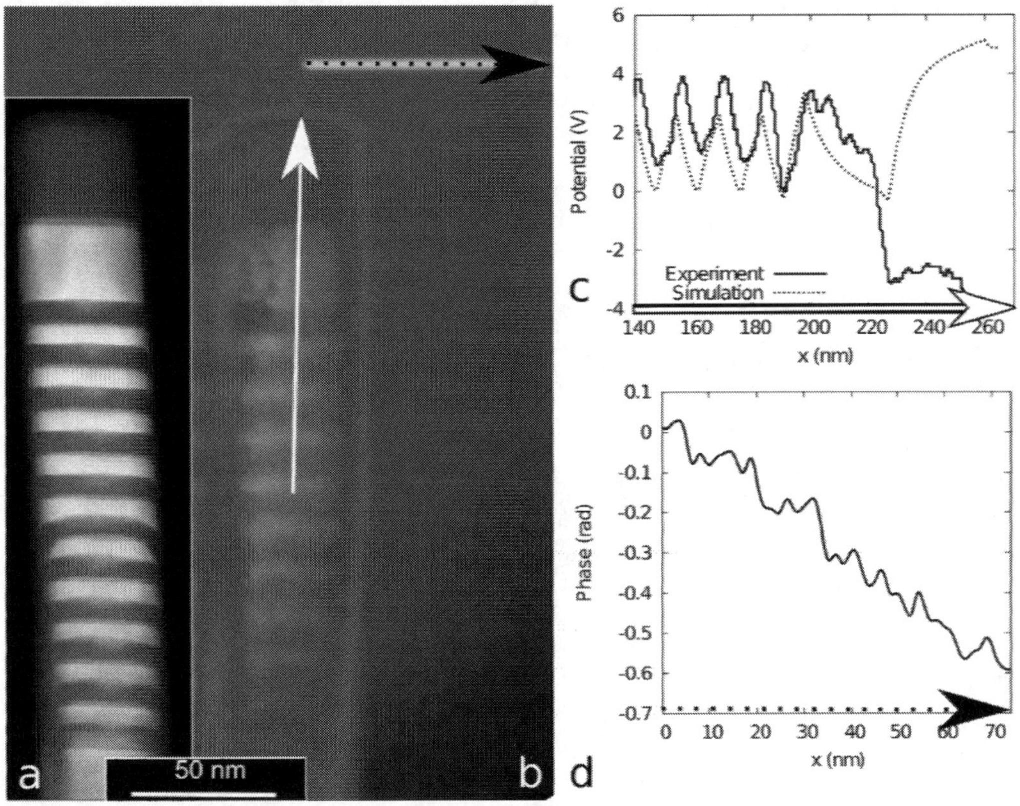

Figure 2. (a) HAADF STEM image of a GaN/AlN NW heterostructure. (b) Phase image of the same NW. (c) Potential profile obtained from the phase along the vertical white arrow compared with two simulated profiles one including a surface charge density of -2×10^{13} electron charges cm^{-2} and one without surface charges. (d) Phase profile in the vacuum obtained along the horizontal black dotted arrow.

For the studied NW, the expected change in potential when going from a GaN to AlN region due only to the variation in MIP would be around 0.8 V, if we assume the NW diameter of 44 nm is roughly equal to the thickness. The experimentally measured potential change at the GaN/AlN heterointerfaces is much larger than the calculated 0.8 V, indicating that indeed the measurement is sensitive to the electric field. However, in the studied NWs there are composition gradients at the interfaces, creating also a gradient in the MIP. Therefore, we cannot provide a quantitative assessment of the interface charge in this structure, but quantitative measurements are feasible in structures with abrupt interfaces and homogeneous alloy composition.

Making a qualitative analysis, the experimental profile is comparable to the simulated profiles and consistent with the N-polarity of the NW. Interestingly, the internal field is so strong that the surface charge does not have much influence on the projected potential, even though the value of the surface charge used in the simulation is higher than experimentally measured [20]. On the other hand, the measurement at the AlN top region presents a deviation from simulations, which might be due to charging of the NW under the electron beam. To investigate charging issues, we display in Fig. 2d a phase profile in the vacuum at the side of the NW along the black horizontal dotted arrow in Fig 2b. Going away from the NW the phase decreases, indicating a positive charging of the NW [21]. This could explain the lowering of the potential in the AlN top region.

3. Conclusion

In summary we have shown first results of holography on GaN/AlN NW heterostructures, compared to theoretical calculations. We confirm the N-polarity of the NW and observe a positive charging of the sample under the electron beam. Comparison with simulated profiles does not allow setting a value on the surface charge of the NW. We have described some of the challenges that have to be met to allow quantitative field mapping in nitride NW heterostructures.

Financial support from the French FMN-SMINGUE 2011, the French CNRS and CEA METSA network, the ANR-2011-NANO-027 "UVLamp" project, the ANR 12 JS10 00201 "COSMOS" project, the EU ERC-StG "TERAGAN" (#278428) project, and the Société Française des Microscopies (SFμ) for a travel grant is acknowledged. We thank Adeline Grenier for providing the Si(111) patterned pillars as the growth substrate.

References

[1] Cherns D, Jiao CG 2001 *Phys. Rev.Lett.* **87** 205504
[2] Mc Carthney MR, Ponce FA, Cai J, Bour DP 2000 *Appl. Phys. Lett.* **76** 3055
[3] Cai J, Ponce FA 2002 *Appl. Phys. Lett.* **91** 9856
[4] Fischer AM, Sun KW, Ponce FA, Songmuang R, Monroy E 2012 *Appl. Phys. Expr.* **5** 025001
[5] Yoshizawa M, Kikuchi A, Mori M, Fujita N, Kishino K 1997 *Jpn. J. Appl. Phys.* **36** L459
[6] Sanchez-Garcia MA, Calleja E, Monroy E, Sanchez FJ, Calle F, Muñoz E, Beresford R 1998 *J. Cryst. Growth* **183** 23
[7] Furtmayr F, Vielemeyer M, Stutzmann M, Laufer A, Meyer BK et al 2008 *J. Appl. Phys.* **104** 074309
[8] Ristic J, Rivera C, Calleja E, Fernàndez-Garrido S et al 2005 *Phys. Rev. B* **72** 85330
[9] Calarco R, Meijers RJ, Debnath RK, Stoica T, Sutter E, Lu H 2007 *Nano Lett.* **7** 2248
[10] Songmuang R, Ben T, Daudin B, González D, Monroy E 2010 *Nanotechnology* **21**, 295605
[11] Carnevale SD, Yang J, Phillips PJ, Mills MJ, Myers RC 2011 *Nano Lett.* **11** 866
[12] Furtmayr F, Teubert J, Becker P, Conesa-Boj S, Morante J R, Chernikov A, Schäfer S, Chatterjee S, Arbiol J, Eickhoff M 2011 *Phys. Rev. B* **84** 205303
[13] Songmuang R, Landré O, Daudin B 2008 *Appl. Phys. Lett.* **81** 251902
[14] den Hertog MI, Schmid H, Cooper D, Rouviere JL, Björk MT, Riel H, Rivallin P, Karg S, Riess W 2009 *Nano Lett.* **9** 3837
[15] den Hertog M I, PhD thesis, http://tel.archives-ouvertes.fr/tel-00493934/en/
[16] den Hertog M I, González-Posada F, Songmuang R, Rouviere J L, Fournier T, Fernandez B, Monroy E 2012 *Nano Lett.* **12** 5691
[17] Kruse P, Schowalter M, Lamoen D, Rosenauer A, Gerthsen D 2006 *Ultramicroscopy* **106** 105
[18] Birner S, Zibold T, Andlauer T, Kubis T, Sabathil M, Trellakis A, Vogl P, 2007 *IEEE Trans. Electron Dev.* **54** (9) 2137
[19] Kandaswamy P K, Guillot F, Bellet-Amalric E, Monroy E, et al 2008 *J. Appl. Phys.* **104** 093501
[20] Bertelli M, Löptien P, Wenderoth M, Rizzi A, Ulbrich R G, Righi M C, Ferretti A, Martin-Samos L, Bertoni C M, Catellani A 2009 *Phys. Rev. B* **80** 1115324
[21] den Hertog M I, Rouviere J L, Schmid H, Cooper D et al 2010 *J. Phys.: Conf. Series* **209** 012027

Multislice calculations for quantitative HAADF STEM analysis of germanium diffusion in strained silicon

L Favre
IM2NP, UMR 7334 CNRS, Aix Marseille Université,
Faculté des Sciences de Saint-Jérôme, F-13397 Marseille, France
E-mail: luc.favre@im2np.fr

Abstract. For decades, microelectronic industry has reduced components size while increasing their performance and reliability. Actual systems rely on multilayer nanostructures that require a perfect control of composition at atomic scales. In this study, we focus on detection of germanium contaminants in a thin silicon layer, using High Angle Annular Dark Field (HAADF) with Scanning Transmission Electron Microscopy. We proceeded to multislice calculations in order to extract quantitative information from HAADF measurements. Models consist of Ge substitutional impurities introduced in 10 nm thick Si layers. We analysed atomic column HAADF intensities along ⟨100⟩ zone axis, varying Ge impurities amount and position in an atomic column. Results show that for a reduced number of Ge atoms ($n < 9$), column intensity varies linearly with n. However depending on Ge atoms positions, a column containing n impurities can display a higher signal than a column containing $n + 1$ impurities. Therefore, a direct correlation between atomic column intensities and their Ge atoms content cannot be established without additional assumptions or complementary measurement.

1. Introduction

Strained silicon is intensively used in microelectronic industry to keep improving transistor characteristics [1]. For instance, Si channels of complementary metal-oxide-semiconductor transistors are grown on relaxed Si_{1-x}-Ge_x layers. The stress induced by lattice mismatch increases carriers mobility and limits short channel effects [2–4]. It results in an enhancement of the drain current, a key parameter to gain switching speed. This technique is now applied to one or two dimensional nano-objects, to combine stress effects with quantum confinements. A new components generation emerges, like high-performance field effect transistors or solar cells, based on band-gap engineering [5–7]. Calculations show that an atomic layer-by-layer growth of alternating Si and Ge atomic planes with a specific sequence would lead to a direct band gap Si-Ge film [8–10]. Such a material represents a great interest for optoelectronic applications.

Considering devices dimensions, a perfect control of layer composition on short scales is required. Small fluctuations would drastically change their properties. Yet, many physical processes like diffusion during multilayers growth, can deteriorate interface sharpness. Scanning Transmission Electron Microscopy (STEM) with High Angle Annular Dark Field (HAADF) is a relevant technique for quantitative detection of impurities. Several studies proceeded to detection of dopant [11,12] and chemical mapping in semiconducting materials [13].

In this paper, we focus on Ge atoms detection in a thin silicon film. Quantitative calculations have been performed in order to analyse image intensity as a function of sample composition. We varied substitutional Ge atom number and position in the Si lattice. We report simulation results and discuss quantitative information that can be extracted from HAADF measurements on this kind of materials.

2. Simulations setup

Quantitative HAADF STEM simulations have been carried out using the QSTEM software [14], employing the frozen lattice calculations. Results have been averaged over 30 frozen phonon configurations. Microscope characteristics were fixed, considering a 200 keV aberration-corrected probe: $C_3 = 4$ µm, $C_c = 1.4$ mm and energy spread $dE = 1.4$ eV. The probe-forming aperture's half-angle has been set to 18 mrad, inner- and outer-angle of the annular detector to 60 and 160 mrad, respectively. Electron beam was focused in the middle of the layer i.e. $\Delta f = -5$ nm.

The layer model consists of a bulk 3x3 Si supercell, 10.5 nm thick. As the sample is observed along the ⟨100⟩ zone axis, each atomic column displayed in simulations (see **Figure 1**) contains 20 atoms. Ge atoms have been introduced as substitutional impurities into the Si lattice. For each simulation, all Ge atoms are located in the same atomic column. Considering the small amount of impurities introduced, we neglected potential lattice relaxation.

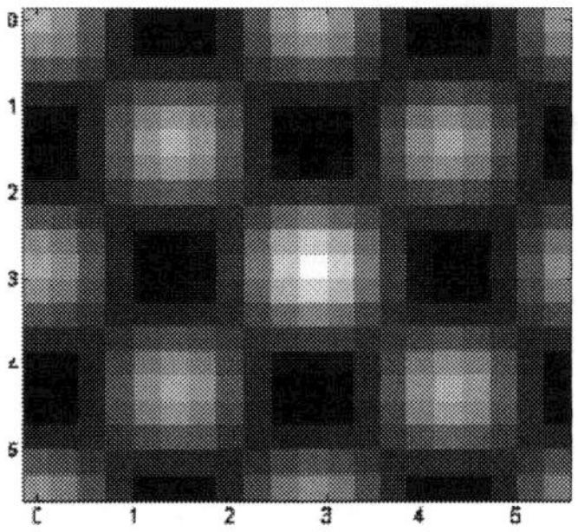

Figure 1. HAADF simulation of a Si lattice containing 1 Ge substitutionnal atom in the middle column. Specimen observed along ⟨100⟩.

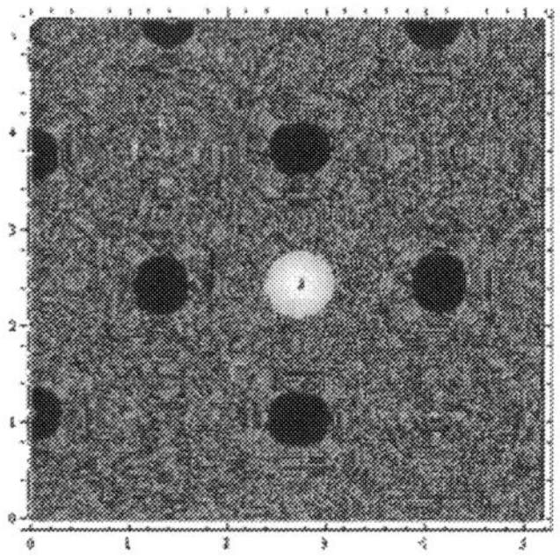

Figure 2. Fit of the simulation displayed in Figure 1 using Gaussian functions.

Using method of last squares, image intensity was then fitted with a sum of Gaussian functions corresponding to the 13 atomic columns contained in the simulated image:

$$I_{\text{tot}}(x, y) = I_0 + \sum_{i=1}^{13} \frac{a_i}{\left(2\pi s_i^2\right)^{1/2}} \exp\left[-\frac{(x - x_i)^2 + (y - y_i)^2}{2s_i^2} \right] \quad (1)$$

x_i and y_i refer to the column coordinates, $a_i(2\pi s_i^2)^{-1/2}$ and s_i to their amplitude and standard deviation, respectively (see **Figure 2**). I_0 was set to zero, as no noise was introduced during simulations. In the following, amplitude of the atomic column containing Ge impurities was reported as a function of simulation conditions, using pure silicon column amplitude as amplitude reference.

3. Results

3.1. Column intensity as a function of Ge atom number

In a first step, column intensity containing Ge impurities was investigated as a function of n, the number of germanium atoms. Theses atoms are located in the middle of the column, in successive planes. Calculations show that for $n \leq 8$, a linear increase of intensities by about 20 %, per additional atom impurity, is observed (see **Figure 3**, red circles). For higher Ge contents ($9 \leq n \leq 20$), additional Ge atoms are located far from the beam focused point, and progressively fill the column. As a consequence, the intensity variation is no longer linear and saturation arises for $n > 13$.

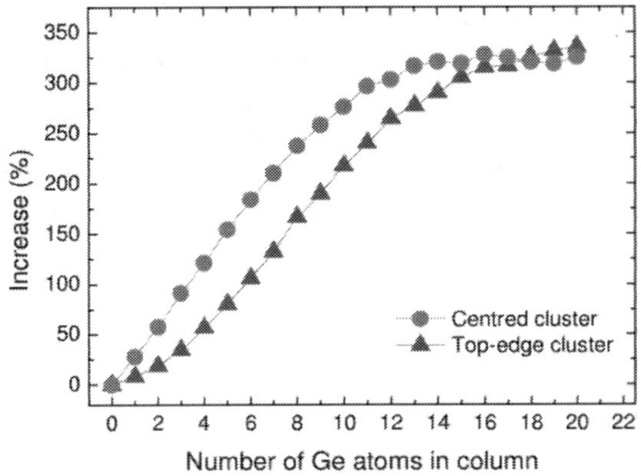

Figure 3. Column intensity as a function of *n*, number of Ge substitutionnal atoms. Ge atoms are inserted in successive planes, located in the centre of the column (circles) or at top column edge (triangles).

Using the same model but for Ge atoms located close to the specimen top, intensity evolution presents the same behaviour (cf **Figure 3**, blue triangles). A linear response is obtained for $3 \leq n \leq 11$, with a 20 % increase. This shift suggests that the intensity also depends on Ge atom positions.

3.2. Column intensity as a function of Ge atom position

For a single Ge substitutional atom, the intensity can vary from 8 %, if located at top or bottom edge of the sample, to 29 % if placed close to the centre. These variations are directly linked to the microscope focus conditions: signal is maximal when impurities are located at the focus point, i.e. the middle of the layer.

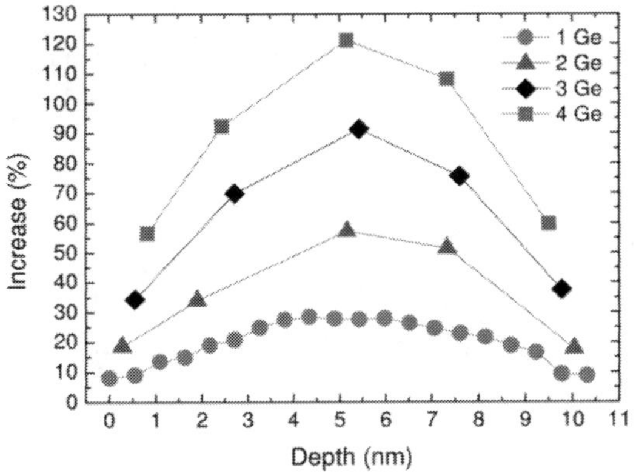

Figure 4. Column intensity increase as a function of Ge cluster position in the column. Ge cluster consists of 1 (circles), 2 (triangles), 3 (diamonds) or 4 (squares) Ge atoms placed in successive planes.

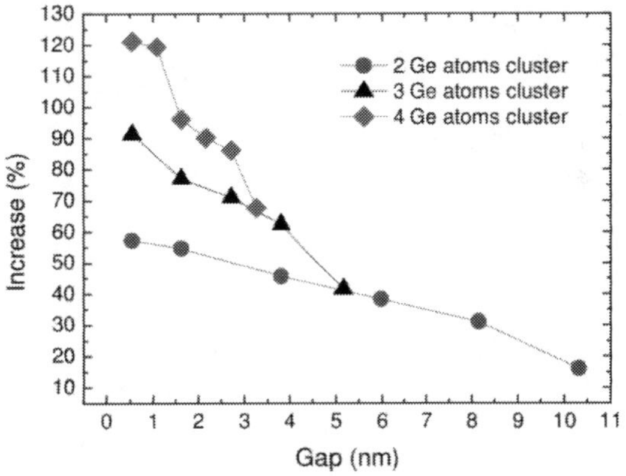

Figure 5. Column intensity increase as a function the gap between Ge substitutional atoms. Ge atoms are located in the same column, regularly spaced and symmetrically distributed around the column centre. Circles, triangle and diamonds correspond to a column containing 2, 3 or 4 Ge atoms, respectively.

Ge atoms clusters consisting of $n = 1$ to 4 adjacent Ge atoms aligned in a single column have been modelled, varying cluster position in the column. Results are displayed in Figure 4. Again, column intensity is maximal when the cluster is located at the microscope focus point, i.e. in the middle of the layer. It drastically decreases, by more than 50 %, when a cluster reaches layer surface. As a consequence, a column containing an *n*-Ge cluster located in the middle of the column can display a

Journal of Physics: Conference Series **471** (2013) 012020 doi:10.1088/1742-6596/471/1/012020

higher intensity than a column for which an $(n+1)$-Ge cluster is located in the top or the bottom of the column. Calculations show that a column containing n atoms can even display a signal higher than a column with $n+2$ Ge atoms for $n \geq 4$.

This observation is confirmed by another set of atom position distributions. This time, Ge atoms are not located on adjacent positions, but are regularly spaced and symmetrically distributed around the column centre, varying the gap between them. As can be seen in Figure 5, column intensity decreases rapidly with gap and, again, signal arising from a column containing $n+1$ Ge atoms can be weaker than from a column containing n Ge atoms.

4. Conclusion

In this study we investigate the possibility for HAADF to be used as a convenient tool for characterising simultaneously sample structure and composition. We focus on the mainstream materials used in microelectronics and optoelectronics industries: silicon and germanium.

The models consider Ge substitutional atoms in a Si lattice. Calculations show that a linear increase of column signal intensity is observed when Ge atoms are located close to the electron probe focus point. As a consequence, the linear regime can be observed only in very thin specimens. For thick ones, a direct correlation between columns' intensity and their composition cannot be established. Complementary experiments would be necessary. Other STEM techniques like EDS would hardly combine high spatial resolution and chemical sensitivity. Silicon and germanium have relatively low and close atomic numbers which represents a major challenge for chemical quantitative investigation at atomic scale.

References

[1] Ang K-W, Tung C-H, Balasubramanian N, Samudra G S and Yeo Y-C 2007 *IEEE Electron Device Lett.* **28** 609 –612

[2] Paul DJ 2004 *Semicond. Sci. Technol.* **19** R75–R108

[3] Capasso F 1987 *Science* **235** 172–6

[4] Schäffler F 1997 *Semicond. Sci. Technol.* **12** 1515–49

[5] Lauhon LJ, Gudiksen MS, Wang D and Lieber CM 2002 *Nature* **420** 57–61

[6] Xiang J, Lu W, Hu Y, Wu Y, Yan H and Lieber CM 2006 *Nature* **441** 489–93

[7] Kelzenberg MD, Boettcher SW, Petykiewicz JA, Turner-Evans DB, Putnam MC, Warren EL, Spurgeon JM, Briggs RM, Lewis NS and Atwater HA 2010 *Nat Mater* **9** 239–44

[8] Froyen S, Wood DM and Zunger A 1987 *Phys. Rev. B* **36** 4547–50

[9] Turton RJ and Jaros M 1990 *Mater. Sci. Eng. B* **7** 37–42

[10] D' Avezac M, Luo J-W, Chanier T and Zunger A 2012 *Phys. Rev. Lett.* **108** 027401

[11] Voyles PM, Muller DA, Grazul JL, Citrin PH and Gossmann H-J L 2002 *Nature* **416** 826–9

[12] Oshima Y, Hashimoto Y, Tanishiro Y, Takayanagi K, Sawada H, Kaneyama T, Kondo Y, Hashikawa N and Asayama K 2010 *Phys. Rev. B* **81** 035317

[13] Rosenauer A, Mehrtens T, Müller K, Gries K, Schowalter M, Venkata Satyam P, Bley S, Tessarek C, Hommel D, Sebald K, Seyfried M, Gutowski J, Avramescu A, Engl K and Lutgen S 2011 *Ultramicroscopy* **111** 1316–27

[14] Koch CT 2002 Determination of core structure periodicity and point defect density along dislocations ProQuest Diss. Thesis PhD-Ariz. State Univ. 2002 Publ. Number AAI3042580 ISBN 9780493562612 Source Diss. Abstr. Int. Vol. 63-02 Sect. B Page 0846 214 P

Journal of Physics: Conference Series **471** (2013) 012021 doi:10.1088/1742-6596/471/1/012021

Cathodoluminescence and TEM investigations of structural and optical properties of AlGaN on epitaxial laterally overgrown AlN/sapphire templates

U Zeimer[1], A Mogilatenko[1,2], V Kueller[1], A Knauer[1] and M Weyers[1]

[1]Ferdinand-Braun-Institut, Leibniz-Institut für Höchstfrequenztechnik, Gustav-Kirchhoff-Str. 4, 12489 Berlin, Germany
[2]Humboldt-Universität zu Berlin, Newtonstr. 15, 12489 Berlin, Germany

ute.zeimer@fbh-berlin.de

Abstract. Surface steps as high as 15 nm on up to 10 µm thick AlN layers grown on patterned AlN/sapphire templates play a major role for the structural and optical properties of $Al_xGa_{1-x}N$ layers with $x \geq 0.5$ grown subsequently by metalorganic vapour phase epitaxy. The higher the Ga content in these layers is, the stronger is the influence of the surface morphology on their properties. For $x = 0.5$ not only periodic inhomogeneities in the Al content due to growth of Ga-rich facets are observed by cathodoluminescence, but these facets give rise to additional dislocation formation as discovered by annular dark-field scanning transmission electron microscopy. For $Al_xGa_{1-x}N$ layers with $x = 0.8$ the difference in Al content between facets and surrounding material is much smaller. Therefore, the threading dislocation density (TDD) is only defined by the TDD in the underlying epitaxially laterally overgrown (ELO) AlN layer. This way high quality $Al_{0.8}Ga_{0.2}N$ with a thickness up to 1.5 µm and a TDD $\leq 5x10^8$ cm^{-2} was obtained.

1. Introduction

Light emitting diodes (LEDs) with wavelengths in the ultraviolet (UV)-B and -C regions have the potential to replace UV lamps used for water purification, photo-chemical curing and many other applications due to their higher efficiency and longer lifetime. For effective light extraction through the sapphire substrate $Al_xGa_{1-x}N$ buffer layers with Al content ranging from 0.5 to 0.8 are required which are transparent to the UV light emitted from the active regions of such LEDs. However, when $Al_xGa_{1-x}N$ layers are grown on planar AlN/sapphire templates they have usually a threading dislocation density (TDD) in the range of > 10^{10}cm^{-2}. This very high TDD reduces the LED efficiency due to nonradiative recombination at dislocations. Therefore several groups developed different approaches to reduce the TDD [1 - 3].

We use lateral epitaxial overgrowth (ELO) over a stripe patterned AlN/sapphire template by metalorganic vapour phase epitaxy (MOVPE) for the reduction of the TDD. First, a thick ELO AlN layer is grown to reach coalescence and a smooth surface which is subsequently overgrown with $Al_xGa_{1-x}N$. In an earlier paper we studied the influence of substrate miscut and etch depth into the AlN/sapphire template on the TDD and the surface morphology of the ELO AlN. We found that the latter plays the key role for the homogeneity of the subsequently grown $Al_xGa_{1-x}N$ layers [4].

In this paper we concentrate on the interplay of the optical properties of $Al_xGa_{1-x}N$ layers studied using 80 K cathodoluminescence (CL) and their defect structure studied by annular dark-field scanning transmission electron microscopy (ADF-STEM), both in cross-sectional geometry.

2. Experimental

500 nm thick AlN layers were grown at 1400°C on c-plane sapphire substrates by MOVPE. A stripe pattern running parallel to the [1-100]AlN direction with 1.5 µm wide grooves and a pitch of 3 µm was processed using plasma etching. Then, 7 to 10 µm thick AlN layers were grown on the patterned AlN/sapphire template at 1400°C to obtain coalescence as well as smooth and defect reduced AlN. This thick ELO AlN layer was overgrown by a 25 nm thick AlN layer followed by a 1 - 3 µm thick $Al_xGa_{1-x}N$ layer, both deposited at 1250°C. The Al content in the $Al_xGa_{1-x}N$ layer was varied between $x = 0.5$ and $x = 0.8$. For the investigation of the structural and optical properties of this layer we used atomic force microscopy (AFM), high resolution X-ray diffraction (HRXRD), CL at 80K using a Gatan Mono-CL4 system on a Zeiss Ultra+ field emission scanning electron microscope (SEM) and ADF-STEM using a JEOL 2200FS transmission electron microscope.

3. Results and discussion

In our earlier paper [4] we have shown that the surface morphology of ELO AlN strongly depends on the miscut direction of the sapphire substrate. A miscut of only 0.25° towards the *m*-direction of the sapphire substrate leads to an ELO AlN surface with 15 nm high periodic steps and a pitch of 3 µm. Such high steps have also been observed by other groups [5]. In contrast, a miscut of 0.25° towards the *a*-direction together with a reduced etching depth leads to a very smooth ELO AlN surface without any steps. On the other hand, the TDD in ELO AlN is not affected by the miscut direction of the template and can be reduced from > $2x10^{10}cm^{-2}$ to below $8x10^{8}cm^{-2}$ [6]. The TDD reduction occurs initially due to bending of dislocation lines to the free surface of the AlN wings as well as further dislocation annihilation during the growth of thick layers. This low TDD is the starting point for the growth of the $Al_xGa_{1-x}N$ buffer layers.

3.1. $Al_xGa_{1-x}N$ with $x = 0.5$

Using our MOVPE conditions, 3.5 µm thick $Al_{0.5}Ga_{0.5}N$ layers with a smooth surface morphology could be grown only on an underlying ELO AlN layer exhibiting surface steps, whereas on atomically flat ELO AlN 3D nucleation occurred leading to a very rough surface. The $Al_{0.5}Ga_{0.5}N$ layer grown on the stepped ELO AlN showed no cracks despite its high layer thickness. HRXRD reciprocal space maps of 002 and 105 reflections showed this layer to be fully relaxed. In fig. 1, a cross-sectional backscattered electron (BSE) image is shown together with monochromatic 80K CL images obtained at two peaks at 271 nm and 284 nm observed in the CL sum spectrum from the same sample area. The intensity distribution in the two CL images (fig. 1b, c) is strongly anticorrelated. We attribute this behaviour to a periodic change in the Al content x of the $Al_xGa_{1-x}N$ layer. This compositional inhomogeneity can be explained by facetted $Al_xGa_{1-x}N$ growth starting at the surface steps of the ELO AlN with a much higher Ga content on the side facets. A detailed explanation for this behaviour can be found elsewhere [4]. From the corresponding wavelengths the Al content was calculated to be $x = 0.45$ on the facets and $x = 0.54$ in the surrounding material.

The consequences of this rather large difference in composition for the defect evolution have been studied in cross-section by ADF-STEM. Fig. 2 shows the distribution of TDs across the interface between the thick ELO AlN, grown at 1400°C, the thin AlN layer grown at 1250°C (called in the following text "LT-AlN") and $Al_xGa_{1-x}N$ with a nominal x-value of 0.5 also grown at 1250°C.

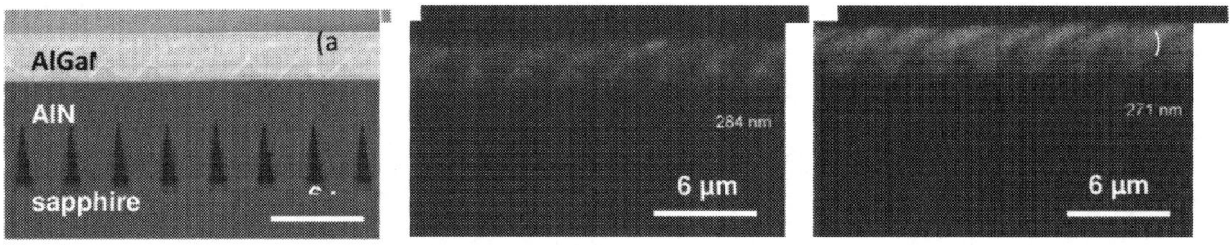

Fig. 1 Cross-section image of a 3.5µm thick $Al_{0.5}Ga_{0.5}N$ layer grown on ELO AlN: (a) BSE image, (b) and (c) monochromatic 80 K CL images

Besides TDs originating from ELO AlN and penetrating through the ELO AlN/LT-AlN interface into the $Al_{0.5}Ga_{0.5}N$ layer, some dislocations in the ELO AlN are observed that bend near the ELO

AlN/LT-AlN interface and propagate parallel to the interface within the ELO AlN. Also, at the LT-AlN/$Al_{0.5}Ga_{0.5}N$ interface new dislocations are created. This was expected taking into account the full relaxation of the $Al_{0.5}Ga_{0.5}N$ layer on ELO AlN. One of the steps on the ELO-AlN surface is also seen. Coalescence of the Ga-rich region growing on the side facet of the surface step with the surrounding $Al_{0.54}Ga_{46}N$ growing on the planar surface results in formation of an inclined grain boundary. This grain boundary gives rise to dislocations with their line directions inclined with respect to the "normal" TDs. Some of those dislocations annihilate with TDs during proceeding growth. From the half width of rocking curves measured at the 002 and 302 reflections of the $Al_{0.5}Ga_{0.5}N$ layer a TDD of about $4x10^9$ cm^{-2} is estimated. This value is by a factor of 5 higher than the TDD observed in the ELO AlN. In an $Al_{0.5}Ga_{0.5}N$ layer grown in the same run on a planar AlN/sapphire template the TDD is in the range of $3x10^{10}$ cm^{-2}. Therefore, the growth of $Al_{0.5}Ga_{0.5}N$ on ELO AlN results in a lower defect density as well as in a higher CL intensity (not shown here) compared to growth on planar substrates. Nevertheless, the inhomogeneity of the Al-concentration should be taken into account when using it as a buffer layer material for LEDs.

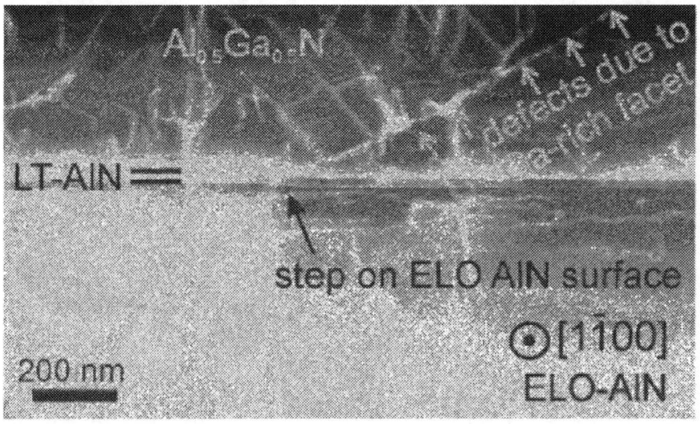

Fig. 2 Cross-sectional ADF-STEM image of the ELO AlN/LT-AlN/$Al_{0.5}Ga_{0.5}N$ interface region (black arrow points to a step on the ELO AlN surface)

3.2 $Al_xGa_{1-x}N$ with $x = 0.8$

For LEDs emitting in the UV-C region around a wavelength of 240 nm a higher Al content in the $Al_xGa_{1-x}N$ buffer layer, i.e. $x = 0.8$, is required. For this purpose an about 1 μm thick $Al_{0.8}Ga_{0.2}N$ layer was grown on ELO AlN with surface steps. In the CL spectrum of the cross-section area shown in fig. 3 only one peak has been observed. On the other hand, periodic inhomogeneities in Al content are still observed in the BSE image as well as in the monochromatic CL images. Due to the low Ga content only one Ga-rich facet develops at each step on the ELO AlN surface. From a Gaussian deconvolution of the CL peak using two wavelengths the difference in the Al content between the facets and the surrounding material has been estimated to be no larger than $\Delta x = 0.04$. HRXRD reciprocal space maps of the 002 and 105 reflections revealed a small relaxation below 5%, but ADF-STEM imaging did not show any formation of additional dislocations in the $Al_{0.8}Ga_{0.2}N$ layer (fig. 4).

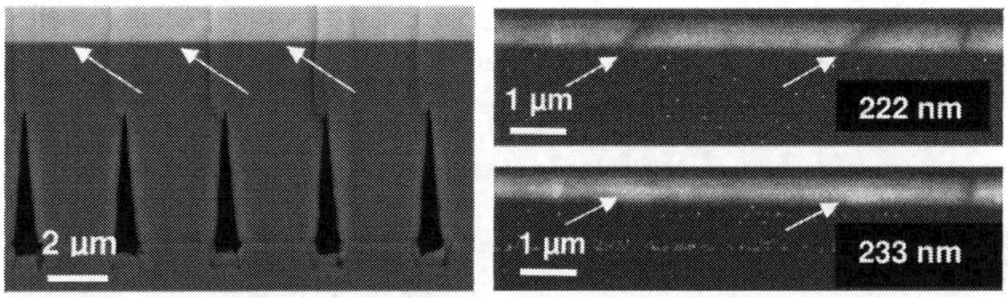

Fig. 3 Cross-section of an $Al_{0.8}Ga_{0.2}N$ layer: (a) BSE-image, (b), (c) 80 K CL images, arrows point to surface steps

However, we observe a certain TD line inclination in $Al_{0.8}Ga_{0.2}N$ whereas TDs in ELO AlN exhibit vertical dislocation lines. Under propagation into the LT-AlN and $Al_{0.8}Ga_{0.2}N$ layers the dislocation line direction changes. Notice that in fig. 4 (obtained in the [1-100]AlN zone axis) some TD lines seem to maintain their vertical directions. However, our tilting experiments proved that this is a projection effect and that all TDs are inclined toward the <1-100>AlN directions. It has been shown previously that dislocation line inclination in $Al_xGa_{1-x}N$ layers grown on planar AlN provides relaxation of compressive strain, since the projected length of an inclined dislocation acts as a misfit dislocation segment [7]. In particular, for $Al_{0.8}Ga_{0.2}N$ layers average projected inclination angles in the range of 6° to 10° were observed. Interestingly, here we observe a much higher projected inclination angle of about 20° in $Al_{0.8}Ga_{0.2}N$ on ELO AlN corresponding to a real inclination angle of 30°. Compared to the $Al_{0.8}Ga_{0.2}N$ growth on planar AlN/sapphire the distinct difference is the lower TDD in ELO AlN. Consequently, to achieve compressive strain relaxation a larger TD line inclination resulting in a longer projected length of TD lines would be required.

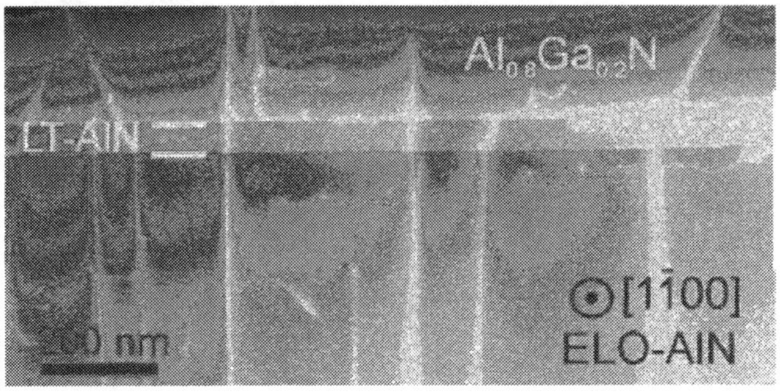

Fig. 4 Cross-sectional ADF-STEM image of the ELO AlN/LT-AlN/$Al_{0.8}Ga_{0.2}N$ interface region

As visible, TD line inclination starts already at the ELO AlN/LT-AlN interface. This observation suggests that a certain Ga incorporation into the LT-AlN layer might take place. It is known that at the growth temperature of 1250°C some residual Ga atoms from the reactor environment can be incorporated into the LT-AlN. Using energy dispersive X-ray spectroscopy we have proved parasitic Ga incorporation into LT-AlN resulting in an actual Al content x of about 0.9 in LT-AlN.

4. Conclusions

$Al_xGa_{1-x}N$ layers with $x \geq 0.5$ can successfully be grown on ELO AlN/sapphire templates. The defect structure in the layers depends on the TDD of the ELO substrate and the Al content x. In the case of $x = 0.5$ additional dislocation creation is observed both at the LT-AlN/ $Al_{0.5}Ga_{0.5}N$ interface and at the Ga-rich facets formed at the surface steps of ELO AlN substrates. In the case of high Al content, i.e. $x = 0.8$, $Al_xGa_{1-x}N$ layers can be grown with the same TDD as in the underlying ELO AlN, although a 30° bending of the TDs toward the <1-100> directions was observed at the ELO AlN/LT-AlN interface.

Acknowledgement

This work has been partially supported by the German Federal Ministry of Education and Research (BMBF) (03WKBT01C) and the German Research Foundation (DFG) within the Collaborative Research Center 787.

References

[1] Jiang H, Egawa T, Hao M, and Liu Y, 2005 *Appl. Phys. Lett.* **87** 241911
[2] Imura M, Nakano K, Fujimoto N, Okada N, Balakrishnan K, Iwaya M, Kamiyama S, Amano H, Akasaki I, Noro T, Takagi T, and Bandoh A, 2006 *Jpn. J. Appl. Phys.* **45** 8639-8643
[3] Iida K, Kawashima T, Iwaya M, Kamiyama S, Amano H, Akasaki I, and Bandoh A, 2007 *J. Cryst. Growth* **298** 265 – 267
[4] Zeimer U, Kueller V, Knauer A, Mogilatenko A, Weyers M, and Kneissl M, 2013 *J. Cryst. Growth* **377** 32-36
[5] Jain R, Sun W, Yang J, Shatalov M, Hu X, Sattu A, Lunev A, Deng J, Shturm I, Bilenko Y, Gaska R, and Shur M S, 2008 *Appl. Phys. Lett.* **93** 051113
[6] Kueller V, Knauer A, Zeimer U, Rodriguez H, Mogilatenko A, Kneissl M, and Weyers M, 2011 *phys. stat. sol. (c)* **8** 2022-2024
[7] Follstaedt D M, Lee S R, Allerman A A, Floro J A, 2009 *J. Appl. Phys.* **105** 083507

Journal of Physics: Conference Series **471** (2013) 012022 doi:10.1088/1742-6596/471/1/012022

Characterization of $L2_1$ order in Co_2FeSi thin films on GaAs

B Jenichen[1], T Hentschel[1], J Herfort[1], X Kong[1], A Trampert[1] and I Zizak[2]

[1]Paul-Drude-Institut fuer Festkoerperelektronik, Berlin, Germany
[2]Helmholtz-Zentrum fuer Materialien und Energie, Berlin, Germany

bernd.jenichen@pdi-berlin.

Abstract. Co_2FeSi/GaAs(110) and Co_2FeSi/GaAs(-1-1-1)B hybrid structures were grown by molecular-beam epitaxy (MBE) and characterized by transmission electron microscopy (TEM) and X-ray diffraction (XRD). The films contain inhomogeneous distributions of ordered $L2_1$ and B2 phases. The average stoichiometry could be determined by XRD for calibration of the MBE sources. Diffusion processes lead to inhomogeneities, influencing long-range order. An average $L2_1$ ordering of up to 65% was measured by grazing-incidence XRD. Lateral inhomogeneities of the spatial distribution of long-range order in Co_2FeSi were imaged using dark-field TEM with superlattice reflections and shown to correspond to variations of the Co/Fe ratio.

1. Introduction

Device concepts based on the spin rather than the charge of the electron are introduced in the field of spintronics. Heusler alloys can be useful for sources of spin injection as a first step for the fabrication of spintronic devices [1]. The Heusler alloy Co_2FeSi has some outstanding properties: it is a ferromagnetic half-metal with a Curie temperature larger than 1100 K and a magnetic moment of $6\mu_B$. Thus, high spin injection efficiency is expected in Co_2FeSi/GaAs hybrid structures. The lattice parameter (0.5658 nm) matches that of GaAs (0.5653 nm), i.e. it can be grown epitaxially on GaAs without the formation of misfit dislocations. In order to obtain a maximum spin injection efficiency fully ordered Co_2FeSi films and perfect ferromagnet/semiconductor (FM/SC) interfaces (IF) are needed. Figure 1 shows a schematic view of the cubic $L2_1$ structure of Co_2FeSi and the corresponding electron diffraction pattern of a Co_2FeSi epitaxial film with the incoming electron beam along the <110> direction. The fundamental reflections form a rectangular pattern and are clearly distinguished by their higher intensity. They are not sensitive to disorder. However, the remaining reflections of lower intensity are superlattice (SL) reflections, which may be used for characterization of the long-range order inside the epitaxial film [2].

For the present work Co_2FeSi/GaAs hybrid structures were grown by molecular-beam epitaxy (MBE) and investigated by transmission electron microscopy (TEM) in connection with electron energy loss spectroscopy (EELS) and grazing incidence diffraction (GID) of X-rays in order to characterize the phase stability of the FM/SC IF and the structural properties of the Co_2FeSi film. The average stoichiometry can be calibrated by lattice parameter measurements. Although the films are lattice matched to the substrate, they contain inhomogeneous distributions of ordered $L2_1$ and B2 phases. Average $L2_1$ ordering is measured by GID via comparison of fundamental and SL reflections using an information depth smaller than the film thickness. The as-grown Co_2FeSi films are highly but not fully ordered. In a first approximation the degree of average $L2_1$ ordering is increasing with growth temperature. The lateral distribution of ordering is imaged by dark-field TEM using the corresponding SL reflections.

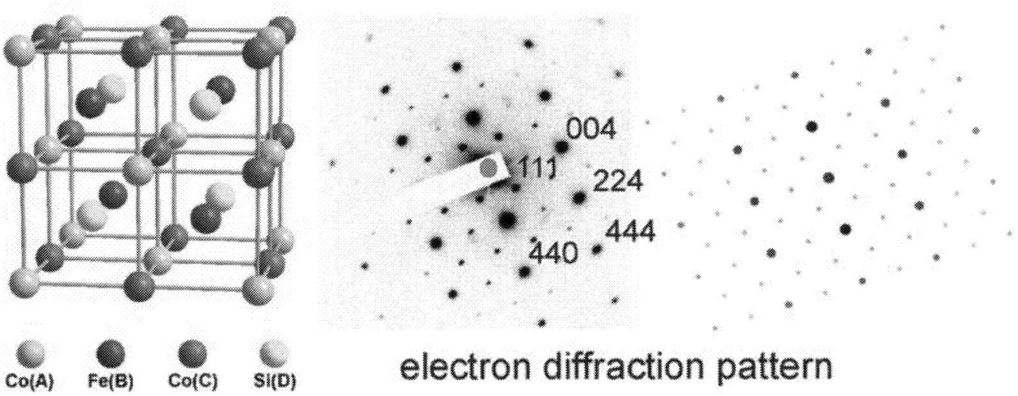

Figure 1. Schematic view of the cubic $L2_1$ structure of Co_2FeSi and the corresponding electron diffraction pattern (experiment and calculation using JEMS) of a thin epitaxial Co_2FeSi film grown on a GaAs (110) substrate (incident electron beam directed along <110>).

2. Experiment

The Co_2FeSi films were grown on GaAs(110) and on GaAs(-1-1-1)B substrates by MBE [2,3]. The calibration of the fluxes was described in [4]. The growth rate was near 0.1 nm min^{-1}. The substrate temperature during MBE growth, T_S, was varied between 100 and 350 °C. The nominal Co_2FeSi film thicknesses were 40 nm on GaAs(110) and 15 nm on GaAs(-1-1-1)B. The samples were investigated by dark-field (DF) and high-resolution (HR) TEM. For that purpose cross-sectional TEM specimens were prepared by mechanical lapping and polishing, followed by argon ion milling according to standard techniques. TEM images were acquired with a JEOL 3010 microscope operating at 300 kV. Electron energy loss spectroscopy (EELS) measurements were performed in the TEM with a spot size of 10 nm. The Co/Fe ratio was determined by the analysis of the Fe-$L_{2,3}$ and Co-$L_{2,3}$ edges after background subtraction. Since the analysis was carried out without standards only the lateral changes of the composition ratio were determined in line scans. We determined the average long-range order using a comparison of the integrated intensities of SL and fundamental reflections [5] measured by GID of X-rays. We used the 111, 222, and 220 reflections with synchrotron radiation (energy 6900 eV, wavelength 0.179687 nm) at the beamline KMC2 of the electron storage ring BESSY II of the Helmholtz-Zentrum Berlin.

3. Results and discussion

A perfectly ordered Co_2FeSi lattice is needed to obtain the extraordinary properties of this half-metallic Heusler alloy. In order to determine the long-range order by XRD or TEM we can distinguish different types of diffraction peaks: the 220 and the 444 reflections are fundamental (i.e. not sensitive to disorder) whereas the 333, 222 and the 111 reflections are SL reflections [2,5,6]. The 222 reflection arises when at least the CsCl-type B2 order is present in the Co_2FeSi lattice, whereas the 111 or the 333 reflection can be found only in regions of $L2_1$ order. Co_2FeSi/GaAs(-1-1-1)B samples are convenient for investigation of SL reflections by TEM, however SL reflections cannot be investigated by GID for this sample orientation. Figure 2 shows the diffraction pattern and corresponding DF micrographs of a nominally 15 nm thick Co_2FeSi film grown on GaAs(-1-1-1)B. The micrograph of the fundamental 444 reflection is more homogenous whereas the 222 and 333 SL reflections exhibit typical inhomogeneities demonstrating laterally disordered regions. Such inhomogeneities were shown by EELS to correspond to local deviations of stoichiometry (Fe/Co ratio) [6].

Dynamical effects prevented quantitative analysis of the DF TEM micrographs. However, in the Co_2FeSi/GaAs(110) samples we found the SL reflections well oriented for GID measurements, i.e. their diffracting netplanes are perpendicular to the surface. GID has the advantage of a limited information depth for incidence (and/or exit) angles below the critical angle. In our case, this information depth is smaller than the film thickness of 40 nm, i.e. using GID we measure the region of the film near the surface and exclude an influence of the substrate. Dynamical effects can be avoided in GID. Figure 3 displays the three types of maxima of $\omega/2\Theta$-scans of a Co_2FeSi film grown on GaAs(110) at a substrate temperature $T_S = 250°C$.

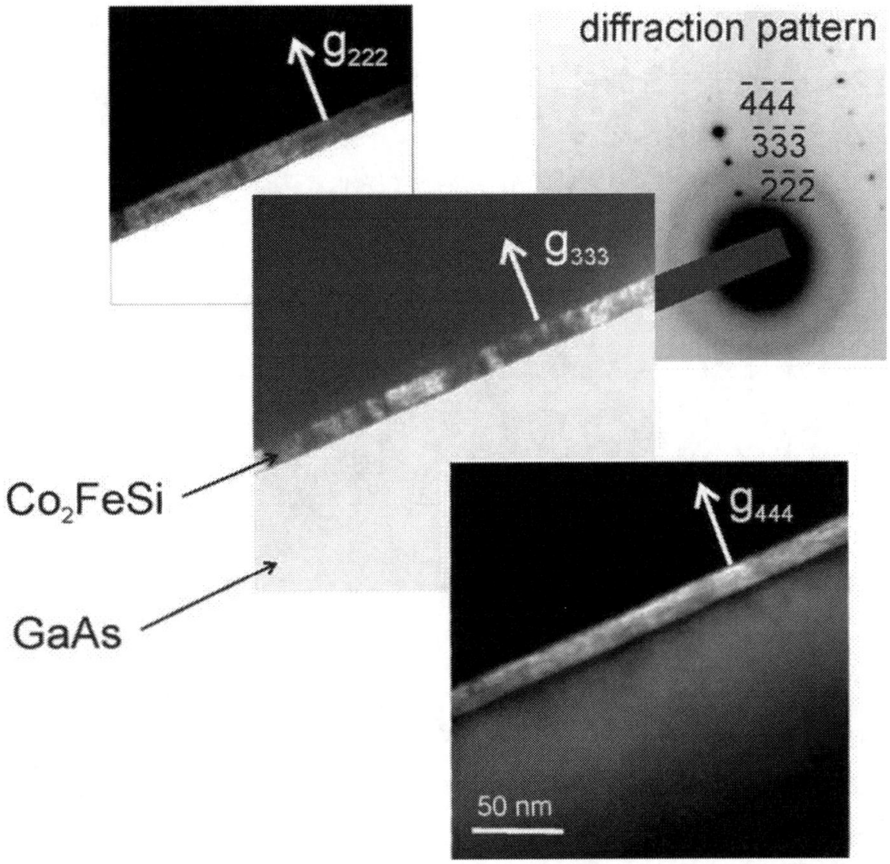

Figure 2. Electron diffraction pattern and dark-field TEM micrographs demonstrating the use of the SL maxima 222 and 333 for imaging the lateral distribution of ordered regions in a Co_2FeSi film on GaAs(-1-1-1)B substrate ($T_S = 275$ °C).

On such a basis, we determined the average ordering in three of our samples grown on GaAs(110) at substrate temperatures $T_S = 100$ °C, $T_S = 200$ °C and $T_S = 250$ °C (see Table 1). The averaging was performed over the whole information depth of the X-rays, which is however smaller than the film thickness, and laterally over an area of several mm^2. The region near the FM/SC IF had to be excluded in order to avoid contributions of the GaAs lattice. The result of about 65% $L2_1$ ordering for T_S = 200°C and T_S = 250°C seems reasonable for thin films although for bulk material tempered at high temperatures for long durations, a higher degree of order was reported. Such annealing is not possible for nanometer thick films because of accompanying diffusion processes, which lead to non-stoichiometry near the FM/SC IF. On the other hand, the growth at higher substrate temperatures leads to an increased order in the as-grown film as shown in Table 1 [2, 6]. Often, HRTEM micrographs are taken as evidence for the high structural quality of Heusler alloy films. However, in this mode of operation of the TEM many reflections interfere. The SL reflections are of low intensity compared to the fundamental ones. One has to pay attention to minor changes in the interference contrast in order to observe effects due to inter-diffusion and ordering.

Table 1. Percentages of average long-range order S_{B2} and S_{L21} of Co_2FeSi films grown on a GaAs(110) substrate at different substrate temperatures T_S. Regions of B2 and $L2_1$ order may overlap.

T_S (°C)	S_{B2} (%)	S_{L21} (%)	*Error* (%)
100	32	48	±2
200	61	65	±2
250	38	63	±2

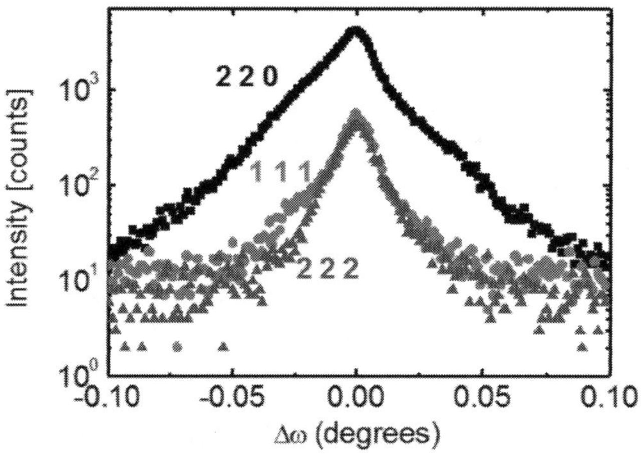

Figure 3. Grazing incidence X-ray diffraction peaks of a 40 nm thick Co_2FeSi film on GaAs(110) grown at T_S = 250°C. The measurements are performed at an incidence angle below the critical angle of Co_2FeSi. Therefore the information depth is below 10 nm. The fundamental 220 reflection (squares) and the 222 (triangles) and 111 (circles) SL reflections are given.

References

[1] Zutic I, Fabian J and Sarma S D 2004 *Rev. Mod. Phys.* **76** 323
[2] Jenichen B, Herfort J, Kumakura K and Trampert A 2010 *J. Phys. D* **43** 285404
[3] Hentschel T, Jenichen B, Trampert A, Kumakura K and Herfort J 2010 *J. Phys. D* **45** 055002
[4] Hashimoto M, Herfort J, Schönherr H P and Ploog K H 2005 *Appl. Phys. Lett.* **87** 102506
[5] Takamura Y, Nakane R, Sugahara S 2009 *J. Appl. Phys.* **105** 07109
[6] Jenichen B, Herfort J, Hentschel T, Nikulin A, Kong X, Trampert A and Zizak I 2012 *Phys. Rev. B* **86** 075319

Journal of Physics: Conference Series **471** (2013) 012023 doi:10.1088/1742-6596/471/1/012023

High resolution imaging in cross-section of a metal-oxide-semiconductor field-effect-transistor using super-higher-order nonlinear dielectric microscopy

N Chinone, K Yamasue, K Honda and Y Cho

Research Institute of Electrical Communication, Tohoku University, 2-1-1, Katahira, Aoba-ku, Sendai 980-8577, Japan

chinone@riec.tohoku.ac.jp

Abstract. Scanning nonlinear dielectric microscopy (SNDM) can evaluate carrier or charge distribution in semiconductor devices. High sensitivity to capacitance variation enables SNDM to measure the super-high-order (higher than 3rd) derivative of local capacitance-voltage (C-V) characteristics directly under the tip ($d^nC/dV^n, n=3,4,\cdots$). We demonstrate improvement of carrier density resolution by measurement of $d^nC/dV^n, n=1,2,3,4$ (super-higher-order method) in the cross-sectional observation of metal-oxide-semiconductor field-effect-transistor.

1. Introduction

The requirement for small, high-speed, and high-function large scale integrated circuits (LSIs) and large scale memories has been promoting miniaturization of transistors in these devices [1]. Widely known as Moore's law, the number of electric devices in LSIs has been exponentially increased with exponential decrease of device scale [1]. According to prediction by International Technology Roadmap of Semiconductors, this trend will continue and the required gate length of transistor will go down to less than 10 nm within the next 10 years [2]. Commercially, flash memories containing transistors whose scale is less than 20 nm are already available [3]. In research, a transistor with 3 nm of gate length has been achieved [4]. However, highly advanced miniaturization enhances the difficulty of suppressing the fluctuation of device properties arising from random spatial dopant distribution [5]. In addition, the trend of low voltage operation devices [2] enhances the influence of the fluctuation. Therefore, a technique for analysing small variation of carrier density with high spatial resolution is indispensable for developing present and future device performance.

Scanning probe microscopies (SPMs) are promising technique for measuring nanometer-scale devices with high spatial resolution. As a purpose of carrier concentration analysis in semiconductor devices, for example, scanning spreading resistance microscopy (SSRM) [6], scanning capacitance microscopy (SCM) [7], and scanning nonlinear dielectric microscopy (SNDM) [8] have been used. Among these SPMs, SNDM can analyse carrier or charge distribution thorough local capacitance-voltage (C-V) measurement. In previous studies, carrier distribution in metal-oxide-semiconductor field-effect-transistor (MOSFET) and stored charge distribution in flash memory were clearly visualized [9–11]. For SNDM, detectable minimum capacitance variation is typically 10^{-22} F [12]. This high sensitivity enables SNDM to acquire not only the 1st order response of capacitance to external voltage ($\mathrm{d}C_\mathrm{s}/\mathrm{d}V$), but also the n-th order response ($\mathrm{d}^nC_\mathrm{s}/\mathrm{d}V^n$), where $n \geq 2$ [13, 14]. We here call the term $\mathrm{d}^2C_\mathrm{s}/\mathrm{d}V^2$ higher-order term and the term $\mathrm{d}^nC_\mathrm{s}/\mathrm{d}V^n$ ($n \geq 3$) super-higher-order term. The higher-order and super-higher-order terms are more sensitive to capacitance variation compared to $\mathrm{d}C_\mathrm{s}/\mathrm{d}V$ term. Measurement of up to super-higher-order terms (super-higher-order method) [14] enables us to detect small variations of C-V characteristics caused by small carrier density variation.

Previously, we reported that super-higher-order imaging could reveal more detailed structure of a MOSFET compared to 1st and 2nd order images [14]. In this paper, we focus on carrier density in the dopant area of a MOSFET and demonstrate that the super-higher-order method is able to clarify slight spatial variation of carrier density.

2. Principle of SNDM

SNDM measures capacitance variation under the sharp conductive tip through frequency modulation and demodulation [8]. Figure 1 shows a schematic diagram of SNDM which is based on contact-mode atomic force microscopy (AFM). The probe of SNDM consists of a metal-coated cantilever tip and an active LC oscillator, whose oscillating frequency f is determined by tip-sample capacitance C_s, static capacitance C_0 (including built-in and stray capacitance), and built-in inductance L. When the C_s has nonlinear dependence on the external voltage V applied between sample and grounded tip, the f is modulated by V. When $V = V_\mathrm{DC} + V_\mathrm{p}\cos\omega_\mathrm{p}t$, the variation of C_s is approximately described as follows [14].

$$\Delta C_\mathrm{s} \approx \left.\frac{\mathrm{d}C_\mathrm{s}}{\mathrm{d}V}\right|_{V_\mathrm{DC}} V_\mathrm{p}\cos\omega_\mathrm{p}t + \frac{1}{4}\left.\frac{\mathrm{d}^2C_\mathrm{s}}{\mathrm{d}V^2}\right|_{V_\mathrm{DC}} V_\mathrm{p}{}^2\cos 2\omega_\mathrm{p}t$$

$$+\frac{1}{24}\left.\frac{\mathrm{d}^3C_\mathrm{s}}{\mathrm{d}V^3}\right|_{V_\mathrm{DC}} V_\mathrm{p}{}^3\cos 3\omega_\mathrm{p}t + \frac{1}{192}\left.\frac{\mathrm{d}^4C_\mathrm{s}}{\mathrm{d}V^4}\right|_{V_\mathrm{DC}} V_\mathrm{p}{}^4\cos 4\omega_\mathrm{p}t + \cdots \quad (1)$$

Therefore, by measuring the n-th order harmonic ($n\omega_\mathrm{p}$), n-th order capacitance response $\mathrm{d}^nC_\mathrm{s}/\mathrm{d}V^n$ is measured. The amplitude and phase of n-th harmonic extracted by lock-in amplifiers represent the magnitude and sign of n-th derivative of C-V characteristics. The super-higher-order SNDM acquires not only $1\omega_\mathrm{p}$ image, but also $2\omega_\mathrm{p}$, $3\omega_\mathrm{p}$, and $4\omega_\mathrm{p}$ images.

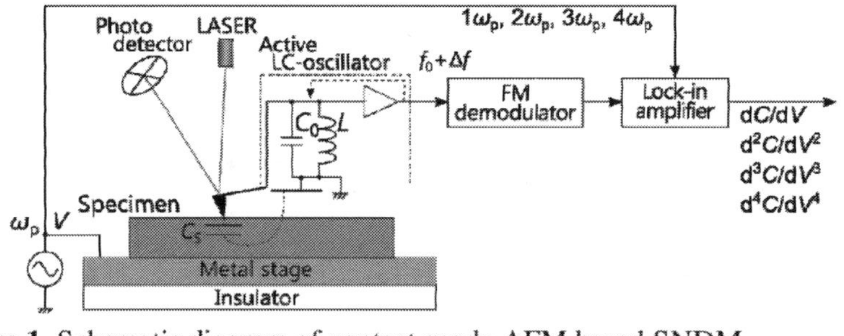

Figure 1. Schematic diagram of contact-mode AFM based SNDM.

3. Experimental result and discussion

In order to demonstrate the usefulness of the super-higher-order method, we measured the cross-section of an n-channel MOSFET whose channel length is 80 nm. We used a commercial AFM (SPA300HV, Seiko Instrument Inc.) for maintaining the tip-sample contacting force and simultaneous acquisition of topography. The cantilever was a metal (Pt-Ir) coated conductive cantilever tip with the nominal tip radius of 25 nm and spring constant of 2.8 N/m (EFM, nanosensors). In order to obtain super-higher-order harmonics, we developed a high sensitive SNDM probe, whose oscillating frequency is about 4 GHz. For demodulating the output of the probe and for extracting amplitude and phase of each harmonic, we used a commercial modulation analyser (MS616B, Anritsu) and digital lock-in amplifier (LI5640, NF corp.). The measurement was done in air, and other conditions were as follows: $V_p = 3\,\text{V}, V_{DC} = -1.5\,\text{V}, f_p = \omega_p / 2\pi = 23\,\text{kHz}$.

Figures 2 (a)–(d) show the acquired SNDM images of 1–4 ω_p [14]. These four SNDM images (Figs. 2 (a)–(d)) reveal the MOSFET structure; gate (G) and source/drain (S/D). The S/D region is visualized as dark in Fig. 2 (a) where the dC_s/dV is negative since the S/D region is n-type. We can see that images of d^3C_s/dV^3 and d^4C_s/dV^4 (super-higher-order ones) visualize the MOSFET in more detail than ones of dC_s/dV and d^2C_s/dV^2 [14].

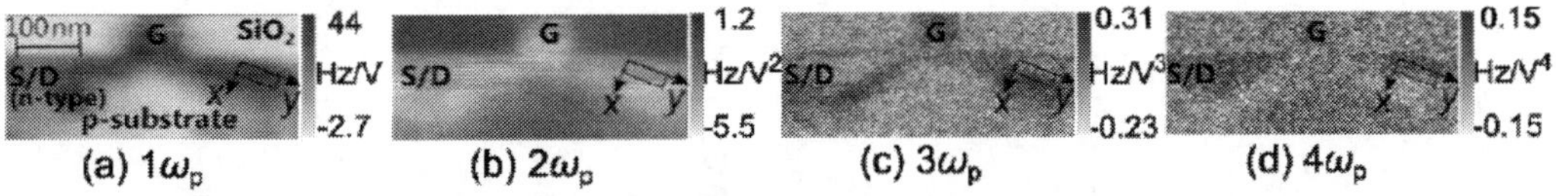

Figure 2. Acquired images of $d^nC/dV^n, n = 1, 2, 3, 4$. Reprinted with permission from Appl. Phys. Lett. **101**, 213112 (2012). Copyright 2012 American Institute of Physics

In order to discuss the acquired images in more detail, we extracted lineprofiles from the marked areas in four SNDM images along the *x*-direction. Since the signal-to-noise ratio was not sufficient, the lineprofiles were averaged in the *y*-direction. The averaged lineprofiles are shown in Figs. 3 (a)–(d), which correspond to positional dependencies of dC_s/dV – d^4C_s/dV^4 signals. The dC_s/dV signal kept a constant negative value and the d^2C_s/dV^2 signal kept an almost constant positive value. In contrast, the super-higher-order signals (d^3C_s/dV^3 and d^4C_s/dV^4) had strong dependence on position along the *x*-axis. Along the *x*-axis, the d^3C_s/dV^3 signal varied from nearly 0 to a negative value and the d^4C_s/dV^4 signal varied from a negative value to 0.

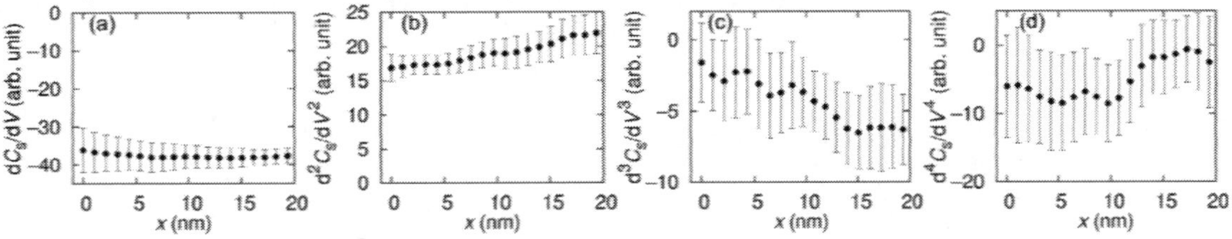

Figure 3. The averaged lineprofiles extracted from areas marked in Figs. 2 (b)–(e). Each graph represents the positional dependence along the *x*-axis of $d^nC_s/dV^n, n = 1, 2, 3, 4$.

To explain these signal variations, we considered how the d^nC_s/dV^n signals behave depending on the small change of carrier density. Table 1 (a)–(d) shows schematic graphs illustrating the relationship between d^nC_s/dV^n and stage voltage V in two cases: low carrier density and high density. The illustrated graphs in these two carrier density conditions are very close since the absolute difference in density is quite small. The graphs in the low carrier density case are a little shrunk along V-axis compared to graphs in the high carrier density case. When the carrier density slightly decreases, the signals from dC_s/dV to d^4C_s/dV^4 at $V = V_{DC}$ vary in the following manner. The d^3C_s/dV^3 varies from ~0 to negative and d^4C_s/dV^4 varies from negative to ~0. On the other hand, variation of

dC_s/dV and d^2C_s/dV^2 are not clear. Comparing the schematic behaviour of each derivative and the behaviour with acquired data shown in Figs. 3 (a)–(d), we can understand that the carrier density decreased along the *x*-axis. This result indicates that the super-higher-order method has the potential for measuring small spatial deviations of carrier density that cannot be clarified by conventional local dC_s/dV measurement.

Table 1: Schematic graphs illustrating the relationship between d^nC_s/dV^n $(n = 1, 2, 3, 4)$ and stage voltage V in the n-type region corresponding to (A) high carrier density and (B) low carrier density.

	Carrier density	(a) dC_s/dV	(b) d^2C_s/dV^2	(c) d^3C_s/dV^3	(d) d^4C_s/dV^4
(A)	High				
(B)	Low				
		const. (negative)	const. (positive)	0→negative	negative→0

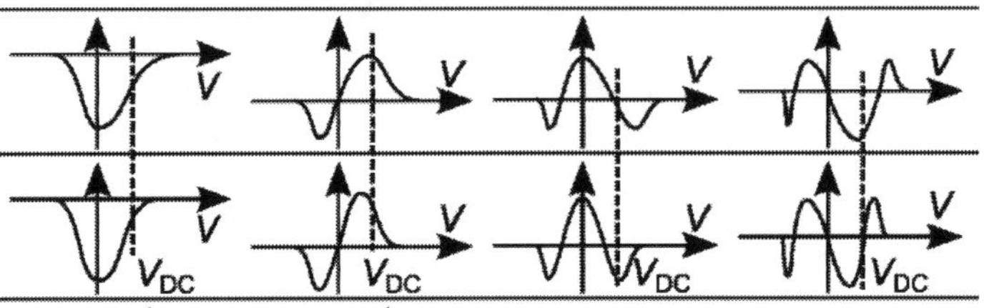

4. Conclusion

We demonstrated that the super-higher-order method can reveal small variations of carrier density that cannot be detected by conventional local dC_s/dV measurement through the cross-sectional imaging of a MOSFET. This method will be useful for precise carrier density analysis in semiconductor devices.

Acknowledgement

We would like to acknowledge Masao Uno of Murata Manufacturing Co. Ltd. and Toshihiko Iwai of Research Institute of Electric Communication, Tohoku University, for supporting the development of the high-sensitive SNDM probe. This work is partly supported by a Grant-in-Aid Scientific Research S (23226008) from the Japan Society for the Promotion of Science.

References

[1] Schwierz F *et al* 2010 *Nanometer CMOS* (Pan Stanford Publishing Pte. Ltd., Singapore)
[2] *International Technology Roadmap for Semiconductors*, 2011 edition, Process Integration, Devices, and Structures (Semiconductor Industry Association, 2011)
[3] Shibata N *et al* 2012 *International Solid State Circuit Conference* 25.1
[4] Migita S, Morita Y, Masahara M and Ota H 2012 *International Electron Devices Meeting* 8.6
[5] Hoeneisen B and Mead C A 1972 *Solid-State Electron.* **15** 819
[6] Shafai C, Thomson J D and SimardNormandin M 1994 *J. Vac. Sci. Technol. B* **12** 378
[7] Williams C C *et al* 1989 *Appl. Phys. Lett.* **55** 1662
[8] Cho Y, Kirihara A, and Saeki T 1996 *Rev. Sci. Instrum.* **67** 2297
[9] Masahara M et al 2004 *Appl. Phys. Lett.* **85** 4139
[10] Ishikawa K, Honda K and Cho Y *et al* 2006 *Nanotechnology* **17** S1
[11] Honda K, Hashimoto S and Cho Y 2005 *Nanotechnology* **16** S90
[12] Cho Y 2011 *J. Mater. Res.* **26** 2007
[13] Cho Y and Ohara K 2001 *Appl. Phys. Lett.* **79** 3842
[14] Chinone N, Yamasue K, Honda K and Cho Y 2012 *Appl. Phys. Lett.* **101** 213112

STEM strain analysis at sub-nanometre scale using milli-second frames from a direct electron read-out CCD camera

K Müller[1], H Ryll[2], I Ordavo[2], M Schowalter[1], J Zweck[3], H Soltau[2], S Ihle[4], L Strüder[4], K Volz[5], P Potapov[6] and A Rosenauer[1]

[1] Universität Bremen, Otto-Hahn-Allee 1, 28359 Bremen, Germany
[2] PNSensor GmbH, Römerstraße 28, 80803 München, Germany
[3] Universität Regensburg, Universitätsstraße 31, 93040 Regensburg, Germany
[4] Max-Planck-Institut Halbleiterlabor, Otto-Hahn-Ring 6, 81739 München, Germany
[5] Philipps Universität Marburg, Hans-Meerwein-Straße, 35032 Marburg, Germany

E-mail: mueller@ifp.uni-bremen.de

Abstract. We report on strain analysis by nano-beam electron diffraction with a spatial resolution of 0.5nm and a strain precision in the $4\text{-}7\cdot 10^{-4}$ range. Series of up to 160000 CBED patterns have been acquired in STEM mode with a semi-convergence angle of the incident probe of 2.6mrad, which enhances the spatial resolution by a factor of 5 compared to nearly parallel illumination. Firstly, we summarise 3 different algorithms to detect CBED disc positions accurately: selective edge detection and circle fitting, radial gradient maximisation and cross-correlation with masks. They yield equivalent strain profiles in growth direction for a stack of 5 $In_xGa_{1-x}N_yAs_{1-y}$/GaAs layers with tensile and compressive strain. Secondly, we use a direct electron read-out pnCCD detector with ultrafast readout hardware and a quantum efficiency close to 1 both to show that the same strain profiles are obtained at 200 times higher readout rates of 1kHz and to enhance strain precision to $3.5\cdot 10^{-4}$ by recording the weak 008 disc.

1. Introduction

In transmission electron microscopy (TEM), lattice strain leaves its fingerprint in positions, amplitudes and phases of diffracted beams. Since especially the latter usually depend on lens aberrations as well as on thickness, orientation, and chemical composition of the specimen, phase-exploiting strain measurement via high-resolution [1] or holographic TEM methods [2, 3] can suffer from artefacts due to gradients of these parameters. In contrast, *positions* of diffracted beams are related to lattice plane spacings simply via Bragg's law. In combination with scanning TEM (STEM), strain analysis by nano-beam electron diffraction (SANBED) has become an attractive method with probe diameters of 2.7 nm for parallel beam electron diffraction (PBED [4]) and 0.5 nm for convergent beam electron diffraction (CBED [5]), while strain precision in the 10^{-4} range is comparable. Currently, one severe drawback of SANBED is the limited speed of the CCD detector as a large number of CBED patterns (typically $\sim 10^3$) is to be acquired during the scan. After a brief summary of dedicated algorithms for accurate CBED disc position detection, this article focuses on SANBED using a pnCCD [6, 7, 8] detector which is a low-noise, high-sensitivity and scintillator-free direct electron detection system with kHz readout rate.

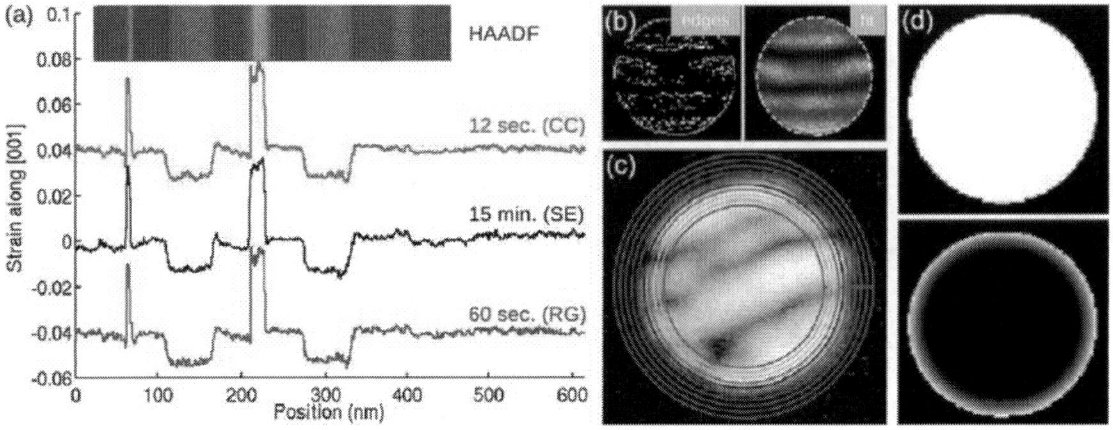

Figure 1. (a) Compressive and tensile strain profiles along [001] measured in a stack of 5 $In_xGa_{1-x}N_yAs_{1-y}$/GaAs layers. The profiles correspond to the 3 algorithms SE, RG and CC with durations of the full evaluation of 800 diffraction patterns also given. (b) 004 disc: initial edges (left) and final circle fit (right) for SE. (c) Rings for inner (coloured) and outer (blue) averages for RG. (d) Full (top) and weighted ring mask (bottom) for CC.

2. Summary of the development of the SANBED method

By focusing the probe with a semi-convergence angle of 2.6 mrad, which we did using an FEI Titan microscope operated in STEM mode at 300 kV, the probe diameter becomes a factor of 5 smaller than in PBED. On the other hand, this requires robust disc position recognition algorithms to cope with the intensity distribution in the CBED reflections which is not only strongly varying inside one disc but also changes drastically among different CBED patterns of one series due to local changes of both thickness and orientation of the specimen. In this study, the sample was 100±10 nm thick, measured in Ref. [5] by high-angle annular dark field (HAADF) STEM.

The first algorithm developed is called *selective edge detection and circle fitting* (SE). Here, Prewitt's edge detection method is applied to raw images yielding edge pixels throughout the full disc as depicted on the left in Fig. 1b. Iteratively, circles are fitted to these edges where the edge pixel with the largest difference from a fit is ignored in subsequent iterations, so that the subset of edge pixels on the disc border finally remains for the last fit on the right in Fig. 1b.

The second algorithm, *radial gradient maximisation* (RG), detects the disc border by maximising the difference between 2 sums: one is calculated from rotational averages inside the disc, one from the outside as shown in Fig. 1c by coloured and blue rings, respectively. During maximisation, position and/or radius of the rings are changed. To stress the disc border, a weighting scheme can be applied to the inner rings as indicated by the different colours.

In a third option called *cross-correlation with masks* (CC), we assume that the CBED disc is similar to an artificial mask so that the relative shift can be calculated by cross-correlation. For the disc-shaped mask in Fig. 1d this is rarely the case due to the inner disc structure in experiment. However, correlating with the weighted ring-shaped mask below stresses the influence of the disc edge and is very robust against such artefacts.

All 3 algorithms have been applied to a sequence of 5 $In_xGa_{1-x}N_yAs_{1-y}$/GaAs layers visible in the HAADF STEM image in Fig. 1a. Given disc positions from SE, RG or CC for all diffraction patterns of a series, strain along [001] is calculated from

$$\varepsilon_{[001]} = \frac{R^0_{004}}{R^0_{004} + \Delta_{004}} \quad (1) \qquad \text{or from} \qquad \varepsilon_{[001]} = \frac{\sin\theta^B_{004}}{\sin(\theta^B_{004} + \frac{1}{2}\Delta\theta_{004})} \quad (2)\ .$$

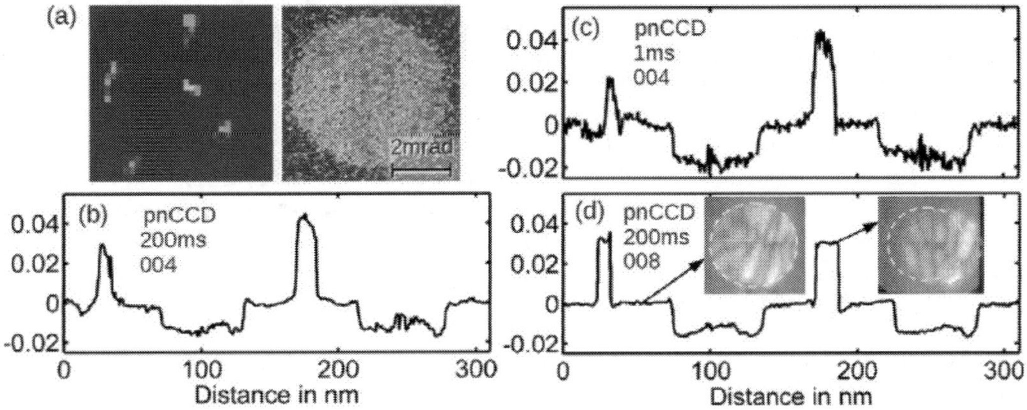

Figure 2. (a) Single electron events at 300 kV (left) and the 004 reflection (right) recorded with the direct electron detector pnCCD. (b) Strain profile as in Fig. 1 but calculated with RG using the 004 disc from pnCCD frames with 200ms frame time. (c) Same as in (b) but with 1 ms frame time. (d) Strain profile calculated with RG from pnCCD frames with 200 ms frame time using the 008 reflection. The two insets demonstrate the high strain sensitivity of 008.

Eq. (1) is valid if both primary and beam 004 are recorded, R^0_{004} being the average distance between both discs in unstrained GaAs and Δ_{004} being the deviation from R^0_{004}. In Eq. (2), only one reflection is needed (here: 004) where the Bragg angle θ^B_{004} in unstrained GaAs is taken from literature and angular deviations $\Delta\theta_{004}$ are measured from the disc movement.

The black (middle) graph in Fig. 1a shows the strain profile through the layers in the inset, measured by SE from the 004 reflection in 800 diffraction patterns. It demonstrates that strain in a complex structure with alternating tensile/compressive strain can be determined accurately over a range of more than 500 nm with a precision of $7.3 \cdot 10^{-4}$ being the standard deviation in the GaAs buffer. Best results from the RG and CC approaches have been obtained from logarithmically scaled intensities. The blue (bottom) and red (top) graphs in Fig. 1a show that RG and CC reproduce the SE result with the advantage that the evaluation speeds up by a factor of 15 (RG) or 75 (CC), respectively.

3. Direct electron detection

To take advantage of the SANBED technique in STEM, detectors with a high signal-to-noise ratio, high detection quantum efficiency (DQE) and high-speed readout are needed, especially with respect to 2-dimensional strain mapping. Whereas a conventional scintillator-based CCD camera (Gatan UltraScan 2000) with frame times of 0.5 s has been used to develop pattern recognition algorithms above (Fig. 1), we investigated the performance of a 264^2 pixel, scintillator-free CCD camera, mainly consisting of a pnCCD chip for electron detection and ultrafast readout hardware with technical details described elsewhere [6, 7, 8]. Mounted on an FEI Titan microscope operated at 300 kV at a camera length of 943 mm, we conducted 2 types of experiments:

3.1. Strain analysis with 1 kHz readout rate

The specimen from Fig. 1a has been investigated along a path in growth direction sampled in steps of 0.5 nm with a STEM probe dwell time of 200 ms. Only the 004 disc has been recorded comprising 125 pixels in diameter so that strain was calculated from Eq. (2). A readout rate of 1 kHz was chosen which yields 200 frames of the 004 reflection per scan point so as to check strain results for different frame times afterwards by integrating over 1-200 frames. It is important to notice that, usually, direct electron detectors for TEM are tuned to high readout rates because this enables single event processing of the traces each relativistic electron causes in the chip as depicted left in Fig. 2a. On the contrary, we nearly saturated the detector in this study to get a complete image of the 004 disc already in a 1 ms frame as shown on the right in Fig. 2a, where single events are impossible to separate. However, as this point spread is isotropic, which we confirmed by Fourier transform of low-dose images [9], it indeed blurs the CBED discs but nevertheless the disc position should remain unchanged.

Because of these substantial differences to scintillator-based CCD images, we first evaluated the direct electron data with comparable frame times of 200 ms (i.e. integration over 200 frames) with the RG algorithm to compare strain precision and -profiles. The result is shown in Fig. 2b which not only reproduces the profiles of Fig. 1a, but also exhibits the same precision of $7.0 \cdot 10^{-4}$. Secondly, we calculated the strain profile in Fig. 2c by using only 1 frame per scan point. Again, the same result is obtained, however, with a lower precision of $1.3 \cdot 10^{-3}$.

3.2. Strain analysis using weak high-order reflections

Finally, we rather aimed at high precision than at high speed and exploited the camera's excellent DQE and signal-to-noise ratio by recording the comparably weak 008 disc with frame times of 200 ms. According to Bragg's law, this reflection is twice as sensitive to lattice parameter changes as 004. This is demonstrated by the two insets in Fig. 2d showing an 008 frame of an unstrained and a strained region, respectively. Consequently, the strain profile in Fig. 2d, obtained with the RG algorithm and Eq. (2) with indices replaced by 008, is characterised by very low noise compared to all other previous measurements. Remarkable precisions in unstrained GaAs have been determined to $3.5\text{-}5.2 \cdot 10^{-4}$.

4. Conclusions

Conceptionally, STEM strain analysis by nano-beam electron diffraction (SANBED) is able to push the spatial resolution of nano-beam-diffraction to the sub-nm range with dedicated algorithms for automated CBED disc position measurement while strain precision is in the range of $4\text{-}7 \cdot 10^{-4}$. In view of the large amount of data recorded in SANBED, cross-correlation with masks is in principle capable of in-situ strain evaluation at the TEM. As to hardware requirements, direct electron detectors such as pnCCD devices with an ultrafast readout unit provide the possibility to either high-speed strain mapping with kHz-readout for 2-dimensional strain analysis, or high-precision strain mapping with high-order reflections. Future tasks could focus on analogous experiments with delay-line detectors, in-situ implementation of SANBED or simulation studies concerning the actual spatial resolution of SANBED at interfaces where regions with different lattice parameters contribute to one diffraction pattern.

References

[1] Bierwolf R, Hohenstein M, Phillipp F, Brandt O, Crook G and Ploog K 1993 *Ultramicroscopy* **49** 273–285

[2] Hÿtch M, Houdellier F, Hüe F and Snoeck E 2008 *Nature* **453** 1086–1089

[3] Koch C T, Özdöl V B and van Aken P A 2010 *Appl. Phys. Lett.* **96** 091901

[4] Béché A, Rouvière J L, Clément L and Hartmann J M 2009 *Appl. Phys. Lett.* **95** 123114

[5] Müller K, Rosenauer A, Schowalter M, Zweck J, Fritz R and Volz K 2012 *Microsc. Microanal.* **18** 995–1009

[6] Strüder L et al. 1997 *Rev. Sci. Instrum.* **68** 4271–4274

[7] Hartmann R, Buttler W, Gorke H, Herrmann S, Holl P, Meidinger N, Soltau H and Strüder L 2006 *Nucl. Instrum. Methods Phys. Res., Sect. A* **568** 118–123

[8] Ordavo I et al. 2011 *Nucl. Instrum. Methods Phys. Res., Sect. A* **654** 250–257

[9] Müller K, Ryll H, Ordavo I, Ihle S, Strüder L, Volz K, Zweck J, Soltau H and Rosenauer A 2012 *Appl. Phys. Lett.* **101** 212110.

Journal of Physics: Conference Series **471** (2013) 012025 doi:10.1088/1742-6596/471/1/012025

Indium Nitride and Indium Gallium Nitride layers grown on nanorods

R F Webster[1], D Cherns[1], L E Goff[2], S V Novikov[2], C T Foxon[2], A M Fischer[3] and F A Ponce[3]

[1]Department of Physics, University of Bristol, Tyndall Avenue, Bristol BS8 1TL, UK
[2]School of Physics and Astronomy, University of Nottingham, Nottingham NG7 2RD, UK
[3]Department of Physics, Arizona State University, Tempe, AZ, US

E-mail: richard.webster@bristol.ac.uk

Abstract. Molecular beam epitaxy has been used to grow InN layers on both Si and SiC substrates and $In_{0.5}Ga_{0.5}N$ layers on Si substrates using a nanorod precursor array. Transmission electron microscopy (TEM) studies show that nanorods grown first under N-rich conditions, and then under more metal-rich conditions to promote lateral growth are free of dislocations until coalescence occurs. At coalescence, dislocations are introduced at grain boundaries. These are predominantly twist boundaries, with better epitaxial alignment seen on SiC substrates. The lateral growth of $In_{0.5}Ga_{0.5}N$ is shown to be cubic, tentatively ascribed to the growth of basal plane stacking faults at the start of the lateral growth and the low growth temperatures used.

1. Introduction

In order to develop high efficiency solar cells the choice of material becomes very important. The majority of current multi-junction solar cells combine different materials systems that are lattice-matched and each with a different band gap range. This has advantages over a single junction solar cell as band gaps of the materials used may be tuned to increase the amount of absorption at different regions of the solar spectrum.

$In_xGa_{1-x}N$ has been proposed as a single materials system for a multi-junction solar cell as the band gap varies from 0.7eV (x=0) to 3.4eV (x=1) which spans the whole visible spectrum. Previous attempts to grow $In_xGa_{1-x}N$ solar cells had low efficiencies [1] when compared to theoretical models [2] as films grown hadve suffered from high defect densities, and possibly phase separation at high In contents. Molecular beam epitaxy (MBE) is an effective method of growing high In content alloys of reasonable quality, but requires low growth temperatures as the In-N bond is quite weak. This means that at temperatures greater than 450°C indium desorbs more readily leading to variations in indium content [3].

One way to overcome defects is to start growth as nanorods before coalescence. Previous work on GaN grown on sapphire has shown that it is possible to first grow nanorods which are defect free, with dislocations only forming as a continuous film is grown on top of the rods [4].

This paper demonstrates the ability to grow InN and $In_xGa_{1-x}N$ nanorods on silicon (Si) and silicon carbide (SiC) and suggests growth mechanisms. By varying the growth conditions it is shown that coalescence of nanorods result in overgrowths which have grains of the order 500nm wide which span multiple nanorods.

2. Experimental Method

InN and $In_xGa_{1-x}N$ nanorods have been grown directly on Si substrates by plasma assisted MBE in a Varian ModGen II system. Nanorods were grown for 5hrs at temperatures between 350-400°C in N-rich conditions. For lateral growth the In and Ga fluxes were increased for a growth time of 3hrs. InN samples were grown on SiC using similar conditions in a custom built MBE system with a RF plasma nitrogen source and a K-cell for the indium flux.

TEM was performed on a Philips EM430 at 200kV using imaging, selected area diffraction (SAD), convergent beam electron diffraction (CBED) and energy dispersive X-ray spectroscopy (EDX) techniques. Cross section and Plan view samples were prepared for TEM by mechanical polishing by diamond lapping films to less than 100μm thickness, then ion thinned using a Gatan PIPS, first at 5kV until a small hole began to form and finishing with a 15 minute polish at 3kV to reduce surface damage.

Figure 1. Shows cross-section TEM of N-rich growth of (a) InN on Si, (b) InN on SiC and (c) $In_xGa_{x-1}N$ (x=~0.5) on Si.

3. Results

Figures 1(a) and 1(b) show TEM of InN growth on Si and SiC grown under N-rich conditions at 400°C. On Si (Figure 1(a)) the InN nanorods grown have lengths of 250±30nm and have a degree of lateral growth. These nanorods have no threading dislocations, with the only visible contrast being bend contours. In comparison the TEM of InN nanorods grown on Si-polar SiC (Figure 1(b)) show a dense forest of short nanorods ~600nm long which have some lateral growth and some nanorods which grow 100nm longer. CBED has shown that nanorods of both lengths have In-polarity; consequently the difference in length has been attributed to competitive growth. Figure 1(c) shows TEM of $In_xGa_{1-x}N$ nanorods grown on Si substrates. These were grown in N-rich conditions at a temperature of 400°C. They show evidence of some tapering as seen in the InN and do not have dislocations.

Plan view TEM of a sample grown on Si first under N-rich conditions to promote nanorod growth, and then under In-rich growth to promote lateral growth, is shown in Figure 2(a). Coalescence of the InN nanorods occurred, forming a nearly continuous film. However, significant rotational misalignments of the nanorods about the c-axis prior to coalescence caused the formation of large and small angle grain boundaries. Figure 2(b) shows a cross-section TEM image of InN grown on SiC with the same growth processes as on Si, which resulted in coalescence of the nanorods occurring right down to the substrate. In some areas where coalescence has occurred screw and mixed dislocations are generated, however some grains have no dislocations present and are 500nm wide which indicates that a good epitaxy of the nanorods was achieved before coalescence on the SiC substrates. $In_xGa_{1-x}N$ coalescence is shown in figure 2(c). It is clear here that the lateral overgrowth has led to coalescence, creating large defect-free grains wider than 500nm which span multiple nanorods.

Figure 2. TEM of N-rich growth followed by metal-rich growth. (a) Plan view of InN on Si (b) cross-section of InN on SiC (c) cross-section of $In_xGa_{x-1}N$ (x~0.5) on Si.

Further investigation of the overgrown films is shown in Figure 3. A [110] SAD pattern confirms that the overgrowth region is cubic $In_xGa_{1-x}N$ grown on hexagonal nanorods. Also visible in Figure 3 are multiple stacking faults at the base of the overgrowth region which extend to the edge of the crystals. Towards the top of the lateral growth region, where the cubic material is present, there are {111} stacking faults inclined at 70° to the basal plane after which the phase reverts to hexagonal. In Figure 4, EDX has been taken from the base and the top of a nanorod grown in N-rich conditions. At the bottom of the nanorod the indium content is 9%, which increases to 55% at the top of the nanorod.

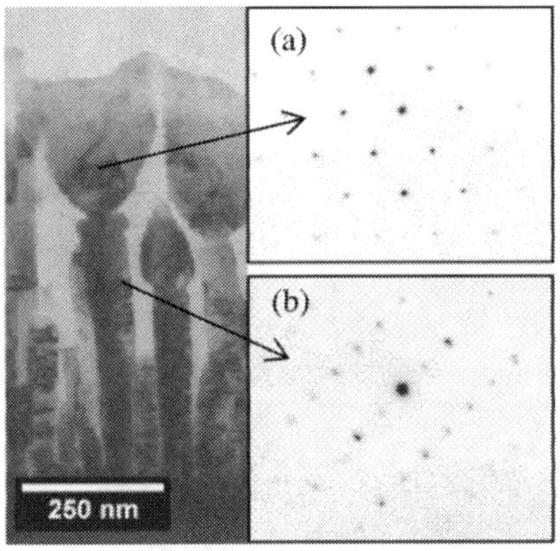

Figure 3. TEM of an $In_{0.5}Ga_{0.5}N$ nanorod grown on Si, first with N-rich conditions then metal rich conditions with insets (a) a [110] SAD pattern and (b) a [11$\bar{2}$0] SAD pattern.

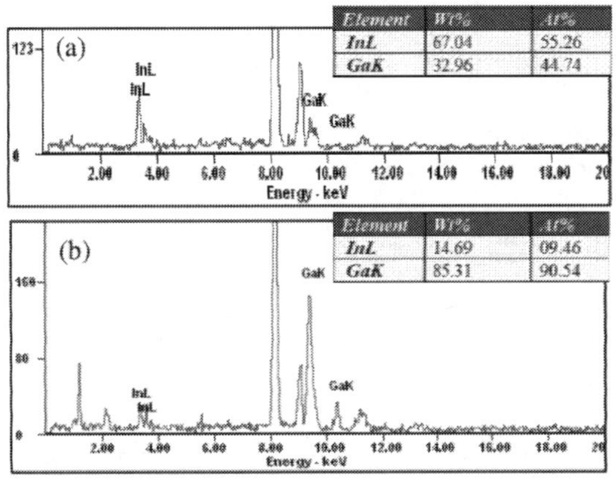

Figure 4. EDX taken from a nanorod grown under N-rich conditions. (a) EDX from the top of a nanorod has a composition *x*=0.55 and (b) EDX from the bottom of a nanorod has a composition *x*=0.09.

4. Discussion

When considering the growth of InN, the general trend is for the nanorods to taper as the growth progresses. This was seen on both Si and SiC substrates. One possible growth mechanism is that at temperatures as high as 400°C, the In atoms have enough energy to desorb during the start of growth increasing the concentration of In atoms at the growth surface. As the growth progresses, the increased indium concentrations promote more lateral growth, which results in tapering of the nanorods.

The growth of $In_xGa_{1-x}N$ is slightly more complicated by the addition of Ga atoms. However at temperatures as low as 400°C it may be assumed that once a Ga atom adsorbs it is not likely to desorb, when compared to In atoms. In this case a similar model may be proposed as for the InN. As the growth begins, very few In atoms adsorb without subsequently desorbing, and when they desorb can be replaced by a Ga atom. As the growth continues, the concentration of In increases, which increases the likelihood of a desorbed In atom being replaced by a second In atom. This results in Ga-rich material at the start of the growth and more In-rich material towards the end of the growth. This has been confirmed by the EDX in figure 4(b) which shows an increase of In content as the growth progresses and is in agreement with other work on $In_xGa_{1-x}N$ nanorods [5].

The coalesced regions with cubic regions may grow due to the stacking faults, which extend to the edges of the crystal, locally changing the stacking sequence from hexagonal (...ABAB...) to a cubic sequence (...ABCABC...), thus providing a cubic substrate for subsequent growth. Other work has shown it was possible to grow cubic GaN at low temperatures (~680°C)[6], this could possibly explain why the cubic growth continues.

5. Conclusion

The growth of $In_xGa_{1-x}N$ nanorods over the composition range x=0 to x=0.5 has been demonstrated. Over this range the nanorods grown show no evidence of dislocations, only bend contours. After lateral growth, defects form due to rotational misalignment of the nanorods. This is more prevalent on Si than on SiC where InN nanorods show better alignment. $In_xGa_{1-x}N$ nanorods show a segregation of the In content, with In-rich material at the tops of the rods. In the lateral growth of $In_{0.5}Ga_{0.5}N$ material, we have confirmed the presence of cubic zinc blende material which is seeded by basal plane stacking faults and continued due to low growth temperatures, before reverting to the hexagonal phase.

Acknowledgements

This work is supported by EPSCR grant EP/I035501/1.

References

[1] R Dahal, B Pantha, J Li, J Y Lin, and H X Jiang 2009 *Appl. Phys. Lett.* **94** 063505
[2] X Zhang, X Wang, H Xiao, C Yang, J Ran, C Wang, Q Hou, and J Li 2007 *J. Phys. D: Appl. Phys.* **40** 7335-8
[3] H Komaki et al 2007 *J. Cryst. Growth.* **301–302** 473-477
[4] D Cherns, L Meshi, I J Griffiths, S Khongphetsak, S V Novikov, N Farley, R P Campion, C T Foxon 2008 *Appl. Phys. Lett.* **92** 121902
[5] T Kehagias 2010 *Physica E: Low-dimensional Systems and Nanostructures* **42** 2197–2202
[6] S V Novikov, N M Stanton, R P Campion, R D Morris, H L Geen, C T Foxon and A J Kent 2008 *Semicond. Sci. Technol.* **23** 015018

Advanced TEM Characterization for the Development of 28-14nm nodes based on fully-depleted Silicon-on-Insulator Technology

G Servanton, L Clement, K Lepinay, F Lorut, R Pantel, A Pofelski and N Bicais

STMicroelectronics, 850 rue Jean Monnet, F-38926 Crolles, France

E-mail: germain.servanton@st.com

Abstract. The growing demand for wireless multimedia applications (smartphones, tablets, digital cameras) requires the development of devices combining both high speed performances and low power consumption. A recent technological breakthrough making a good compromise between these two antagonist conditions has been proposed: the 28-14nm CMOS transistor generations based on a fully-depleted Silicon-on-Insulator (FD-SOI) performed on a thin Si film of 5-6nm. In this paper, we propose to review the TEM characterization challenges that are essential for the development of extremely power-efficient System on Chip (SoC).

1. Introduction

The semiconductor industry has continuously been driven by the scaling down of transistor dimensions. Tomorrow's nomad applications will require higher speed performances with lower power consumption compared to today's devices. To fulfill these conditions, STMicroelectronics decided to develop the 28-14 nm technological nodes based on fully-depleted Silicon-on-Insulator (FD-SOI). Such a 28 FD-SOI chip, showing all metal levels till Al pads, is presented in the SEM cross-section from Figure 1. FD-SOI technology provides a unique value for extremely power-efficient chips compared to bulk MOS technologies. Actually, transistors run at higher frequencies (30%), are more power efficient (reduced leakage, thus preserved battery lifetime) and are much simpler for the manufacturing process compared to FinFETs [1]. A scanning TEM (STEM) Energy Dispersive X-ray (EDX) 3D reconstruction of a 28 FD-SOI transistor, presented in figure 1, shows the classical transistor parts (source, drain, gate, and channel) and the FD-SOI main process bricks (SiO_2 box and thin Si film). Advanced physical characterization is mandatory to develop FD-SOI technology, in particular to adjust critical process steps. For information purposes, in 2012, the average number of TEM analyses for 28-14 FD-SOI reached ~20 TEM studies per week with a response time of ~ 2.5 days. Specific issues are addressed through these analyses and will be detailed in this paper. Emphasis will be put on: the TEM sample preparation challenges using Focused Ion Beam (FIB), the need for quantified chemical characterization (STEM EDX) and strain/stress measurements by Nano Beam Electron Diffraction (NBED) or Dark Field Electron Holography (DFEH).

Figure 1. *(left)* SEM image showing a 28 nm FD-SOI chip from MOS to Al pads. *(right)* Tilted STEM EDX reconstruction of a 28 nm FD-SOI transistor

2. Characterization Challenges and limitations for 28-14nm FDSOI

2.1. Dedicated 28-14nm FD-SOI sample preparation

TEM lamellae of semiconductor devices are mainly prepared using a Focused Ion Beam (FIB) system, so as to ensure site specific process characterization. In this in-situ sample preparation method, a tab is extracted from the piece of wafer at target location. This tab is section milled to shape the lamella, from top layers to silicon substrate. For 28 FD-SOI devices, TEM lamellae have to be thinned down to typically less than 70nm, to not get overlapping structures, and even smaller for the smallest devices (e.g. SRAM cells). With the even smaller dimensions of the device, the top-down section approach appears not to be the most appropriate direction of observation; in-plane observation brings additional information about dimensions of the devices, while allowing relaxed TEM lamellae dimensions. If cross-sectional TEM observation is required, e.g. to control stacked layers of the process, reduced dimensions of the devices impose even thinner lamellae. This implies the use of low accelerating voltage for the FIB, to minimize the amorphous/damage layer; in case of the conventional top-down approach, this would lead to strong curtaining effects (coming from metal layers). Therefore, inversed sample prep is preferred for front-end device observation: the tab is flipped so that substrate is on top and metal layers at the bottom.

2.2. TEM/STEM morphological studies

In order to ensure a good scalability of SOI technology, the SiO_2 box and thin SiGe film thicknesses must be precisely monitored. Actually, for electrostatic control purpose, very low Si film thickness variations may be tolerated along the entire wafer. HR-TEM and STEM analysis have been performed in order to evaluate thin Si film roughness. These two imaging techniques are complementary since TEM enables high spatial resolution while STEM helps for Z-contrast imaging, as described in Figure 2.

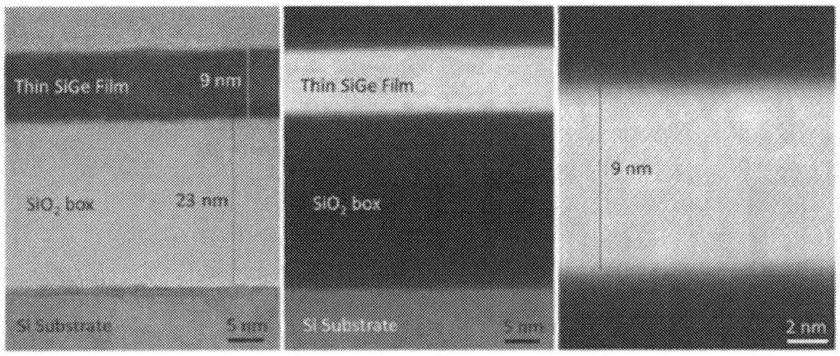

Figure 2. HRTEM *(left)* and HRSTEM *(centre)* images of SiO_2 box and thin SiGe film. *(right).* HRSTEM zoom showing SiGe lattice from thin Si film in *p*MOS 14nm FD-SOI.

2.3. STEM EDX chemical characterization

For all the FD-SOI elemental steps development, morphology and composition of the MOS transistors and structures had to be controlled at high resolution. The following TEM and STEM EDX analyses were carried out at 120 keV using Tecnai Osiris combining high brightness Field Emission Gun (FEG) and four EDX Silicon Drift Detectors (SDD). Figures 3, 4, 5 show analytical TEM analyses for the development of 28 and 14 nm FD-SOI technologies. Figure 3 shows STEM-EDX analysis of 28nm FD-SOI when source/drain epitaxy problems were encountered, leading to local defects. The initial interface showed silicon oxidation of a few nanometers, and finally this problem was solved by cleaning optimization.

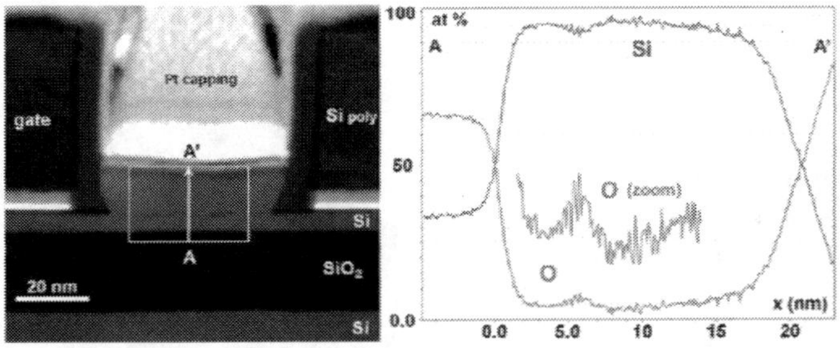

Figure 3. STEM ADF image and quantified STEM EDX profiles (Si, O) revealing an epitaxy issue during the raised source drain process of 28 nm FD-SOI technology

Figure 4 shows TEM morphology and STEM-EDX dopant process control of *n*MOS 14nm FD-SOI. Arsenic doping from the gates is mapped and atom segregation can be measured at nanometer scale.

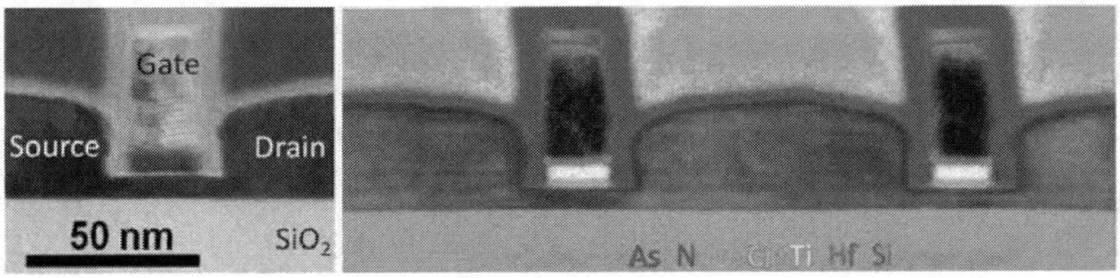

Figure 4. TEM bright field *(left)* and STEM-EDX elemental map *(right)* showing the dopant distribution in gate (As) 14 nm *n*MOS FD-SOI. *See online version for colour code.*

Figure 5 shows the complexity of 14nm *p*MOS FD-SOI alloys integration for channel and source/drain. The channel's SiGe composition should be homogeneous and was targeted at a concentration of 15% Ge. The SiGe raised sources/drains, after Ni(Pt) silicidation are made of quaternary alloys (Si, Ge, Ni, Pt) which must be accurately controlled.

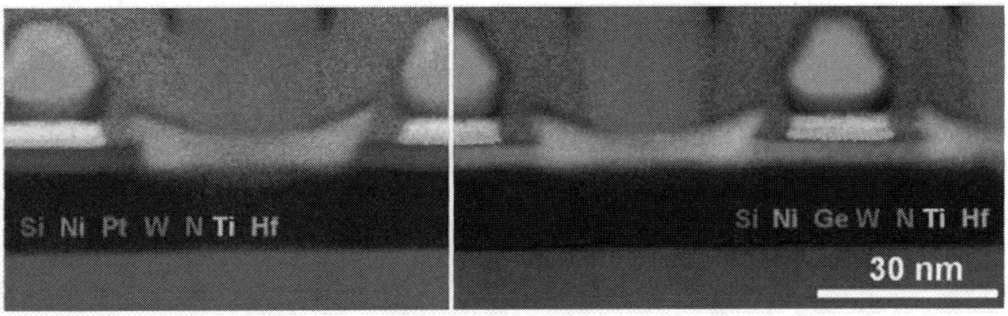

Figure 5. STEM EDX maps showing *p*MOS FD-SOI final structure. *See online version for colour code.*

2.4. NBED & DFEH strain characterization

Nowadays, the objectives regarding the performance cannot be achieved without strain engineering and therefore without strain characterization. Two techniques are mostly used in our laboratory: Nano Beam Electron Diffraction (NBED) and Dark Field Electron Holography (DFEH). The FD-SOI technology brings some specific challenges making a classic extension from bulk technology not always possible for strain techniques. The thickness of the film and its misorientation compared to the bulk Si are the two major constraints increasing considerably the complexity in the technique applications and results interpretation. For example, the tilt needed to bring relative good diffraction conditions in DFEH makes all interfaces blur (around 1-2nm). The reliability of the technique can be questioned on a 5nm thick film. For NBED, the probe size is equal to the thickness of the film, therefore very limited information can be extracted from those results. The NBED results taken along a SiGe film made by condensation are presented in figure 6(b). It is possible to see that the film is kept strained after the process in the middle of the area, and major relaxation is observed on its edge.

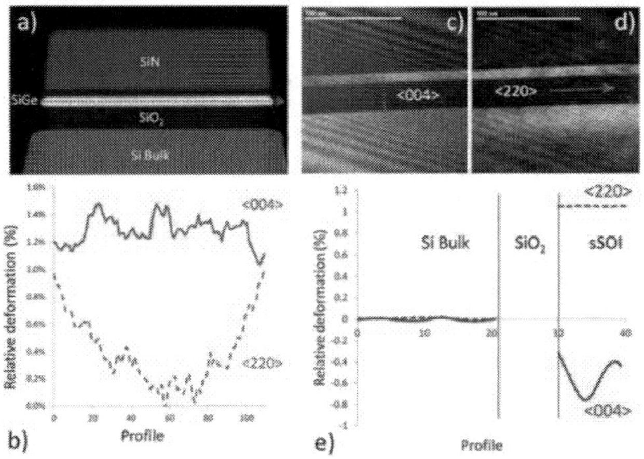

Figure 6 a) High Angle Annular Dark Field (HAADF) STEM picture of the SiGe film with corresponding NBED line profile (dashed). b) NBED relative deformation results from <220> & <004> directions along the SiGe film showing major relaxation at SiGe film edges. c) <004> and d) <220> dark-field holograms acquired in strained sSOI. e) DFEH relative deformation results showing a positive deformation in 220 direction and a negative deformation in 004 (results processed with HoloDark)

To overcome the issues cited above, some relaxed samples can be used to characterize the strain depending on the process steps. For example, the first steps of process use a thicker film (around 9-10nm) which enables more reliable strain characterization. Figures 6 c & 6d present DFEH results on a full wafer tensile strained SOI (sSOI) sample, which are clearly showing in Figure 6 (e) a positive deformation in the 220 direction and a negative one in the 004 direction (results obtained on a F20 TECNAI with Dual Lens configuration [2]). FD-SOI precise strain measurements show high challenges; however, a qualitative approach is an important source of information for process development in order to determine some configuration to keep the strain in the film or anticipate a loss of performance in some devices.

3. Conclusions

This paper demonstrates that advanced TEM characterizations are required and even unavoidable to ensure the development of new transistor generations such as FDSOI technologies. At first, the sample preparation is essential and FIB-based preparations have been optimized. Then, high resolution TEM and STEM are essential to precisely monitor each deposited layer. Finally, chemical analysis (EDX) coupled with strain measurements (NBED, DFEH) information are more and more routinely performed to get a deeper understanding of these materials physics.

References

[1] Skotnicki T et al. 2008 *IEEE Transactions on Electron Devices* **55**, 96-130
[2] Wang Y Y, Li J Domenicucci A and Bruley J 2013 *Ultramicroscopy* **124**, 117-129

Journal of Physics: Conference Series **471** (2013) 012027 doi:10.1088/1742-6596/471/1/012027

Defect analysis of a silicon nanowire transistor by X-ray energy dispersive spectroscopy technique in a STEM: 2D mappings and tomography

K Lepinay[1, 2], F Lorut[1], A Pofelski[1], R Coquand[1, 3], R Pantel[1] and T Epicier[2]

[1]STMicroelectronics, 850 rue Jean Monnet, F-38926 Crolles, France
[2]MATEIS, umr CNRS 5510, INSA de Lyon, 20 Av. A. Einstein, 69621 Villeurbanne, France
[3]CEA-LETI, MINATEC, 17 rue des Martyrs, F-38054 Grenoble, France

E-mail: kevin.lepinay@st.com

Abstract. The robust development of exotic devices strongly relies on physical characterization to better understand the stacked layers architecture and the resulting morphology. In such a context, a silicon nanowire, 200nm long and of around 15nm of diameter, encapsulated by a high-k metal gate stack (TiN/HfSiON) of few nanometers per layer, is characterized by 2D and 3D X-ray energy dispersive spectroscopy in a STEM to evaluate the process flow methodology.

1. Process description

It is commonly acknowledged that the end of the CMOS roadmap is likely to be reached with multi-gate devices, such as gate-all-around nanowires. We proposed the fabrication of such devices using an innovative integration scheme. Based on the Silicon-On-Nothing (SON) process, sacrificial SiGe followed by future channel Si layers are epitaxially grown on 300mm wafer. After nanowire patterning (300nm long pattern, with a top-width down to 15nm), a selective SiGe dry etch is performed. Thanks to wider source/drain regions, remaining SiGe pads at the edges sustain the suspended top Si nanowire. This process has to be optimized to avoid remaining SiGe below the top Si channel, but over-etch must be controlled to make sure SiGe pads remain to sustain the top Si channel.

Once released, the nanowire is covered with Hydrogen Silsesquioxane (HSQ), a spin-coated flowable oxide which has the advantages of being an excellent gap fill material, and which can be used as a non-organic electron-beam resist. We developed a lithographic step in order to form a 40nm-wide trench over the Si nanowire, which is then filled with gate materials (HfSiON, TiN ALD and PolySi). This allows the formation of damascene gate, surrounding the suspended Si nanowires, and HSQ wet etch will lead to a conventional integration to form CMOS devices.

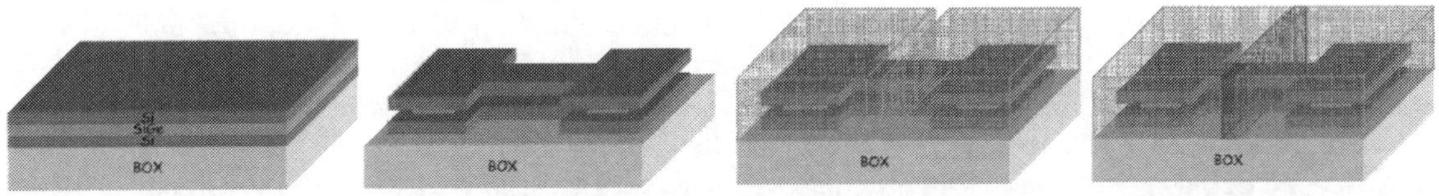

Figure 1: Integration scheme to form Si nanowires with self-aligned gate-all-around structure

Unfortunately, the trench formation in such patterns is highly limited in resolution, and trench clogging can occur, e.g. below the Si nanowires, thus disabling the gate deposition all-around the Si channel. In such small devices, physical characterization is limited to cross-sections normal to the smaller dimension, but observations along the nanowires are hardly realistic.

2. 2D characterisation of the device

Such small devices must be characterized using TEM based techniques since the resolution here is key to reach the information requested. The first classical approach is a 2D analysis on a thin extracted lamella (around 70nm) with a clean FIB sample preparation.

2.1. Nanowire 2D analysis

To analyze the device, energy filtered transmission electron microscopy (EFTEM), scanning transmission electron microcopy (STEM) and X-ray energy dispersive spectroscopy (XEDS) are performed. All experiments below were performed on an FEI OSIRIS microscope operating at 200keV. For the STEM imaging, a 0.5nm probe has been used, and then a 1.2nA current has been configured for the EDX analysis resulting in a probe size around 1nm.

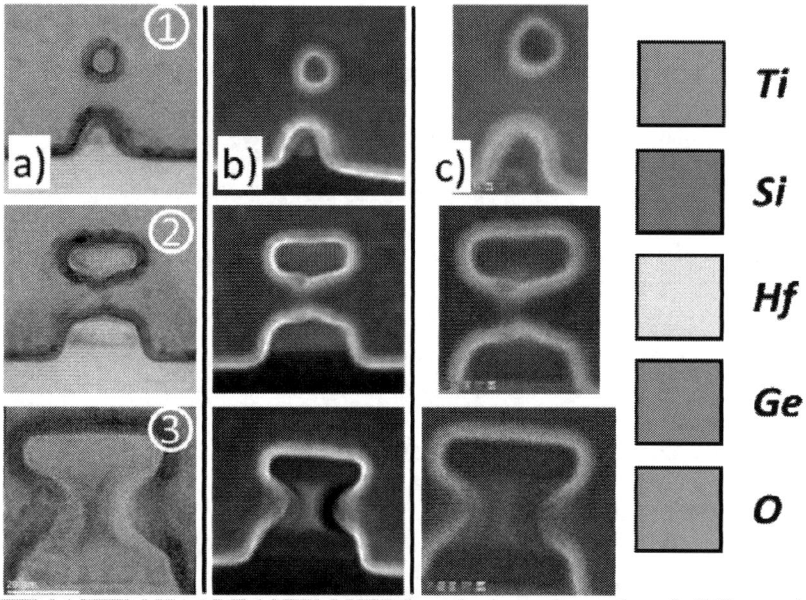

Figure 2: TEM / STEM HAADF / STEM XEDS analysis performed on 3 different devices (See online for colour version)

The goal here is to characterize precisely the SiGe etch and the high-k metal gate (HKMG) deposition since these two process steps could be a source of defects. The shape on device #1 (fig. 2) is very close to the one expected from the process steps with no SiGe residues and good HKMG deposition all around the gate. However, device #3 would highlight a major failure, probably caused by SiGe being not properly etched. The results on device #2 would suggest a minor issue with some SiGe residues left on the bottom of the nanowire. However, the 2D approach encounters the classic projection issue and it is not possible to be fully conclusive here. For example, the SiGe residue found on the second device could come from the SiGe pad where its presence is expected if the lamella is too thick. On the third device, the SiGe is not etched properly but again it is not possible to know if it is a SiGe etch problem or a misalignment of the trench for the HKMG deposition. For example, the lamella could have been extracted on the SiGe pads where its presence is expected. This would suggest an issue in the trench formation not having been well centered between the two pads. Therefore another lamella in the perpendicular direction could be an interesting complement to understand the failure. However, the complexity of the sample preparation will be critical here because of the size of the nanowire. The best option is finally to extract one large sample including the full device and perform a 3D analysis.

3. 3D Chemical characterisation of this device by STEM XEDS Tomography

3.1. STEM XEDS tomography method presentation

To have the full characterisation of this complex device, a 3D STEM XEDS is performed with a needle shaped specimen of around 150nm diameter prepared in the FIB. During the experiment, the sample is analysed over 180° of tilt to avoid missing wedge effects and chemical maps are acquired every 5°. Then, specific elemental mappings are extracted manually from the 36 full mappings. Following this, the alignment parameters are calculated using one elemental stack and applied to others. All these stacks are reconstructed using the SIRT algorithm with 20 iterations.

3.2. Well processed device analysis

For a long acquisition time analysis such as 3D XEDS STEM tomography, the experimental setup has to be tuned to limit electron beam damage. Consequently, acceleration voltage and the probe current are respectively fixed at 120keV and 1.2nA. To get sufficient signal for the acquisition of 750x900 pixels mappings, the dwell time is tuned at 800µs/px leading to a mapping acquisition time of 10min and a global acquisition time of 6 hours. The reconstructed volume is presented in figure 3.

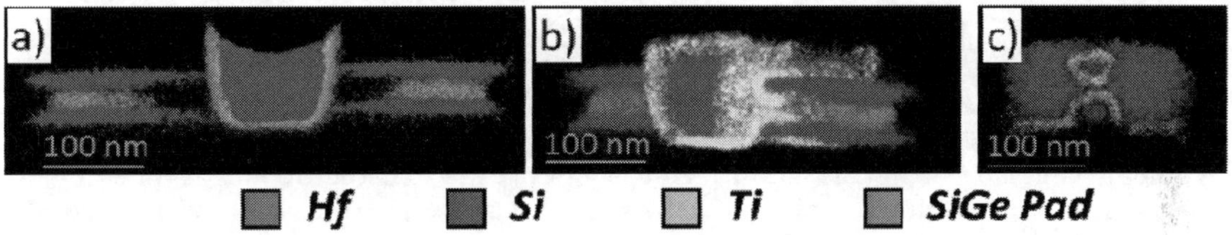

Figure 3: 3D chemical representation of the device. Silicon in Blue, Germanium in Purple, Titanium in Yellow, Hafnium in Red. a) & b) global rendering showing the geometrical complexity of the device. c) Focus on the 10nm Si nanowire and its high-k metal gate deposition all around the nanowire (See online for colour version)

On this well processed device, the SiGe is well etched to retain only the pads. So, the Si nanowire is also perfectly supported. With the front view, the deposition of the high-k metal gate is observed. The HfSiON and TiN layers are surrounding the Si nanowire as expected. These two complex steps are obviously decisive for the device functionality.

3.3. Failed device analysis

To analyse this device, the experimental setup is almost the same as previously for the same reasons. The chemical mappings of 820x880 pixels are acquired with a dwell time of 1000µs leading to a global acquisition time of 7hours. Results are presented in figure 4. The main observations are that the SiGe pads are not aligned with the nanowire direction and the deposition of the high-k metal gate has failed. The first point can be explained by a mask misalignment or etching issue. The second one is a consequence of the first fail. Due to the SiGe pads issue, the suspended Si nanowire seems to be not at its place. Consequently, the HKMG layer deposition can't be successful, as presented in figure 4 b). The 3D representation of this failed device brings more information than a simple 2D analysis to understand the origin of the failure. The localisation of the SiGe pads would have been lost if just a 2D analysis had been performed on this device. However, the resolution of this technique is limited to 3nm. Consequently, a comparison between the 2D results and an extraction of the 3D volume from an equivalent region is realized.

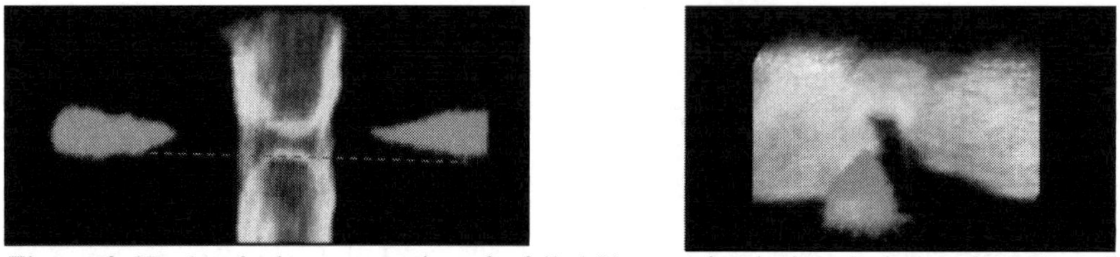

Figure 4: 3D chemical representation of a failed Si nanowire device. Ge in purple, Ti in yellow. a) Top global view of the device. The dashed blue line points out a misalignment between the SiGe pads and the gate. b) View in the direction of the expected Si nanowire: the gate deposition has failed. (See online for colour version)

3.4. 2D vs 3D analyses study

In this study, a STEM XEDS 2D analysis performed on a lamella localized on the gate is compared to the projections of slices, extracted of around 40nm of thickness, in the equivalent region. Figure 5 shows the different results.

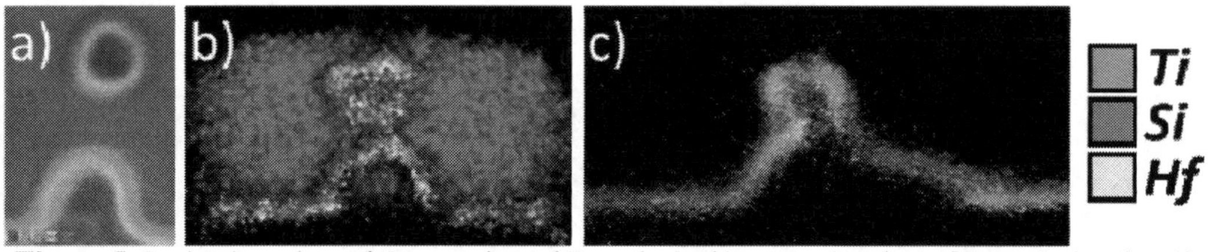

Figure 5: Projection view of the gate in a) 2D analysis, b) extraction from 3D analysis of well processed device and c) extraction from 3D analysis of failed device. (See online for colour version)

As result, the 2D analysis seems to be better resolved. It's linked to the use of a higher magnification and more dwell time on a thinner sample allowing better photon collection. The 3D results have the advantage of being localized more precisely although the layer thicknesses can only be estimated.

4. Conclusion

The contribution of the 3D technique is undeniable for complex devices presenting 3D chemical interest. The 2D can be very useful for a quantitative and more resolved approach but the aim of this tomographic technique is to be more qualitative than quantitative which is still useful. The best advantage is that the analysed volume can be observed in all the directions post acquisition without favouring one as for a 2D analysis. Moreover, since two lamellas could not be extracted on the same device, the probe can include the full 3D information of an isolated failure and identify more precisely its cause.

References

[1] Coquand R et al. *Innovative through-Si 3D lithography for ultimate Self-Aligned Planar Double-Gate and Gate-All-Around nanowire transistors*, to be presented at VLSI technology symposium 2013.
[2] Lepinay K, Lorut F, Pantel R, Epicier T 2013 *Micron* **47** 43

Journal of Physics: Conference Series **471** (2013) 012028 doi:10.1088/1742-6596/471/1/012028

SiGe epitaxially grown in nano-trenches on Si substrate

O Richard, B Vincent, P Favia, P Lagrain and H Bender

Imec, Kapeldreef 75, B-3001 Leuven, Belgium

E-mail: olivier.richard@imec.be

Abstract. SiGe lines epitaxially grown in silicon oxide nano-trenches on Si substrate are characterized by transmission electron microscopy (TEM) based techniques. Due to the high aspect ratio of the trenches most of the extended crystallographic defects are limited to the bottom of the lines. Few of them are still observed at the top of the lines and emerge in the cap above the lines. The SiGe composition is not homogenous in the width of the line and in the cap. This is linked to the formation of {111} facets. This variation of composition is not observed for the lines with an actual width narrower than 20nm.

1. Introduction

In order to increase the electron and hole mobility, co-integration of III-V materials or tensile Si and Ge, respectively, is considered as replacement for the Si channel [1]. The hole mobility in a Ge p-Fin field effect transistor (FinFET) can be even further increased by introducing compressive stress in the Ge channel [2]. This can be achieved by growing the Ge channel epitaxially on top of a SiGe relaxed buffer (SRB).

The aim of this paper is to characterize the SiGe layer epitaxially grown in trenches etched with HCl vapour in shallow trench isolation (STI) structures prepared on (001) silicon wafers. The nominal trench widths are 20-80nm but the actual widths are different, depending on the processing. The nominal width of the lines is mentioned in the rest of this paper except if otherwise specified.

2. Experimental

TEM specimens are prepared parallel and perpendicular to the trenches with the focused ion beam (FIB, Strata, FEI) lift-out technique and characterized with a Tecnai F30 TEM operating at 300kV. A chemical vapour deposited (CVD) glass or an electron beam deposited platinum layer and a FIB induced platinum layer are deposited beforehand on the area of interest in order to protect the specimen during the FIB milling.

Two-beam bright field TEM images are acquired in order to better reveal the presence of extended defects in the SiGe lines.

The composition homogeneity of the SiGe is assessed with energy dispersive x-ray spectroscopy (EDS) chemical analysis and related to the contrast observed in high angle annular dark field scanning TEM (HAADF-STEM) images. The atomic concentration profiles are calculated using the TIA (FEI) software.

3. Results

3.1. SiGe layer grown in shallow trench isolation

Typical TEM images of the nominal 20nm lines perpendicular and parallel to the lines are presented in figures 1a and 1b, respectively. The HCl vapour etch of the trenches results in {113} and {111} facets at the bottom of the trenches in the (001) Si substrate (figure 1a) [3]. The layers exhibiting different contrasts for the parallel orientation (figure 1b) can be explained by the overlap of different materials (e.g. Si-SiGe at the bottom of the trench, SiO_2-SiGe higher in the line) and by the different thicknesses of the materials present in the thickness of the TEM specimen (e.g. SiGe cap vs. SiGe line). Due to the {111} faceting of the bottom of the caps (see paragraph 3.2), voids are present where the caps merge due to the lateral growth of the SiGe on the oxide.

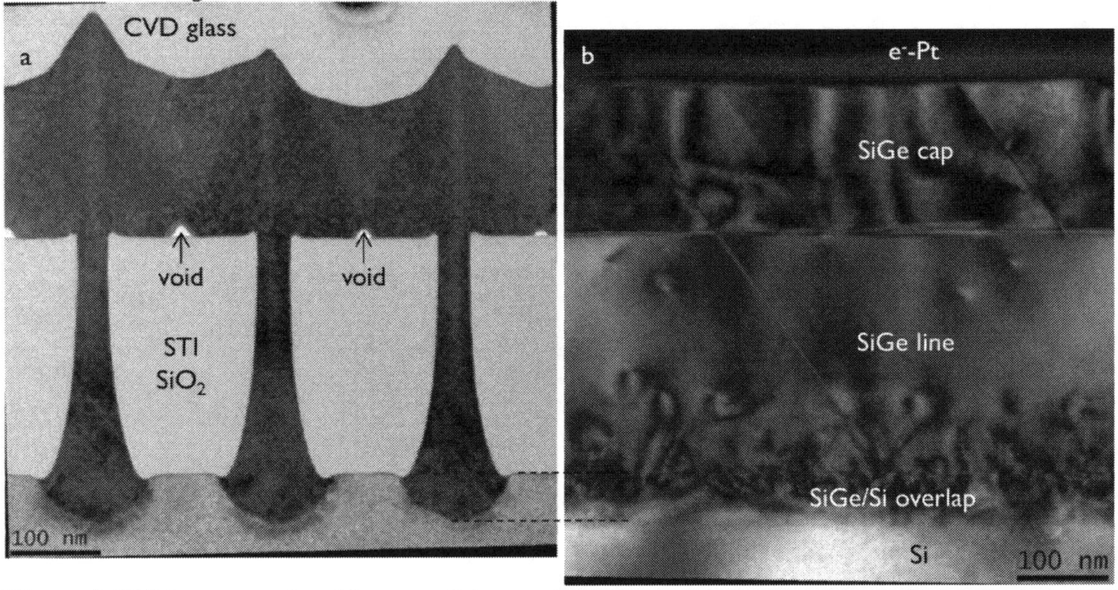

Figure 1. TEM cross-sectional images of the nominal 20nm wide lines perpendicular (a) and parallel (b: BF (220)) to the lines. For the parallel orientation (b) only one 20nm line is present in the thickness of the TEM specimen.

Higher density of {111} defects (stacking faults, nanotwins) and dislocations are observed at the bottom of the lines. Most of these defects start at the SiGe/Si interface and emerge at the sidewalls of the lines leaving the top of the lines with lower defect density. This is induced by the high aspect ratio of the trenches [4]. As can be seen with the trench embedded within the TEM specimen (figure 1b) only few defects propagate through the line and emerge at the top of the cap. Dislocations are also observed higher in the SiGe lines and cap.

Bands exhibiting a brighter contrast are observed in the centre and on the top of the lines in the TEM image (figure 1a). These bands exhibit a darker contrast than the surrounding SiGe in the HAADF-STEM image (figure 2) where the contrast is proportional to about $\langle Z \rangle^2$. It indicates that the SiGe composition is not homogenous in the width of the lines and of the cap.

Energy dispersive x-ray spectroscopy (EDS) is performed across the line (3-4) and across the cap (1-2) as outlined by the white lines on the HAADF-STEM image (figure 2a). The Si and Ge atomic concentration line profiles are calculated using the automatic procedure included in the TIA (FEI) software. The EDS signal is noisy due to the rather short acquisition time (3s), the atomic Si and Ge concentration profiles are therefore also noisy. In the top part of the lines (figure 2c) the Ge content is higher at the sidewalls of the lines (in average $Si_{0.32}Ge_{0.68}$) whereas lower in the central part of the lines (in average $Si_{0.45}Ge_{0.55}$). In the SiO_2 at both sides of the line 100% Si is obtained as only Si and Ge are taken into account for the quantification. In the cap (figure 2b) the Ge content is also lower in the central part above the line (~ $Si_{0.42}Ge_{0.58}$) than on the sides (~$Si_{0.30}Ge_{0.70}$) and then again lower (~$Si_{0.47}Ge_{0.53}$) with faceted frontiers close to {111} in the HAADF-STEM image (figure 2a). The nominal composition of the SiGe layer for these lines is $Si_{0.25}Ge_{0.75}$. It is worth noting that no standard specimen has been used but previous comparisons show that the SiGe quantification is quite accurate (± 5%). The variation of composition is qualitatively in agreement with the variation of contrast observed in the HAADF-STEM image. This composition distribution is not desired since it will create variation of the lattice parameter and strain in the width of the lines.

The variation of Ge content is due to the formation of {111} facets at the edges of the lines compared to the centre of the lines. Due to the higher atomic density of the {111} facets the growth perpendicular to these facets is slower than perpendicular to {001}. The Ge incorporation is then higher on the {111} facets.

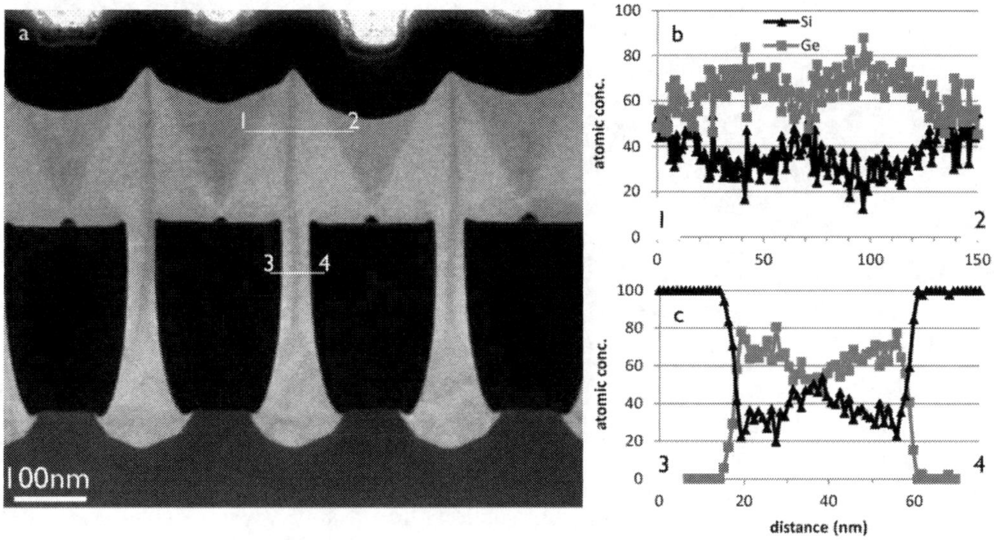

Figure 2. HAADF-STEM image of the 20nm wide lines (a) and Si and Ge atomic concentration profiles calculated from EDS obtained along the lines 1-2 (b) and 3-4 (c).

3.2. SiGe layer grown in shallow trench isolation with Ge marker layers

In order to better observe the correlation between the formation of the {111} facets and the composition variation in the width of the lines the growth of the SiGe layer is interrupted and 5 thin Ge marker layers are grown (figure 3). For the narrowest trenches the Si substrate is less deeply etched than for the wider trenches. It is worth noting that the growth rate of the first SiGe layer on top of the Si is higher for the wider than for the narrower lines.

For the lines wider than nominally 40nm, the 1st Ge marker is {111} faceted at both sides of the lines whereas rounded in the centre. The variation of composition in the width of these lines is clearly observed (figures 3b and c) between the 1st and 2nd marker layer. For the 40nm (actual width: 18nm) line (figure 3a) the first marker layer is rounded and no clear facets are observed at both sides of the lines. No variation of contrast is observed in the width of this line, indicating that the composition of the narrowest lines, the technologically most important ones, is more constant in the width of the lines than for the wider lines.

For all the line widths, the Ge marker layers are clearly {111} faceted in the cap and rounded at the intersection while the outer shape is sharper. For the 40nm lines (figure 3a) the different layers in the cap are rougher, leading to a widening of the Ge marker layers due to the projection of the roughness. Variation of contrast due to the variation of the Ge content is clearly observed in the caps.

The steps on the SiO_2 layer at both sides of the SiGe lines are most likely at the origin of the bottom {111} facets resulting in voids when the caps merge.

It can also be remarked that the growth of the Ge marker layers introduced a perturbation. On top of each Ge marker layer the SiGe area poorer in Ge is wider than below and above. The width of this area increases after each Ge marker layer, corresponding with the enlargement of the rounded top surface.

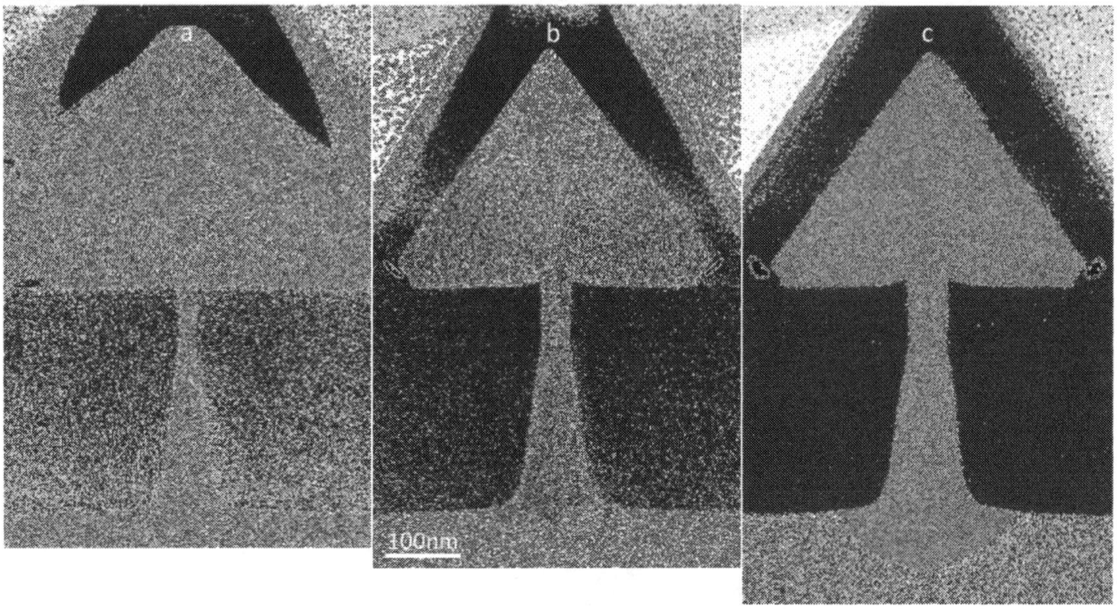

Figure 3. HAADF-STEM images of the nominal 40nm (a), 60nm (b) and 80nm (c) wide lines.

4. Conclusions

20-80nm nominal SiGe lines epitaxially grown in shallow trenches between silicon oxide are characterized by TEM based techniques. Due to the beneficial effect of the high aspect ratio of the trench a lower density of extended defects is observed at the top than at the bottom of the lines. A variation of composition linked to the formation of {111} facets is observed in the HAADF-STEM images and detected by EDS in the width of the lines and in the cap. This undesirable effect is not observed for the lines with an actual width narrower than 20nm.

Acknowledgements

P Van Marcke is acknowledged for the tedious TEM specimen preparation and the imec/EPI group for providing the samples and for numerous discussions.

References

[1] Heyns M et al 2011 *Proc. International Electron Devices Meeting (Washington)* p 299-302

[2] Eneman G et al 2012 *Proc. International Electron Devices Meeting (San Francisco)* p 6-5

[3] Loo R, Caymax M, Meunier-Beillard P, Peytier Y, Holsteyns F, Kubicek S, Verheyen P, Lindsay R and Richard O 2004 *Appl. Surf. Sci.* **224** 63-67

[4] Langdo T A, Leitz C W, Currie M T, Fitzgerald E A, Lochtefeld A and Antoniadis D A 2000 *Appl. Phys. Lett.* **76** 3700-3702

Dynamic segregation of metallic impurities at SiO_2/Si interfaces

A De Luca, A Portavoce, M Texier, N Burle and B Pichaud

IM2NP, UMR7334 CNRS - Aix-Marseille Université, Faculté des Sciences et Techniques, Campus de Saint-Jérôme, Av. Escadrille Normandie Niémen, F-13397 Marseille Cedex, France

E-mail: anthony.deluca@im2np.fr

Abstract. The behaviour of two metallic impurities, iron and tungsten, during oxidation of silicon wafers has been investigated using transmission electron microscopy and atom probe tomography. Metallic contamination has been introduced by implantation of ^{54}Fe at 65 keV and ^{186}W at 150 keV, with a dose of 10^{15} at/cm^2. Oxidation of Fe-contaminated Si wafer results in the precipitation of iron as β-$FeSi_2$ at the SiO_2/Si interface. The presence of these precipitates hinders the oxidation front which forms silicon pyramidal defects. Further oxidation of the precipitates leads to iron-rich cluster formation in the SiO_2 layer, surrounding the pyramids. Dry oxidation of a tungsten-contaminated Si wafer is characterised by the formation of nanometric spherical precipitates in the Si layer. The size and density of these precipitates versus depth follow the as-implanted W concentration profile.

1. Introduction

With the increasing complexity of the design of the newest generation of transistors, unintentional contamination by metallic impurities, even at low concentration levels, has become one of the major problems in the manufacturing of modern microelectronic devices. In particular, the presence of very small amounts of contaminants in the active area of transistors is enough to have deleterious impact on their electrical properties making them defective. It is thus essential to improve our understanding of metallic contaminant behaviour during the processing of microelectronic devices and, more specifically, at the interfaces of CMOS structures.

In this context, this work aims at a study of the mechanisms of diffusion and precipitation or segregation of two metallic elements at the SiO_2/Si interface: iron and tungsten. The parallel study of these two impurities aims to compare and to model the diffusion/precipitation mechanisms of species exhibiting different diffusion behaviours. Indeed, whereas iron diffusion uses an interstitial mechanism and is a fast diffusing species [1], tungsten shows low diffusion rates [1,2] and an unknown diffusion mechanism.

Depending on both the contaminating element and oxidising conditions, various situations have been distinguished. Based on the experimental observations, a qualitative model for the dynamic segregation of metallic impurities in silicon is proposed.

2. Experimental details

In this study, 525 μm-thick P-doped (001) Si wafers (resistivity 2-20 Ω.cm) were contaminated with ^{54}Fe and ^{186}W ions, at a dose of 10^{15} at/cm^2 and with implantation energies respectively of 65keV and 150keV.

In order to study the effect of oxidation temperature and duration, the samples were dry oxidised under O_2 flux for various conditions (900°C for 5, 189, 592 min or 1100°C for 12 and 44 min). The oxidation times were set to grow SiO_2 layers with comparable thicknesses for both temperatures.

Post mortem analyses of the oxidised specimens were done using several characterisation tools, such as Cs-corrected HRTEM (High Resolution Transmission Electron Microscopy), Z-contrast STEM (Scanning Transmission Electron Microscopy) and LP-APT (Laser-Pulsed Atom Probe Tomography). TEM thin foils and APT tips were prepared by focused ion beam (DBFIB) micromachining using an FEI Helios microscope. TEM lamellae were cleaned using the Fischione Nanomill device. In order to facilitate the evaporation of the oxide layer during APT analysis, the tips were prepared by the lift-out preparation method [3] using a specific sample orientation: the wedge was attached to a microtip with its surface tilted by 10°. APT analyses were carried out with a LEAP 3000xHR at a temperature of 80 K, with a laser pulse frequency of 100 kHz, using a laser power of 1.2nJ, corresponding to a I_{Si}^{2+}/I_{Si}^{1+} ratio of 16.5 and a I_W^{2+}/I_W^{1+} ratio of 5.6.

3. Results and discussion

3.1. Fe behaviour during dry oxidation

Dry oxidation of iron-implanted Si wafers led to the formation of iron precipitates at the SiO_2/Si interface. Their diameters ranged from 20 to 50 nm. For the shortest oxidation times, facetted precipitates were located at the SiO_2/Si interface, inside the Si layer. For longer oxidation time, the precipitates got partly incorporated in the SiO_2 layer and became round in shape (figure 1).

The precipitate structure was deduced from the Fourier transform (FT) analysis of HRTEM images. All the precipitates have been identified as being the low-temperature stable silicide phase β-$FeSi_2$ [4, 5]. Except for the shorter oxidation time, analysis of the SiO_2/Si interface also revealed the presence of pyramidal-shaped defects (figure 2), already observed in the literature [4]. These defects consist of silicon huts whose origin was unclear. HAADF-STEM imaging revealed the presence of small iron-rich clusters incorporated in the SiO_2 layer around these pyramids (figure 2).

Figure 1. HRTEM image of a precipitate of β-$FeSi_2$ formed at the SiO_2/Si interface.

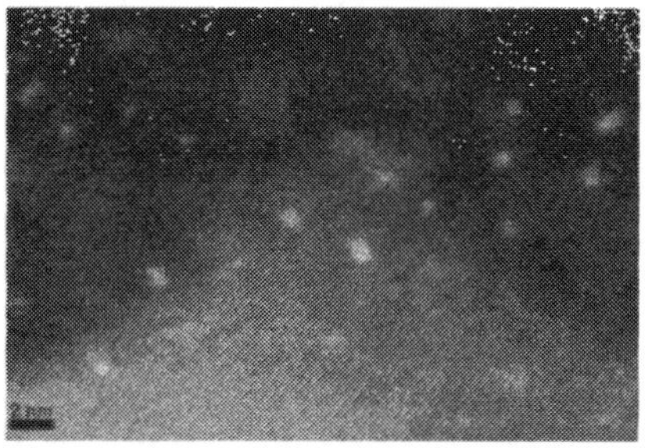

Figure 2. HAADF-STEM image showing iron clusters incorporated in the SiO_2 layer around a silicon pyramidal-shaped defect.

Figure 3 shows a pyramidal-shaped defect surmounted by a precipitate. The observed configuration suggests that the formation of those defects is correlated with the presence of a precipitate which locally hinders the propagation of the oxidation front.

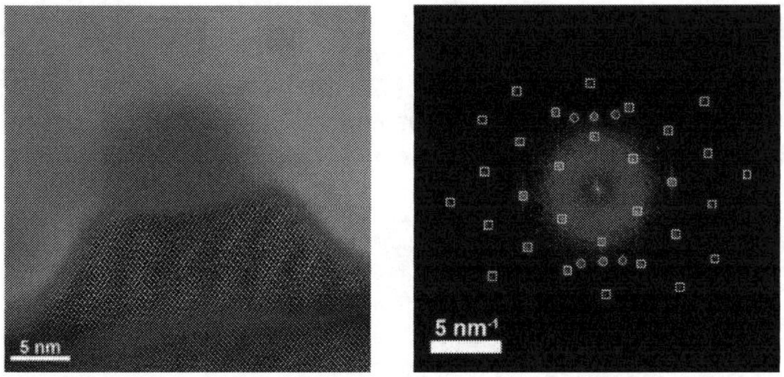

Figure 3. Filtered HRTEM image showing a pyramidal-shaped defect surmounted by a precipitate, and associated Fourier Transform of the original image. Squares indicate selected spots used for filtering.

We thus propose the following model for the formation of these iron-induced defects (figure 4):
i) Iron accumulates at the growing SiO_2/Si interface forming facetted β-$FeSi_2$ precipitates. Oxidation of the precipitate smoothes its surface, leading to a spherical shape.
ii) Assuming that the oxidation rate of iron silicide is significantly lower than the one of silicon, the presence of the precipitate modifies the oxidation front and is responsible for the observed defects at the SiO_2/Si interface.
iii) The complete consumption of the precipitate leaves behind pyramidal-shaped silicon huts at the SiO_2/Si interface that are surrounded by Fe clusters.

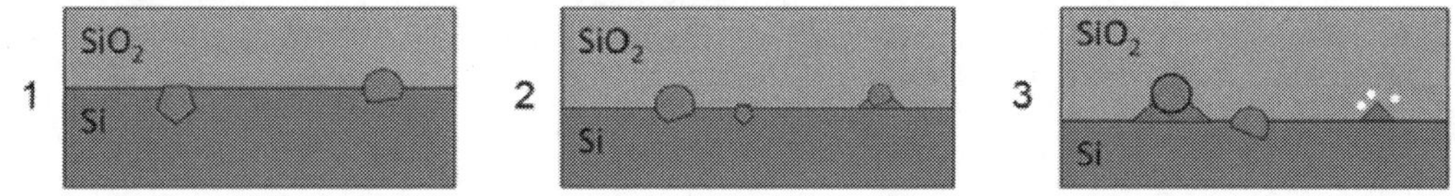

Figure 4. Proposed mechanism for the formation of pyramidal-shaped defects at the SiO_2/Si interface.

3.2. W behaviour during dry oxidation

Dry oxidation of tungsten-implanted Si wafers is characterised by the formation of nanometric spherical precipitates in the Si layer (figure 5). The size and density of these precipitates versus depth follow the as-implanted W concentration profile. APT analysis of the sample oxidised at 900°C for 5 min revealed an asymmetric widening of the W profile toward the growing oxide (figure 5). It thus shows that tungsten atoms can diffuse without forming precipitates at a concentration of ~10^{20} at/cm^3.

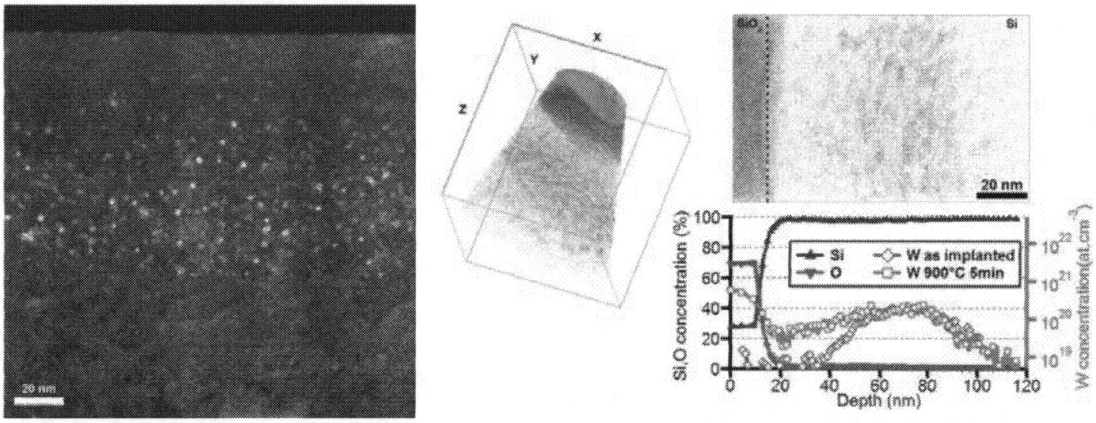

Figure 5. HAADF (left) and APT (right) analysis of the sample oxidised at 900°C for 5min revealed that precipitate density follows the as-implanted tungsten concentration profile. Tungsten atoms are coloured in red, oxygen in blue and nickel in green. Si atoms are not shown. The evaporated volume dimensions are 155×155×180 nm^3 (X×Y×Z).

Structural analysis of precipitates revealed the coexistence of two phases: tetragonal W_5Si_3 and hexagonal WSi_2. These precipitates exhibit topotactic relationships with the silicon matrix (topotaxy: the phenomenon of mutual orientation of two or more crystals of different species resulting from a solid-state transformation [6]). However, various precipitates' orientations have been observed, and no predominant orientation of the precipitates has been evidenced. HAADF imaging also revealed that sub-nanometric W clusters are formed in the SiO_2 layer (figure 6).

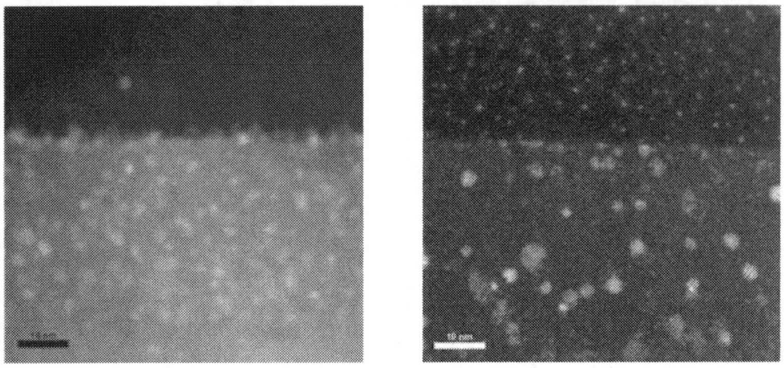

Figure 6. HAADF comparison between samples oxidised at 900°C for 592 min (left) and at 1100 °C for 44min (right).

Size evolution of precipitates in Si and clusters in SiO_2 has been studied as a function of time and temperature. At 900°C, precipitates' size increases with time whereas it doesn't evolve notably at 1100°C, indicating that they probably reach an equilibrium state faster at higher temperature. In addition, it can be noticed that the density of clusters in the SiO_2 layer seems to increase with oxidation temperature.

4. Conclusion

As expected, iron and tungsten exhibit very different behaviours during dry oxidation: iron diffuses fast and forms large β-$FeSi_2$ precipitates at the SiO_2/Si interface whereas tungsten diffusion is limited in length and brings about the formation of nanometric W_5Si_3 and WSi_2 precipitates, with a distribution centred close to the implantation projection range. The Fe-induced pyramidal-shaped defects result from the differences in oxidation rates for iron precipitates and silicon. In both cases, iron and tungsten, oxidation of these precipitates results in the incorporation of metal into the SiO_2 layer as metallic clusters.

Acknowledgement

This work has been funded by the French Ministry of Industry in the frame of the "COMET" project. (Industrial partners: ST Microelectronics, Lfoundry, Ion Beam Services, Rockwood, Biophy Research, TERA Environnement , VEGAtec)

References

[1] Graff K 2000 "Metal Impurities in Silicon Device Fabrication", *Springer Series in Materials Science*, 2nd ed. (Springer, Berlin)
[2] Fujisaki Y et al 1955 *J .Appl. Phys.* **63**, pp2304-2306
[3] Thompson K et al 2007 *Ultramicroscopy* **107**, pp 287-298
[4] Wong-Leung J et al 1998 *J. Appl. Phys* **83**, pp580-584
[5] Oyoshi K et al 2001 *Thin Solid Films* **381**, pp202-208
[6] Bailey SW et al 1977 *Acta Cryst.* **A33**, pp681-684

Journal of Physics: Conference Series **471** (2013) 012030 doi:10.1088/1742-6596/471/1/012030

Atom probe tomography of a commercial light emitting diode

D J Larson[1,*], T J Prosa[1], D Olson[1], W Lefebvre[2], D Lawrence[1], P H Clifton[1] and T F Kelly[1]

[1] CAMECA Instruments, Inc., 5500 Nobel Drive, Madison, WI 53711 USA
[2] GPM UMR CNRS 6634, Université de Rouen, Saint Etienne du Rouvray 76801 FRANCE

Summary: The atomic-scale analysis of a commercial light emitting diode device purchased at retail is demonstrated using a local electrode atom probe. Some of the features are correlated with transmission electron microscopy imaging. Subtle details of the structure that are revealed have potential significance for the design and performance of this device.

1. Introduction

Gallium nitride (GaN) has been successfully used as the fundamental material for a wide range of optoelectronic devices [1]. Undoped GaN emits in the ultra-violet (UV), and alloying it with indium or aluminum produces emission with wavelengths ranging from the green to the deep UV. As a result, GaN-based multiple quantum well (MQW) structures are used as the active regions of commercial light-emitting diodes (LEDs) and laser diodes.

Atom probe tomography (APT) provides unique capabilities for nanoelectronic device characterization including three-dimensional compositional mapping of a relevant volume (>10^6 nm^3), high detection efficiency (>50%), and good sensitivity (<10 ppm) [2]. Although long applied to metals and some semiconductors (see [3] and references contained within), APT has also more recently been applied to semiconducting and insulating materials used in microelectronic devices [4-6]. This change has been enabled by the recent commercial development of laser-pulsed atom probe systems, which were first demonstrated in the 1980s [7]. One application of APT in the microelectronics industry which is less common, however, is competitive analysis and/or failure (FA). In order to facilitate FA of microelectronic structures, a specimen must first be fabricated from a specific, individual device and then satisfactorily survive analysis. As a step on the path toward feasibility for FA using APT, we have collected data from an individual commercial LED originating as a fully packaged device.

2. Experimental

OSRAM 455 nm Golden Dragon® Plus LEDs were purchased from a retail supplier, depackaged, and fabricated into APT-compatible specimens using standard focused-ion-beam (FIB) methods [8-10] with a backside preparation direction [11]. Specimen preparation was carried out in an FEI Novalab dual-beam FIB microscope with an Omniprobe AP200 *in-situ* micromanipulator. Scanning transmission electron microscopy observations were performed along the <01-10> orientation of the wurtzite GaN structure with a JEM ARM 200F operating at 200 kV equipped with a Schottky field emission gun. APT data collection was performed on a CAMECA local electrode atom probe (LEAP ®) 4000X HR operated in a 500 kHz pulsed 355 nm laser mode with a pulse energy of 3-70 fJ (adjusted during the run to maintain a

* David.larson@ametek.com

constant charge-state ratio of 0.5 for Ga^{++}/Ga^{+}). The specimen base temperature was 25 K and the ion detection rate was 3.0% (three ions detected for every 100 laser pulses).

Figure 1 shows a sequence of images during the specimen preparation process from the initial surface of the wafer after depackaging (Figure 1b) to the final state of FIB annular sharpening [12] of the tip (Figure 1d).

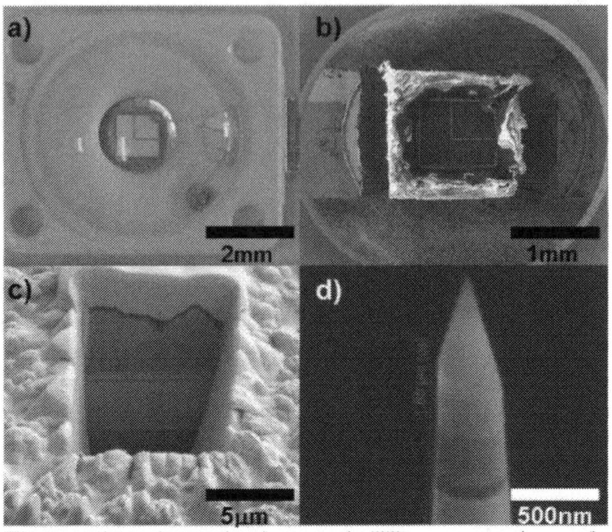

Figure 1. Depackaging and FIB preparation of a commercial GaN device for APT analysis: a) initial LED, b) SEM image of depackaged device, c) FIB-milled trenches, d) the final specimen following annular milling.

3. Results and Discussion

In order to have as much information as possible regarding details of the LED structure, specimens were also prepared and analyzed in both a CAMECA SXFive electron microprobe and a JEM ARM 200F scanning transmission electron microscope (STEM). In the electron microprobe, signals from Al and In were analyzed in order to determine the location of the In MQW region of interest. In the STEM, a high-angle annular-dark-field image was obtained from a portion of the region of interest and is shown in Figure 2. Five repeats of high image intensity are visible in the image. This microstructure would later be determined by APT as InGaN/GaN quantum wells. From Figure 2, the average width of the bright layers (InGaN) and the dark layers (GaN) is measured to be 3.5±0.2 nm and 4.4±0.2 nm, respectively. This information is used to confirm the spatial reconstruction accuracy for APT data.

Following spectral peak identification in the APT data, a three-dimensional atom map may be created, as shown in Figure 3a. The overall approximate concentration of the APT data is 48.3at.%Ga-50at.%N-0.85at.%In-0.75at.%Al-0.14at.%Mg. Figure 3a shows that the data are highly structured with four distinct regions immediately identifiable: 1) a region of Mg doping (to form p-type GaN) near the left side of the structure, 2) an Al-rich region, 3) a region of In-rich MQWs, and 4) an In-based superlattice structure near the right side of the data.

The Mg distribution is visually non-uniform suggesting that some fraction of the dopant may be electrically inactive [13, 14]. The extent of this non-uniformity may be quantified using standard analysis methods [15] which employ an algorithm using a maximum separation distance (D_{max})

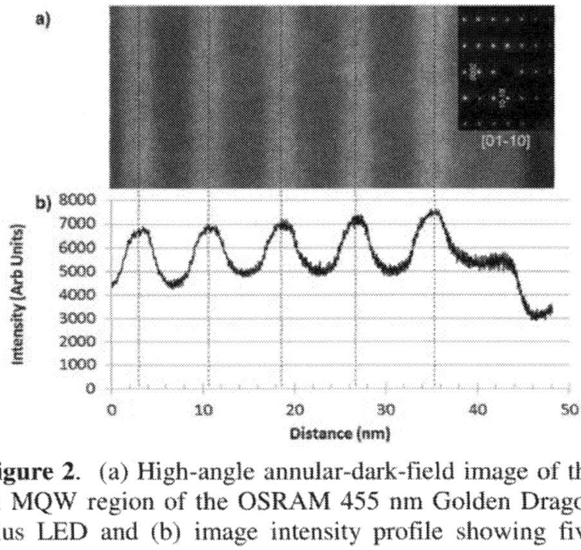

Figure 2. (a) High-angle annular-dark-field image of the In MQW region of the OSRAM 455 nm Golden Dragon Plus LED and (b) image intensity profile showing five layers of high intensity. The average distance between the layers (dashed lines) is 7.9 nm.

between atoms. Using a D_{max} value of 1.5 nm produces 69 clusters identified with a mean concentration of Mg of ~10 at.%. The clusters contain an average 21 Mg atoms with a most likely number of atoms per cluster of five to ten atoms. The average cluster volume is ~20 nm^3 and the average cluster diameter is ~3 nm.

Adjacent to the region of Mg clusters is a region of high Al concentration (Figure 3). The region is an electron blocking layer (EBL), which improves device performance by preventing carrier migration in the wrong direction (which in this case is electrons moving left to right in Figure 3). The concentration of the EBL, as calculated using an isoconcentration surface enclosing the region of space containing at least 5 at.% Al, is found to be 9.68 at.% Al (estimate of standard error 0.02 at.%). The EBL appears asymmetric, with a higher Al concentration (nearly 15 at.%) on the right edge, as shown in Figure 3b.

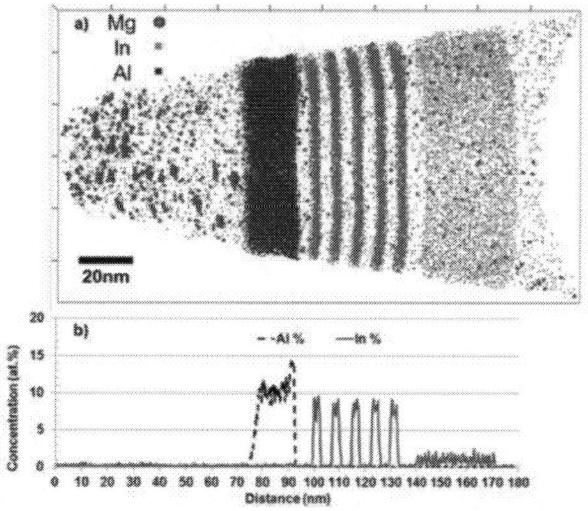

Figure 3. (a) APT atom map of the OSRAM 455 nm Golden Dragon Plus LED showing Mg atoms as large spheres, In atoms as small grey dots and Al atoms as small black dots (for clarity, Ga and N atoms are not shown) and (b) concentration profile suggesting four distinct chemical regions within the dataset.

To the right of the EBL in Figure 3 is the region containing In-rich MQWs. As observed in the TEM image in Figure 2, there are five individual wells. The concentration profile in Figure 3b shows that each well reaches approximately 8at.% In. Enclosing and summing the region of space containing at least 3at.% In and calculating a concentration produces 7.3at.% In (note that this is a somewhat arbitrary measurement as it is summed over multiple quantum wells and also depends on the choice of the 3at.%In isoconcentration surface). Qualitatively the wells do not appear to be asymmetric in concentration or in interface intermixing (Figure 3b). This is confirmed by an estimation of the internal interface roughness [16] based on the 3at.% In isoconcentration surface separating each well from the inter-well GaN region. Comparing the distribution of each lower and upper (proceeding right to left in Figure 3) individual interface measurements, we find that the means are not statistically different, with the mean values of RMS roughness being 0.28 nm and 0.24 nm for the lower and upper interfaces, respectively.

On the far right side of Figure 3 there is another region of non-uniform In concentration. Closer investigation of the APT data from this region reveals a superlattice structure of 21 In-rich layers (Figure 4). These layers are spaced ~1.5 nm from each other. When binned in 0.3 nm bins (to minimize statistical variation) the superlattice In concentrations range from 0.5at.% to 1.5at.%.

4. Conclusion

Many subtle details of the microstructure and composition emerge from

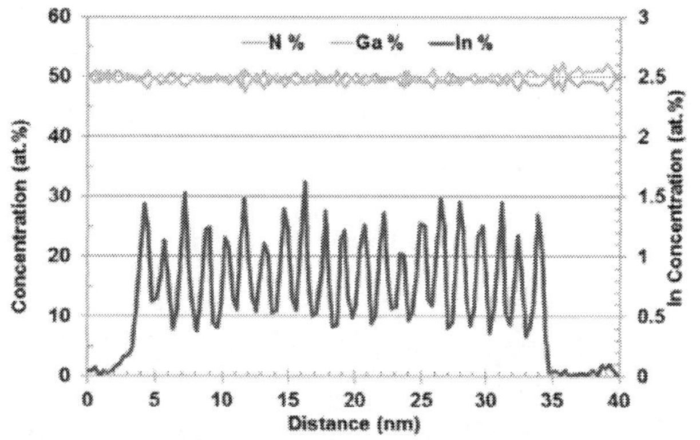

Figure 4. Superlattice structure showing 21 layers having a spacing of ~1.5 nm and In concentrations of ~1at.%.

the LEAP analysis of this commercial device. These details provide crucial insights into the design and operation of the device. Many of these features would not be apparent with other techniques especially the Mg clustering and the In superlattice but are readily observed in a single experiment with the LEAP.

Acknowledgements

The authors would like to thank our colleagues at CAMECA Instruments, Inc., who assisted in assembling the materials presented in this manuscript, including R. Ulfig, B. Geiser, D. Giddings, S. Strennen, J. Olson, D. Lenz, J. Bunton, J. Shepard, T. Payne, E. Strennen and E. Oltman.

References

[1] Schubert E F 2006 *Light Emitting Diodes.* (Cambridge: Cambridge University Press)
[2] Kelly T F and Larson D J 2012 Atom Probe Tomography 2012 *Annual Reviews of Materials Research* **42** 1-31
[3] Miller M K, Cerezo A, Hetherington M G and Smith G D W 1996 *Atom Probe Field Ion Microscopy* (Oxford: Oxford University Press)
[4] Kelly T F, Larson D J, Thompson K, Alvis R L, Bunton J H, Olson J D and Gorman B P 2007 Atom Probe Tomography of Electronic Materials *Annual Review of Materials Research* **37** 681-727
[5] Larson D J, Lawrence D, Lefebvre W, Olson D, Prosa T J, Reinhard D A, Ulfig R M, Clifton P H, Bunton J H, Lenz D, Renaud L, Martin I and Kelly T F 2011 Toward Atom Probe Tomography of Microelectronic Devices *Journal of Physics: Conference Series* **326** 012030
[6] Larson D J, Prosa T J, Lawrence D, Geiser B P, Jones C M and Kelly T F 2011 *Handbook of Instrumentation and Techniques for Semiconductor Nanostructure Characterization,* ed R Haight, *et al.* (London: World Scientific Publishing) 407-77
[7] Kellogg G L and Tsong T T 1980 Pulsed-laser atom-probe field-ion microscopy *Journal of Applied Physics* **51** 1184-94
[8] Larson D J, Foord D T, Petford-Long A K, Cerezo A and Smith G D W 1999 Focused ion-beam specimen preparation for atom probe field-ion microscopy characterization of multilayer film structures *Nanotechnology* **10** 45-50
[9] Miller M K, Russell K F, Thompson K, Alvis R and Larson D J 2007 Review of Atom Probe FIB-Based Specimen Preparation Methods *Microsc. Microanal.* **13** 428-36
[10] Thompson K, Lawrence D J, Larson D J, Olson J D, Kelly T F and Gorman B 2007 In-Situ Site-Specific Specimen Preparation for Atom Probe Tomography *Ultramicroscopy* **107** 131-9
[11] Prosa T J, Lawrence D, Olson D, Larson D J and Marquis E A 2009 Backside Lift-Out Specimen Preparation: Reversing the Analysis Direction in Atom Probe Tomography *Microsc. Microanal.* **15** 298-9
[12] Larson D J, Foord D T, Petford-Long A K, Liew H, Blamire M G, Cerezo A and Smith G D W 1999 Field-Ion Specimen Preparation Using Focused Ion-Beam Milling *Ultramicroscopy* **79** 287-93
[13] Fair R B and Weber G R 1973 Effect of complex formation on diffusion of arsenic in silicon *Journal of Applied Physics* **44** 273-9
[14] Fahey P M, Griffin P B and Plummer J D 1989 Point defects and dopant diffusion in silicon *Reviews of Modern Physics* **61** 289-384
[15] Marquis E A and Hyde J M 2010 Applications of atom-probe tomography to the characterisation of solute behaviour *Materials Science and Engineering R* **69** 37-62
[16] O'Neill R W, Larson D J, Thompson K, Kunicki T C and Geiser B P 2006 Measuring the Roughness of Buried Interfaces in Nanostructures by Local Electrode Atom Probe Analysis *Microsc. Microanal.* **12** 1746-8

Calibration of thickness-dependent *k*-factors for germanium X-ray lines to improve energy-dispersive X-ray spectroscopy of SiGe layers in analytical transmission electron microscopy

Y Qiu[1,2], VH Nguyen[3], A Dobbie[3], M Myronov[3] and T Walther[1*]

[1] Department of Electronic and Electrical Engineering, University of Sheffield, Mappin Building, Mappin Street, Sheffield S1 3JD, UK
[2] LAAS-CNRS, 7 Avenue du Colonel Roche, 31077 Toulouse Cedex 4, France
[3] Department of Physics, University of Warwick, Coventry CV4 7AL, UK

Email: t.walther@sheffield.ac.uk

Abstract. We show that the accuracy of energy-dispersive X-ray spectroscopy can be improved by analysing and comparing multiple lines from the same element. For each line, an effective *k*-factor can be defined that varies as a function of the intensity ratio of multiple lines (e.g. K/L) from the same element. This basically performs an internal self-consistency check in the quantification using differently absorbed X-ray lines, which is in principle equivalent to an absorption correction as a function of specimen thickness but has the practical advantage that the specimen thickness itself does not actually need to be measured.

1. Introduction

Energy-dispersive X-ray spectroscopy (EDXS) in a transmission electron microscope (TEM) offers high spatial resolution but usually suffers from both poor counting statistics due to small collection angles and quantification problems due to self-absorption in thin foil specimens analysed under small take-off angles. A first-order approach for quantification is based on experimentally determined sensitivity factors, called *k*-factors [1]. More accurate quantification of elemental compositions depends on the calculation of the relative ionisation cross-sections of the corresponding elements (*Z*-effect) [2,3] and corrections for absorption (*A*) [4] and for fluorescence (*F*) [5], which can be combined for a given specimen thickness and density in the so-called *ZAF*-correction. While this approximation neglects the top-bottom effect due to the depth of the analysed feature embedded within the thin foil specimen [6], this depth can itself be determined either directly by tilting the specimen through large angles or by means of measuring X-ray intensities as a function of take-off angle during moderate tilting experiments [7] and comparison with corresponding Monte Carlo simulations [8]. It has been shown in previous studies how the conventional thin film approximation for EDXS in a TEM can be improved for a given material system by calibrating *k*-factors based on linear least-squares fitting X-ray line intensities recorded over a range of sample thicknesses and extrapolation to zero thickness [9,10]. For the AlGaN system such a thickness-extrapolation yielded, together with a minor correction for stray X-rays from the substrate, the local Al content with a precision of ~1at% [11,12].

* To whom any correspondence should be addressed.

Here, we show that the absolute sample thickness is not necessary for accurate X-ray quantification if several lines are present for at least one element. Our approach is somewhat different from the method described by Morris et al. [13] who suggested to use K/L intensity ratios to first iteratively estimate the sample thickness and from this to perform a correction for fluorescence and absorption for all relevant lines, a procedure which still relies on absolute thickness values.

2. Experimental

If the X-ray intensity of a certain line is compared to that of Si_K as a standard, then the chemical concentration of an element *j* can be calculated as

$$x_j=(I_j\, k_{j,Si_K}\, a_j/A_j)/(\textstyle\sum_n I_n\, k_{n,Si_K}\, a_n/A_n) \qquad \text{(eqn. 1)}$$

where *I* denotes intensity, *k* the *k*-factor with respect to weight %, *a* the absorption, *A* the atomic weight and the index *n* runs over all elements detected. The *k*-factors in this equation are sensitivity factors that depend on the primary voltage, the X-ray fluorescence yield of the corresponding element and the energy-dependence of the sensitivity of the X-ray detector, compared to that of the standard. The division by atomic weights is only necessary if the *k*-factors refer to weight % (rather than at%). Suitable *k*-factors are usually provided by the manufacturer of the EDXS system, and the user has to choose which lines to include in the quantification. It is important that only one X-ray line per element should be used in the above equation, otherwise the result would be distorted by double weightings. Most microscopists use the hardest X-ray line detectable for each element so that absorption effects are as small as possible. Often, absorption in thin foil TEM specimens is neglected ($a_n=1\ \forall n$). For a precise quantification absorption should always be considered, however, the way standard software includes it is only approximate: by assuming a foil of thickness *t* that is homogeneous in composition along the beam direction. We have developed a method that takes into account all available X-ray lines (K, L etc.) within spectra, thereby performing a self-consistent absorption correction for e.g. K- and L-lines.

We demonstrate this in the following for the SiGe system as an example the results of which have already been applied to quantify the chemical composition of very thin $Si_{1-x}Ge_x$ layers [14]. This model system has the advantage that $k_{Si_K,Si_K}=a_{Si_K}=1$ by definition so that only two further X-ray lines are to be considered: Ge K and Ge L. The K-line is a hard X-ray and shows only very weak absorption, while the L-line is a soft X-ray and heavily absorbed. From above we can derive for the germanium content:

$$x_{Ge_K}=(I_{Ge_K}k_{Ge_K,Si_K}a_{Ge_K}/A_{Ge})/(I_{Si}/A_{Si}+I_{Ge_K}k_{Ge_K,Si_K}a_{Ge_K}/A_{Ge}) \qquad \text{(eqn. 2) for the K-line,}$$

$$x_{Ge_L}=(I_{Ge_L}k_{Ge_L,Si_K}a_{Ge_L}/A_{Ge})\ /(I_{Si}/A_{Si}+I_{Ge_L}k_{Ge_L,Si_K}a_{Ge_L}/A_{Ge}) \qquad \text{(eqn. 3) for the L-line.}$$

which *should* be identical ($x_{Ge_K}=x_{Ge_L}$) if the *k*-factors for both lines were correct. This is certainly not the case in our software package (Oxford Instruments ISIS300), as quantification with Ge K or L yields different apparent concentrations, up to 10at% apart, as shown in figure 3a. Calculating the ratio

$$x_{Ge_{K,L}}/x_{Si_K}=(I_{Ge_{K,L}}k_{Ge_{K,L}}a_{Ge_{K,L}}/A_{Ge})/(I_{Si}/A_{Si}) \qquad \text{(eqn. 4)}$$

for either K- or L-line of Ge and inserting $x_{Si_K}=1-x_{Ge}$ for the binary system, this can be solved for

$$k^*_{Ge_{K,L}} \equiv k_{Ge_{K,L}}a_{Ge_{K,L}}=(x_{Ge_{K,L}}I_{Si}\,A_{Ge})/[(1-x_{Ge})I_{Ge_{K,L}}A_{Si})] \qquad \text{(eqn. 5)}$$

The product of *k*-factor and absorption factor (relative to Si) on the left side of equation 5 basically represents a thickness-dependent *k*-factor [10], which we have experimentally determined for two reference samples of known chemical composition, namely a bulk $Si_{0.54}Ge_{0.46}$ (the chemical composition of which has been confirmed independently as 46.8±0.6 at.% Ge by inductively-coupled plasma optical emission spectrometry using a Spectro Cirus Vision instrument and a nitric acid/HF digest) and a strained layer $Si_{0.45}Ge_{0.55}$ sample (the Ge composition of the layered stack shown in figure 1 has been confirmed independently via strain analysis by X-ray diffraction, and the layer analysed here is the topmost in the stack with x=0.552±0.002). The X-ray spectra were acquired in a JEOL 2010F operated at 197kV and equipped with an Oxford Instruments Si:Li detector with ultrathin window. In the few cases where substantial Cu was detected from the supporting copper ring, a corresponding amount (in at.%) was subtracted from the Si signal to account for stray X-rays (some of which were present despite the use of an analytical stray aperture).

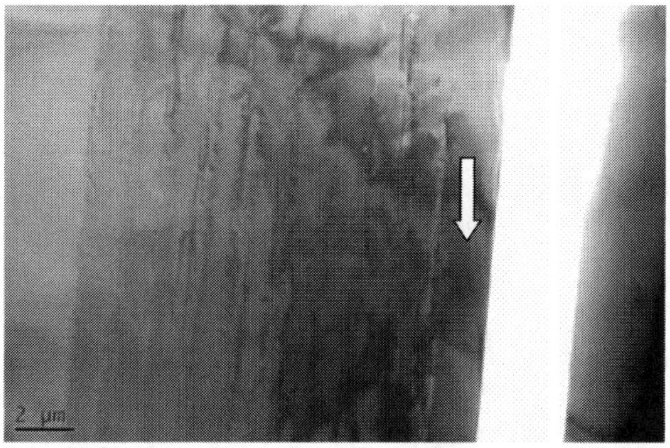

Figure 1. Bright field TEM image of a stack of 10 step-terraced grown SiGe buffer layers grown on Si(001) substrate (visible at the left). The $Si_{0.45}Ge_{0.55}$ layer investigated here was grown last and is marked by an arrow.

In figure 2 we plot for all spectra the I_{Ge_K}/I_{Ge_K} ratio vs. detector dead time, which can be assumed to be proportional to the specimen thickness projected along the electron beam direction [15]. Data points lie on straight lines the slope of which depends on tilt angle, composition (higher Ge content means higher density and thus stronger absorption, i.e. higher slope) and also the beam intensity, however, the intercept with the vertical axis is the same. This gives an estimate of the Ge K/L intensity ratio of 1.3 for a specimen of zero thickness (t=0) which is difficult to achieve in practice. Note that some measurements for small thicknesses led to considerable carbon built-up on the sample surface, thereby absorbing preferentially the softer L-line (hence, larger apparent K/L ratios). These spectra are marked by open symbols and were omitted from fitting.

Figure 3 shows plots of the thickness dependent k-factors from equation (5) as a function of the Ge K/L ratio, i.e. of $k^*_{Ge_{K,L}}$, along with exponential and also parabolic fits, for both Ge K-line (a) and L-line (b) from the same set of spectra taken under various conditions. It can be seen that data from both samples fall onto common lines for both X-ray lines. The "correct" k-factors $k_{Ge_{K,L}}$for ultra-thin samples would then be obtained from an extrapolation to zero thickness. Using the K/L line ratio extrapolated above and the second-order polynomial fits in figure 3, we calculate these k-factors for ideal thin samples as $k_{Ge_K,Si}$= 1.53±0.08 for the Ge K-line and $k_{Ge_L,Si}$= 2.08±0.07 for the Ge L-line. A comparison with the k-factors tabulated in the ISIS300 software in table 1 shows large differences, and in particular $k_{Ge_L} > k_{Ge_K}$, while the tabulated ISIS values would suggest the reverse.

If the ISIS software is used for quantification, without absorption correction the Ge L-line yields somewhat better results than the Ge K-line, however, the spread of the results is large (see figure 4a). While the true compositions lie somewhere between those values reported from K- or L-line quantification, simple interpolation will is inappropriate because the actual values determined at any point will vary with thickness and if a wider thickness range is included the average reported will systematically shift. Even if absorption is included (using an average density of ρ=3.95 g cm^{-3} [16] and assuming a linear interpolation between t=0 for Ge K/L=1.3 and t=1200nm for K/L=2.5 [17]) the discrepancy between quantifications using either the Ge K- or the Ge L-line relative to Si K is only reduced but not eliminated, as demonstrated in table 1.

Relative uncertainties of 3-5% in our extrapolated k-factors should allow us to determine the Ge concentration to $\Delta x=\pm 0.02$ in thin specimens, where the precision is mainly limited by the spread of the K/L ratio determined from the extrapolation to t=0 in figure 2, i.e. ultimately the knowledge of the specimen thickness. Simply averaging over all spectra using these k-factors for ideal thin films without absorption correction then yields consistent values from both K- and L-lines, but the values are too large for higher sample thicknesses if no allowance is made for absorption (see figure 4c)

However, using K- or L-lines with the thickness dependent k^*-factors from figure 3 in which the mean squared deviation of all 55 data points from the fit curves is only $\langle(\Delta k^*)^2\rangle=0.014$ (figure 3a) to 0.039 (figure 3b) now implies mean relative errors in k^* for any spectrum of 1-3% and leads to nearly identical results for both samples *independent of thickness,* as demonstrated in figure 4d, with an rms spread of only $\sqrt{\langle\Delta x^2\rangle}=\pm 0.01$. The scatter observed is mainly due to counting statistics, and the deviation of the average of repeated measurements from the nominal values are typically $\Delta x < 0.005$.

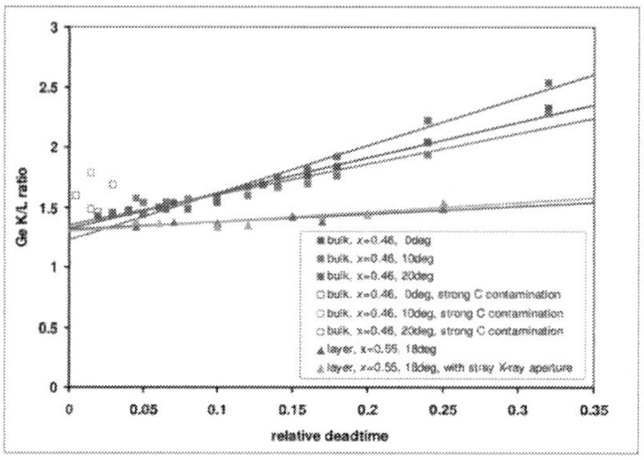

Figure 2. Plot of Ge K/L intensity ratio vs. detector dead time, for two calibration samples of x_{Ge}=0.46 and 0.55 studied at various tilts towards the X-ray detector, along with linear fits. Extrapolation to zero dead time (for an ideal thin sample) yields an intercept of 1.303±0.047 for all five curves. The linear regression coefficients are R^2=0.943±0.059.

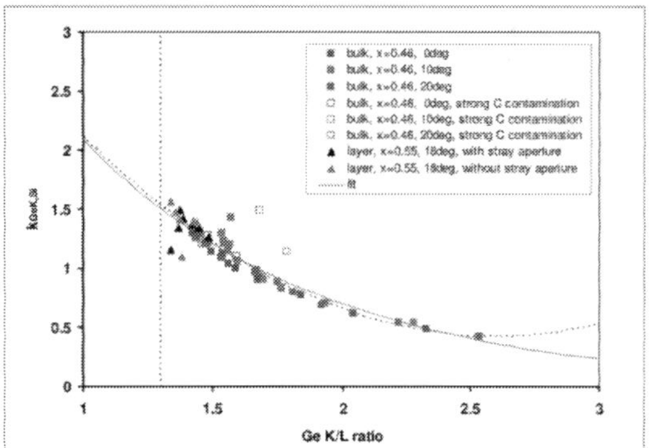

Figure 3a. Plot of the ratio $k^*_{Ge_K}=k_{Ge_K}\,a_{Ge_K}$ vs. Ge K/L intensity ratio for both samples. Solid [dotted] lines indicate exponential [parabolic] fits, respectively, which yield nearly identical values for an intensity ratio K/L=1.303±0.047 for t=0 of k_{Ge_K} =1.501±0.077 [1.529±0.081].

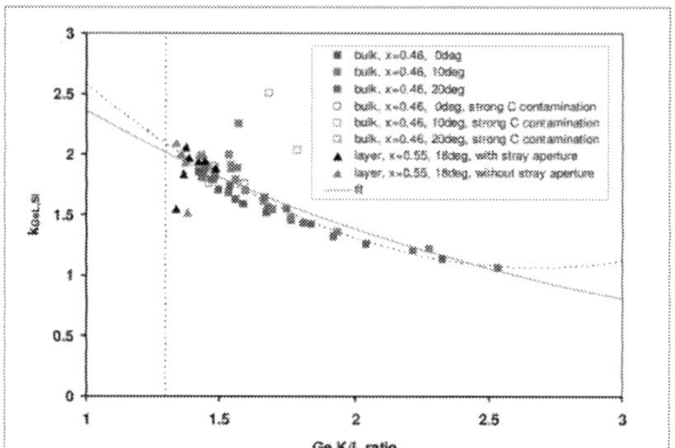

Figure 3b. Plot of the ratio $k^*_{Ge_L}=k_{Ge_L}\,a_{Ge_L}$ vs. Ge K/L intensity ratio for both samples. Solid [dotted] lines indicate exponential [parabolic] fits, respectively, which for t=0 extrapolate to k_{Ge_L} =2.015±0.051 [2.080±0.070].

Table 1. Comparison of k-factors (with respect to Si_K) and reported germanium concentrations

specimen		k-factor		bulk sample (x_{Ge}=0.468)		strained layer (x_{Ge}=0.552)	
method	line	Ge K	Ge L	Ge K	Ge L	Ge K	Ge L
ISIS, thin film approximation (a=1)		1.723	1.651	0.597±0.070	0.462±0.045	0.609±0.024	0.516±0.024
ISIS, with absorption correction for linear increase of thickness from t=0 (K/L=1.3) to t=1200nm (K/L=2.5)		1.723	1.651	0.524±0.033	0.423±0.026	0.584±0.024	0.503±0.025
k*-factor extrapolated to K/L=1.303 from parabolic fit, then used without absorption correction (a=1)		1.53±0.08	2.08±0.07	0.569±0.072	0.519±0.045	0.581±0.024	0.573±0.023
k*-factor for K/L ratio as measured for each spectrum		figure 3a	figure 3b	0.464±0.012	0.461±0.011	0.545±0.010	0.548±0.012

Note: font colour indicates whether numerical results agree with reference within error bars (green) or not (red). When numerical results are correct but error bars >0.02, the error bars are marked yellow.

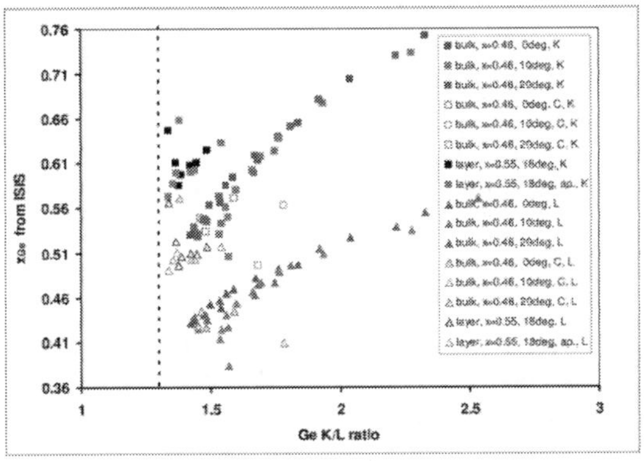

Figure 4a. Plot of x_{Ge} from LINK ISIS300 calculated without absorption correction. Numerical results differ for K and L lines and strongly vary with thickness.

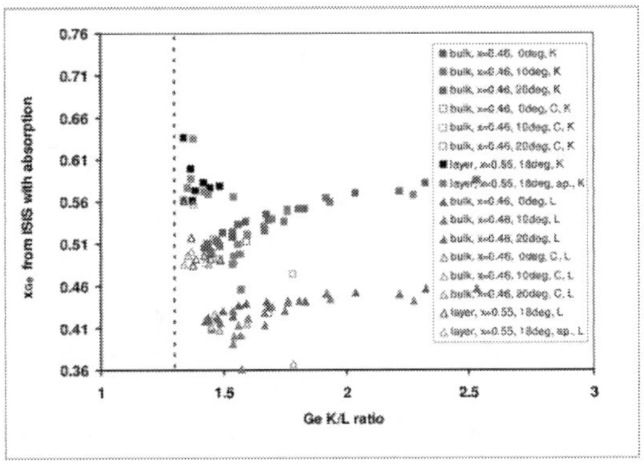

Figure 4b. Plot of x_{Ge} from LINK ISIS300 calculated with absorption correction. Numerical results still differ for K and L lines but vary somewhat less with thickness.

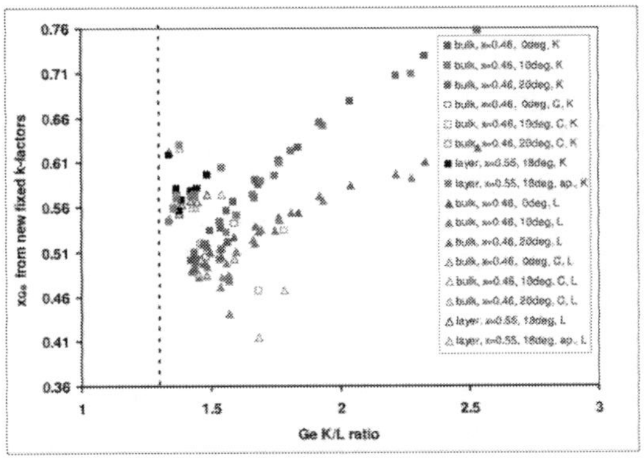

Figure 4c. Plot of x_{Ge} from k-factors extrapolated to t=0 as in [10]. Results from K and L lines now agree better but still depend on thickness, and are too large throughout.

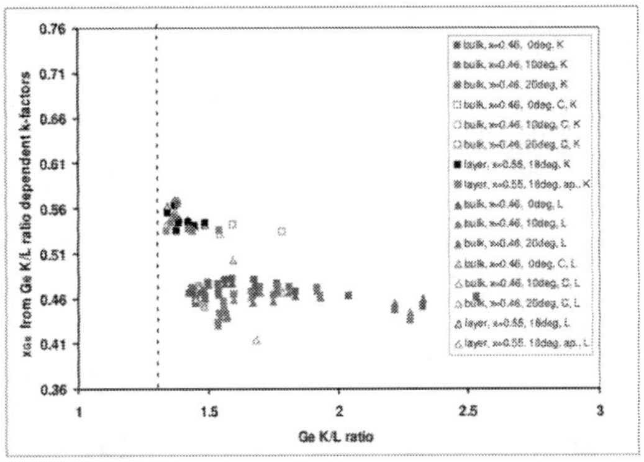

Figure 4d. Plot of output x_{Ge} using improved k*-factors, i.e. $k^*_{Ge_{K,L}}$ from figure 3. Results from K and L lines agree within noise levels and are independent of thickness.

3. Summary

By plotting the *k*-factor in EDXS versus the intensity ratio of two suitable lines (e.g. K/L) it is possible to define an effective *k*-factor (k^*) for each X-ray line which implicitly performs an absorption correction without knowledge of the absolute specimen thickness. For thicker and/or denser specimens and/or smaller take-off angles the absorption of K- and L-lines will be affected differently, and rather than computing this effect numerically for sets of specimen thicknesses, densities and take-off angles we can simply use the effective *k*-factors from calibration curves (such as in figure 2) at the appropriate K/L ratio to account for the different absorption effects. Applying this to two SiGe specimens of known compositions we have shown that we can determine the germanium concentration in a self-consistent way to better than 1at% for a specimen of unknown thickness, at least in the range of $t \leq 1.2\mu m$ which should be sufficient for most experimental cases.

References

[1] Goldstein JI, Costley JL, Lorimer GW and Reed SJB 1977 Quantitative X-ray Analysis in the Electron Microscope *Scanning Electr. Microsc.: Proc. Workshop Anal. Electron Microsc. (IIT Research Institute, Chicago, IL, USA, March 1977,* vol **1***)* ed O Johari, p 315
[2] Bethe HA 1930 *Ann. Phys.* **397**, 325
[3] Duncumb P and Reed SJB 1968 Quantitative Electron Probe Microanalysis *NBS Spec. Publ.* (vol 298) ed KFJ Heinrich, p 133
[4] Philibert J 1963 *Proc. Int. Symp. X-ray Optics and X-ray Microanalysis (Academic Press, New York, USA)* eds HH Pattee, VE Cosslett and E Engström, p 379
[5] Reed SJB 1965 *Brit. J. Appl. Phys.* **16**, 13
[6] Walther T 2007 *Proc. 15th Int. Conf. on Microscopy of Semicond. Mater.*, Cambridge, UK, *Springer Proc. in Physics* **120**, 185
[7] Walther T 2008 *Proc. Electron Microscopy & Analysis Group Conference 2007,* Glasgow, UK, *J. Phys. Conf. Ser.* **126**, 012090
[8] Parri MC and Walther T 2012, *Proc. Electron Microscopy & Analysis Group Conference 2011*, Birmingham, UK, *J. Phys. Conf. Ser.* **371**, 012063
[9] Walther T 2010, *Proc. 16th Int. Conf. on Microscopy of Semicond. Mater.*, Oxford, UK, *J. Phys. Conf. Ser.* **209**, 012029
[10] Walther T 2010, *Proc. Electron Microscopy & Analysis Group Conference 2009*, Sheffield, UK, *J. Phys. Conf. Ser.* **241**, 012016
[11] Amari H, Lari L, Zhang HY, Geelhaar L, Cheze C, Kappers MJ, McAleese C, Humphreys CJ and Walther T 2011, Proc. *17th Int. Conf. on Microscopy of Semicond. Mater.*, Cambridge, UK, *J. Phys. Conf. Ser.* **326**, 012028
[12] Amari H, Kappers MJ, Humphreys CJ, Cheze C and Walther T 2012 *phys. stat. sol. (c)* **9**, 1079
[13] Morris PL, Ball MD and Statham PJ 1979, Proc. EMAG '79, Brighton, UK, *Inst. Phys. Conf. Ser.* **52**, 413
[14] Walther T, Norris DJ, Qiu Y, Dobbie A, Myronov M and Leadley DR 2013 *phys. stat. sol. (a)* **210**, 187
[15] Williams DB and Carter CB 1996 Transmission Electron Microscopy *(Springer, New York)*, p561
[16] Schaffler F 2001, in *Properties of Advanced SemiconductorMaterials GaN, AlN, InN, BN, SiC, SiGe*. Eds. ME Levinshtein, SL Rumyantsev and MS Shur, Wiley, New York, p 149-188
[17] Parri MC, Qiu Y and Walther T 2012 *Proc. 15th Electr. Microsc. Congress, Manchester, UK,* eds DJ Stokes and JL Hutchison, vol **2**, p 693

Journal of Physics: Conference Series **471** (2013) 012032 doi:10.1088/1742-6596/471/1/012032

Towards the structure of rare earth luminescence centres – terbium doped aluminium nitride as an example system

Felix Benz[1,2], Thomas Walther[3] and Horst P Strunk[2]

[2] Institute of Materials Science, University of Stuttgart, Heisenbergstr. 3, 70569 Stuttgart, Germany
[3] Kroto Centre for High Resolution Imaging & Spectroscopy, Dept. Electronic & Eletrical Eng., University of Sheffield, Mappin Street, Sheffield S1 3JD, UK

E-mail: felix.benz@imw.uni-stuttgart.de; fb400@cam.ac.uk

Abstract. Sputter deposited terbium doped aluminium nitride layers were investigated with respect to the distribution and surroundings of the terbium luminescence ions. Semi-empirical calculations indicate that terbium forms complexes consisting of one aluminium vacancy surrounded by three oxygen ions on nitrogen lattice positions and one nitrogen ion that is bound to a terbium ion on a distorted aluminium lattice position. The crystal field splitting of the terbium ions indicate their surroundings to be not tetrahedral but, as anticipated from the determined complex, C_{3v}. Complementary electron microscopic investigations show a random distribution of these complexes within the layer volume. The terbium ions cause a shape of the Tb $M_{5,4}$ edge similar to the shapes in other ionic compounds like Tb_2O_3.

1. Introduction

Rare earth doped semiconductors exhibit unique optical properties such as very narrow emission lines. Possible applications include data transfer [1], biological sensing [2] or light emission in electroluminescent devices [3].

The wavelengths of the emission lines are hardly influenced by the matrix material since the luminescence arises from inner-lying 4f orbitals, which are effectively shielded from the matrix by outer-lying orbitals [4]. Further experimental results suggest that the energy transfer between the rare earth ions is insensitive against the matrix material [5]. Thus one could be tempted to describe the influence of the matrix on a rare earth ion by the dielectric properties and the bandgap of the matrix only and consider this rare earth ion as kind of an independent system in an isotropic medium. This fact holds essentially for the farther surrounding of the ions, say from the next-nearest neighbour shell on outwards. The next-nearest neighbour surroundings, however, and the microstructural position of the rare earth ion with reference to this shell play a mostly beneficial role in rare earth ion luminescence. Firstly, the shell's symmetry influences the excitation of the rare earth ion by matrix excitations and, secondly, it creates a so-called (local) crystal field that breaks the conditions of luminescence transitions, which are in a free ion quantum mechanically forbidden. This contribution presents an attempt to approach these two roles by clarifying the microscopic structure of the centres.

---

[1] To whom any correspondence should be addressed.

Journal of Physics: Conference Series **471** (2013) 012032 doi:10.1088/1742-6596/471/1/012032

2. Experimental procedure

The used terbium doped aluminium nitride samples were produced by reactive DC magnetron sputtering. Two aluminium targets and one terbium target were used in order to adjust the terbium concentration to approximately 1-2 at.-%. A mixture of argon and nitrogen was used as deposition atmosphere. After deposition the samples were annealed at 1000°C under 1 bar pure nitrogen atmosphere for 30 min. This step is necessary for activating the luminescence centres (gain in luminescence intensity approximately 25 times).

Photoluminescence experiments were carried out using a 450 W xenon lamp as excitation source. A first monochromator was used for selecting the desired excitation wavelength, a second one was used for spectral recording the emission using a photomultiplier tube. The intensity of the excitation light was recorded with a silicon photodiode. For these experiments room temperature was used. Additional cathodoluminescence (CL) measurements were recorded at a temperature of 4 K.

Transmission electron microscopic investigations were performed using a JEOL 2010F FEG-TEM microscope operated at 197kV and equipped with scan unit, Gatan Imaging Filter (GIF2000), and X-ray spectrometer (Si:Li, ATW, Oxford Instruments, ISIS300).

3. Simulation of the centre structure

The semi-empirical programme MOPAC2012 [6] together with the appropriate expansion SPARKLE [7] was used for a preliminary investigation of the rare earth centre nature. The PM6 algorithm [8] was used for all calculations. First, an aluminium nitride super-cell consisting of 18 unit cells (72 atoms) was built and optimised. Periodic boundary conditions were used. The obtained lattice constants match with X-ray diffraction experiments. Subsequently, a terbium ion, an aluminium vacancy, and up to three oxygen ions were introduced into the simulation volume. (There are indications that vacancies and oxygen take part in optically active centres in AlN, see *e.g.* refs. [9,10].) Different atomic arrangements were used in order to work towards the energetically lowest configuration. As a result, a configuration is obtained that has the lowest energy of all calculated configurations of approximately 5 eV below a random arrangement of these atoms. It consists of one aluminium vacancy surrounded by the three oxygen ions and one nitrogen ion that itself is bound to the terbium ion. The terbium (1.4 at.-%) and oxygen impurity concentrations (4 at.-%) of our measured samples approximately match the 1:3 relation between terbium and oxygen. Figure 1 shows this complex embedded in the aluminium nitride super-cell. The same result was always obtained for different initial arrangements and different positions in the simulation volume.

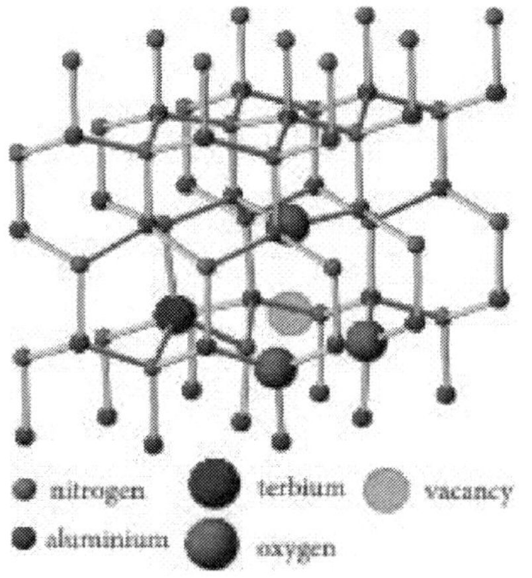

Figure 1. Obtained lowest energetic complex ($E \approx -5$ eV), consisting of an aluminium vacancy surrounded by three oxygen ions and one nitrogen that is itself bound to a terbium ion.

It is remarkable that the local symmetry at the position of the terbium ion in this complex is C_{3v} instead of T_d, the point group for a perfect tetrahedron.

4. Experimental results and discussion

The local symmetry at the position of the rare earth ion influences the number of sublevels into which a state with a certain total angular momentum is split (crystal field splitting). For the state $^7F_5$ of the trivalent terbium ions a splitting into seven states is anticipated for the C_{3v} point group, for T_d four sublevels are expected [11]. The CL spectrum recorded at 4 K (*cf.* Figure 2) shows the emission due to the transition $^5D_4 \rightarrow {}^7F_5$. At such a low temperature it can be expected that the transition occurs from the lowest sublevel of $^5D_4$ only. Therefore the observed splitting is solely due to the splitting of the ground state $^7F_5$. The spectrum shown in Figure 2 confirms the assumption that the local symmetry is lower than T_d, *e.g.* C_{3v} as expected from the simulation shown above.

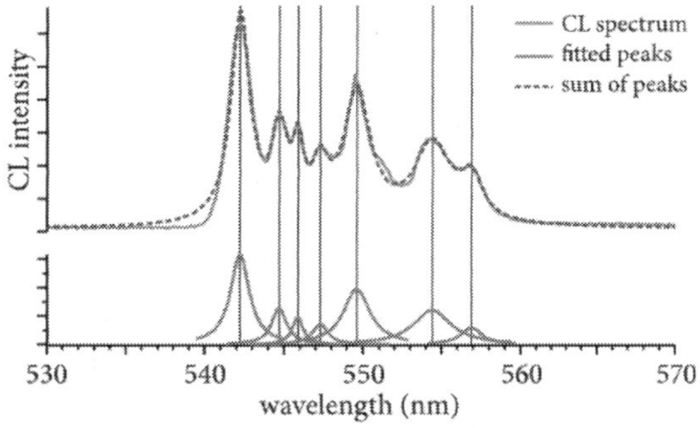

Figure 2. CL spectrum recorded at 4 K of transition $^5D_4 \rightarrow {}^7F_5$: peak split into seven sub-peaks due to crystal field.

Figure 3 shows a bright field image of the investigated layer together with STEM X-ray maps for the elements Si, Al and Tb. The layer has a columnar grain structure with a grain diameter of approximately 20 nm. The X-ray maps indicate that the terbium ions do not occupy specific positions in the microstructure (*e.g.* the grain boundaries, the layer surface, *etc.*). In fact it rather appears that the ions are randomly distributed, although clustering at an atomic scale might occur and not be observable.

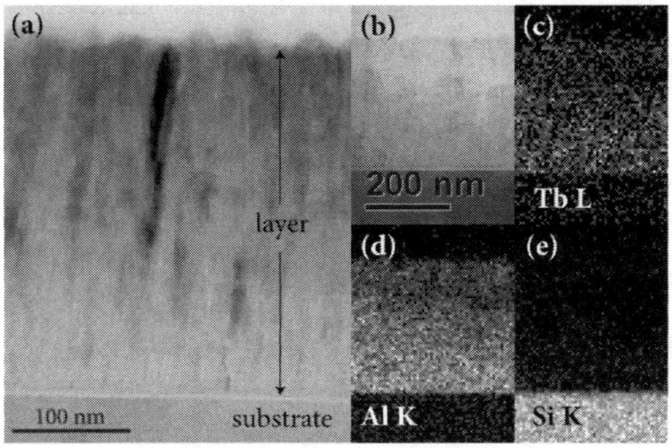

Figure 3. Electron microscopy study of a cross-section in bright field (BF) (a), STEM BF (b) with corresponding X-ray maps of Tb L (c), Al K (d) and Si K (e) at a sampling of 6.4nm/pixel. The sample thickness of the foil tapers off towards the top of the layer, leading to a corresponding contrast gradient.

Additionally, the layer was investigated by electron energy-loss spectroscopy (EELS), see Figure 4. The Tb $M_{5,4}$ edges can be seen at 1240 eV (Tb M_5) and 1259 eV / 1271 eV (Tb M_4). The shapes of the edges are similar to those observed for other ionic terbium compounds like Tb_2O_3 (observed by EELS) [12] or $TbMnO_3$ (observed by XANES) [13], underlining the ionic character of terbium incorporated into AlN.

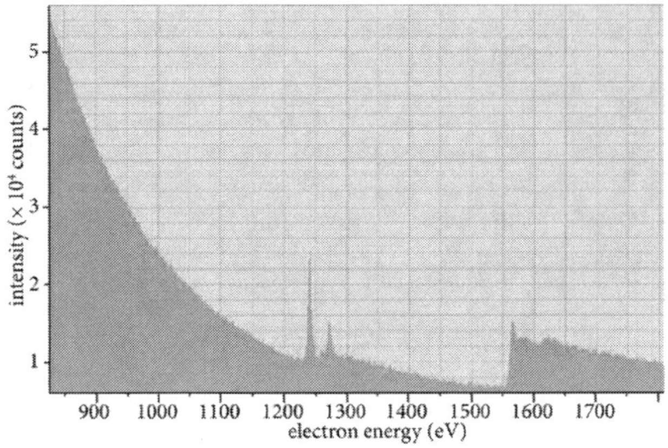

Figure 4. EELS spectrum of AlN:Tb, recorded with an energy offset of 800 eV with 1 eV / channel dispersion and an integration time of 100 s.

5. Conclusion

Both experimental and theoretical results indicate that terbium ions in aluminium nitride form complexes containing one aluminium vacancy, three oxygen ions, and one terbium ion connected to the vacancy through a nitrogen ion. STEM results indicate that the terbium ions are randomly distributed in the microstructure through the whole layer. Furthermore, the EELS spectrum of terbium ions doped into aluminium nitride is presented. The Tb $M_{5,4}$ edge shows discrete peaks at 1240 eV (strong), 1259 eV (weak), and 1271 eV (strong), which is similar to spectra reported for other ionic terbium compounds.

References

[1] J K R Weber, J J Felten, B Cho, and P C Nordine, *Nature* **393**, 769 (1998).
[2] H Jeong, S Lee, G Y Sung, and J Shin, *IEEE Photon. Technol. Lett.* **23**, 88 (2011).
[3] A Nishikawa, T Kawasaki, N Furukawa, Y Terai, and Y Fujiwara, *Appl. Phys. Express* **2**, 071004 (2009).
[4] T Kano in *Phosphor Handbook*, 2nd edition, edited by W M Yen, S Shionoya, and H Yamamoto (CRC Press, Boca Raton, Florida, USA, 2006).
[5] F Benz, J A Guerra, Y Weng, A R Zanatta, R Weingärtner, and H P Strunk, *J. Lumin.* **137**, 73 (2013).
[6] J J Steward (Steward Computational Chemistry), MOPAC2012, http://openmopac.net (2012).
[7] R O Freire and A M Simas, *J. Chem. Theory Comput.* **6**, 2019 (2010).
[8] J J P Steward, *J. Mol. Model.* **13**, 1173 (2007).
[9] R A Youngman and J H Harris, *J. Am. Ceram. Soc.* **73**, 3238 (1990).
[10] B Bastek, F Bertram, J Christen, T Hempel, A Dadgar, and A Krost, *Appl. Phys. Lett.* **95**, 032106 (2009).
[11] B Henderson and R H Bartram, *Crystal-Field Engineering of Solid-State Laser Materials* 1st edition (Cambridge University Press, Cambridge, 2000).
[12] C C Ahn and O L Krivanek, *EELS Atlas* (Gatan Inc. and ASU HREM Facility, 1983).
[13] T R Forrest, *et al.*, *J. Phys.: Condens. Matter* **20**, 422205 (2008).

Journal of Physics: Conference Series **471** (2013) 012033 doi:10.1088/1742-6596/471/1/012033

Transmission electron microscopy study of semi-polar gallium nitride layer grown by hydride-chloride vapour-phase epitaxy on SiC/(001)Si heterostructure

L M Sorokin[1], **A E Kalmykov**[1], **A V Myasoedov**[1], **V N Bessolov**[1], **A V Osipov**[2] **and S A Kukushkin**[2]

[1]Ioffe Physical-Technical Institute, Russian Academy of Sciences, St. Petersburg, 194021 Russia

[2]Institute for Problems of Mechanical Engineering, Russian Academy of Sciences, St. Petersburg, 199178 Russia

aekalm@mail.ioffe.ru

Abstract. The structure of 10-μm-thick GaN layer grown by chloride vapour-phase epitaxy on 1.5-inch SiC/(001)Si templates has been investigated by transmission electron microscopy (TEM). The silicon carbide buffer layer has been fabricated by a new method of solid-phase synthesis. It was found that the GaN layer consists of oriented grains with size of tenths of a micron. The grains have wurtzite structure, and the {0001} GaN planes are oriented parallel to {111} Si, that is, the deviation of the axis *c* of GaN crystallite from the normal to the substrate is about 52°. The revealed epitaxial relationship between substrate and most grains is $(02\bar{2}3)$GaN||(001)3C-Si||(001)Si and $[2\bar{1}\bar{1}0]$GaN||[110]3C-SiC||[110]Si. Some inclusions of sphalerite gallium nitride were also found in the epilayer.

1. Introduction

GaN is a suitable material for high-performance green and ultraviolet light-emitting devices. But the luminous efficiency of the devices is low, because of the presence of electrostatic fields within the active layers which are generated by the spontaneous and piezoelectric polarization along the [0001] axis (*c*-axis) in wurtzite GaN grown along this *c*-axis [1]. This crystallographic orientation corresponds to the natural growth direction of GaN layer deposited on currently available substrates (silicon carbide, sapphire). One way of the field reduction is to grow GaN layers on semi-polar planes. Nonpolar and semi-polar GaN films are most frequently grown by methods of metalorganic chemical vapour deposition and molecular beam epitaxy on *m*-Al_2O_3 [2], $(11\bar{2}0)$4*H*-SiC [3] and (113)Si [4] substrates. Silicon is a promising material for GaN growth: there are the prospects for integration of the gallium-nitride and silicon electronics technologies and the possibility of using inexpensive single crystal silicon substrates, available in large diameters (> 200 mm). Furthermore, Si substrates have reasonable thermal and electrical conductivity. (001)Si substrate orientation is preferred as it is the one used in the silicon industry. Thus, for example, the growth process of semi-polar $(10\bar{1}2)$GaN on (001)Si offcut substrates with 3*C*-SiC buffer layers was realized by metal-organic vapour-phase epitaxy (MOVPE) [5]. Recently, the growth of semi-polar $(1\bar{1}01)$GaN has been attempted on a patterned (001) silicon substrate adopting selective area MOVPE [6]. The growth was initiated on

(111) facets of the Si, which had been prepared by anisotropic etching.

The aim of this study is to investigate orientation and structural quality of GaN layer grown on SiC/(001) Si template by the method of hydride-chloride vapour-phase epitaxy (HVPE).

2. Experimental details

A new method was used for synthesis of SiC buffer layer wherein the Si substrate itself is involved into the chemical reaction and the reaction product grows into the interior of substrate layer [7,8]. This method ensures the relaxation of arising elastic stresses during the growth.

To evaluate the layer microstructure TEM investigation was performed with a PHILIPS EM420 microscope operated at 100 kV. (011)Si oriented cross-sections were prepared by cutting and mechanical polishing followed by Ar^+ ion milling.

3. Results and discussion

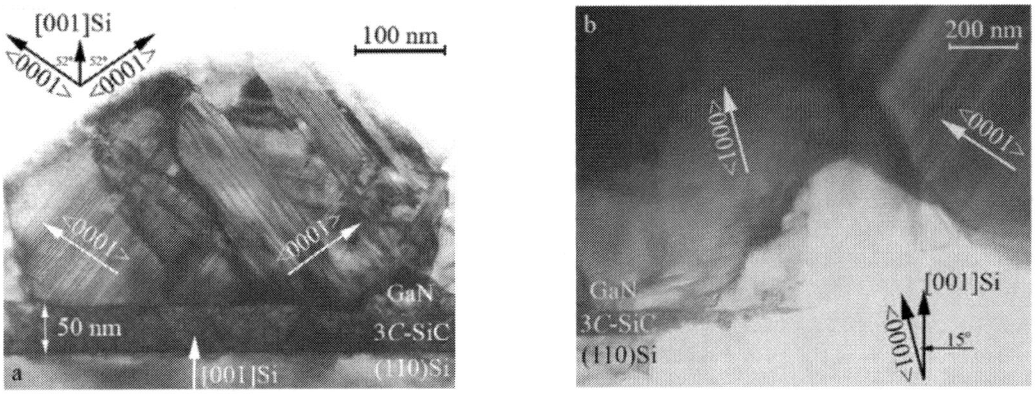

Figure 1. Cross-sectional TEM images of different areas of GaN/SiC/(001)Si heterostrucrure.

TEM study showed that the gallium nitride layer consists of individual grains of different orientations (figure 1). The sizes of grains are in the range of 0.1 - 10 μm. Grains contain basal plane stacking faults with a density varying from grain to grain within 2×10^5 - 3×10^6 cm^{-1}. Figures 2-4 show typical selected area diffraction patterns (DP) taken from the sample.

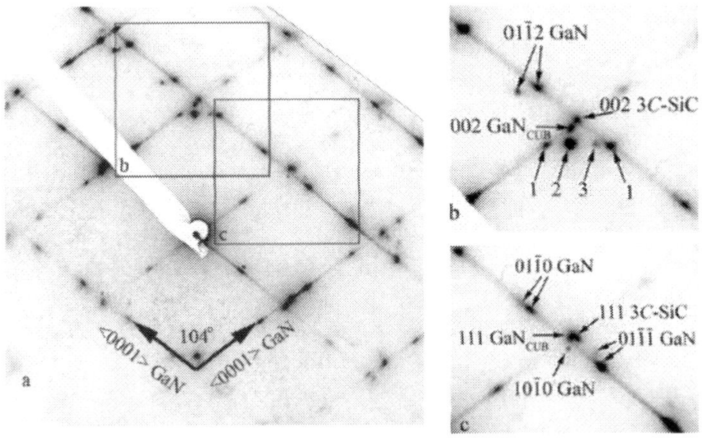

Figure 2. a: DP taken from the area shown in figure 1a; b, c: magnified photo of the DP outlined in figure 2a. In figure 2b the diffraction spots 1 and 3 correspond to wurtzite grains with common *a*-axis. *C*-axis of the grain giving the diffraction spot 1 is parallel to <111>Si. As to the grain giving diffraction spot 2, its *a*-axis does not coincide with [011]3*C*-SiC direction.

There are three groups of diffraction spots in the DPs which are the results of electron scattering by Si substrate, SiC buffer layer and GaN grains, respectively. Most GaN grains have a common zone axis, which coincides with the direction of the incident electron beam: [011]3*C*-SiC/Si is parallel to [$2\bar{1}\bar{1}0$]GaN. DP analysis has shown that the main part of GaN grains have wurtzite structure.

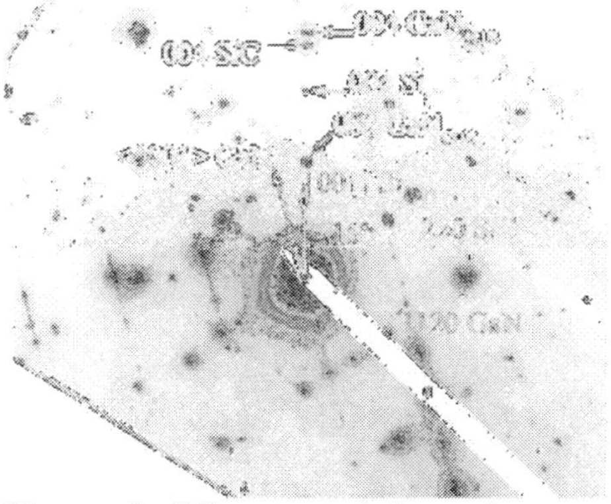

Figure 3. DP recorded from the area represented in figure 1b.

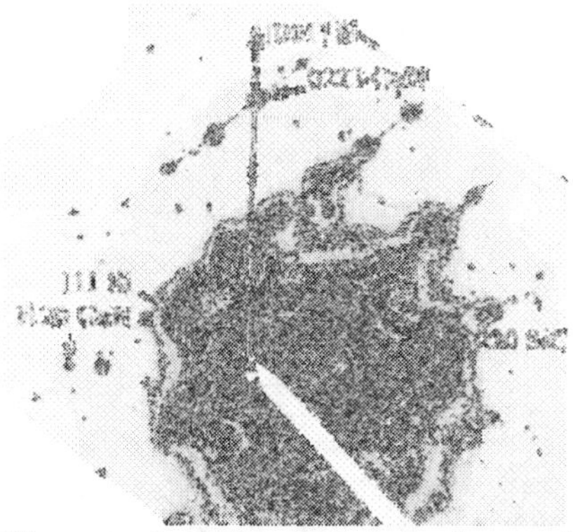

Figure 4. DP recorded from the interface area.

According to figures 2-4 the orientation of *c*-axis of most wurtzite grains coincides with <111> substrate directions and deviates from the normal to the substrate by an angle of 52°. Thus the orientation of the most grains is semi-polar and should be the same as GaN layer orientation in [6] where substrate masking and selective etching were used for (111)Si facets forming.

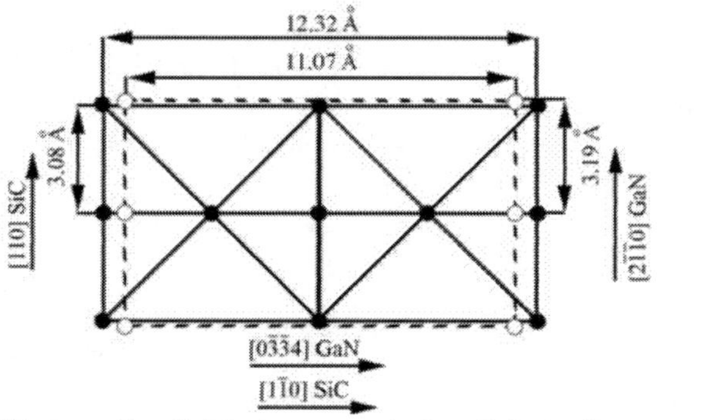

Figure 5. GaN layer and 3*C*-SiC buffer layer matching. The black circles mark SiC lattice points positions and the solid lines show the (001)SiC plane unit cell. The open circles mark GaN lattice points positions and the dashed lines show the ($02\bar{2}3$)GaN plane unit cell.

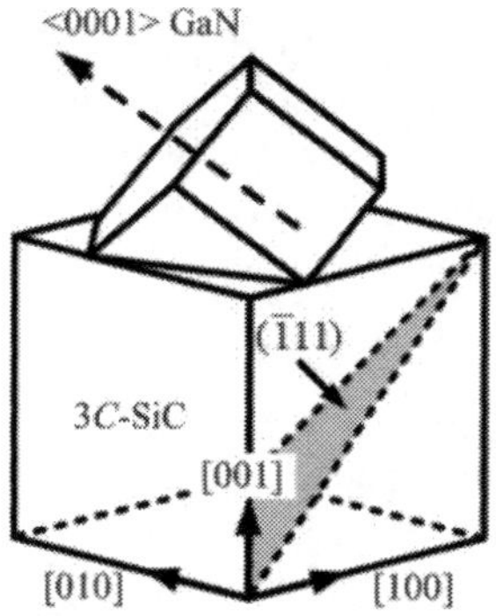

Figure 6. Schematic of grown semi-polar structure <111>SiC||<0001>GaN. The idea of the schematic belongs to [9].

Miller indices of the matching planes of the interface can also be determined by DPs. There is a $02\bar{2}3$ diffraction spot marked in figure 4. Obviously this is the spot with the lowest index which lies on the [001] Si direction in reciprocal space. Accordingly, the matching planes are ($02\bar{2}3$)GaN and (001)SiC, and taking into account that $a = [2\bar{1}\bar{1}0]$ axis coincides with the [110] substrate the following epitaxial relationships were found for these grains: ($02\bar{2}3$)GaN || (001)3*C*-Si || (001)Si and [$2\bar{1}\bar{1}0$]GaN || [110]3*C*-SiC || [110]Si.

The lattice mismatch is only 3.5% in the silicon carbide [110] direction, although it is 10% in the [$1\bar{1}0$] direction. Figures 5 and 6 show the experimental epitaxial orientation of GaN and SiC layers and a schematic of the heterostucture. [110] and $[1\bar{1}0]$ are two equivalent directions in the (001)SiC plane. Therefore there should also be GaN grains with the *c*-axis inclined from the normal of the substrate by rotating around $[1\bar{1}0]$ SiC at 52°. This is confirmed by the presence of $11\bar{2}0$ GaN diffraction spots in the DP, figure 3. Figure 6 shows only one of the four possible orientations of GaN grains relative to the substrate. All possible orientations of GaN grains in the figure can be obtained by rotating the schematic around [001]3*C*-SiC axis through 45°.

The *c*-axis of smaller part of GaN grains deviates from the normal to the substrate at an angle of 15° (figure 3). Similarly, it was found for these grains: $(01\bar{1}7)$GaN||(001)3*C*-Si||(001)Si and $[2\bar{1}\bar{1}0]$GaN||[110]3*C*-SiC || [110]Si. The lattice mismatch in the $[1\bar{1}0]$ SiC direction is 6%.

Occasionally, there are grains in the layer the *a*-axis of which is not parallel to the <011>3*C*-SiC direction. For example, the zone axis of the grain corresponding to diffraction spot 2 in figure 2 is $[\bar{4}\bar{1}53]$.

An adequate regular spot array has been found for each of the identified diffraction spots in the DP (figure 2). This enabled to identify these reflections reliably, which, in turn, allowed us to detect the presence of sphalerite gallium nitride (GaN_{cub}) by DP analysis (figure 2b,c). There are also well resolved 004 GaN_{cub} and 004 SiC diffraction spots in figure 3. Following the results of [5] it can be supposed that the thin sphalerite GaN layer, which may have been initiated by a cubic buffer layer, is initially grown with {111} facets, and then wurtzite GaN film was initiated on the {111} facets of sphalerite GaN. There is some tilt between the GaN_{cub} planes in the film and 3*C*-SiC planes (figures 2, 3). This tilt is due to small difference in lattice constants of GaN_{cub} film and 3*C*-SiC buffer layer.

4. Conclusions

For the first time, semi-polar GaN layer consisting of oriented grains has been grown by HVPE on 1.5-inch Si(001) substrate with SiC interlayer without any masking or etching. The *c*-axis of most wurtzite grains deviates from the normal to the substrate at an angle of 52°. Meanwhile, the GaN film suffers from high basal plane stacking fault density in comparison to the conventional *c*-plane GaN. This result enables us to hope that using offcut substrates will make it possible to fabricate single-crystal semi-polar GaN layers. It should be emphasized that HVPE allows us to grow thick (tens of micrometers) epilayers.

Acknowledgements

This study was supported by the Department of Physical Sciences of the Russian Academy of Sciences (programs "New Materials and Structures"), the Russian Foundation for Basic Research (project nos. 11-02-00496, 12-02-00935-a, 13-08-00074), the St. Petersburg Foundation for Support of Science and Education, the Skolkovo Innovation Centre (registration number 1110023) and in part by Ministry of Education and Science of the Russian Federation. TEM investigation was made on the equipment of the Joint Research Centre «Material science and characterization in advanced technology» (Ioffe Institute, St. Petersburg, Russia).

References

[1] Deguchi T et al. 1999 *Jpn J. Appl. Phys.* **38** L914–L916
[2] Baker T J et al. 2006 *Jpn. Appl. Phys.* 2 45 L154
[3] Wagner B P et al. 2006 *J. Crystal Growth* **290** 504
[4] T. Tanikawa et al. 2008 *Phys. Status Solidi C* **5** 2966
[5] Yoshihisa Abe et al. 2009 *Materials Science Forum* **600-603** 1281-1284
[6] Sawaki N and Honda Y 2011 *China Tech. Sci.* **54**: 38–41.
[7] Kukushkin S A, Osipov A V, Feoktistov N A Patent No 2008102398, filed on 22 January 2008
[8] Kukushkin S A and Osipov A V 2013 *J. Appl. Phys.* **113** 024909
[9] Romanov A E et al. 2006 *J. Appl. Phys.* **100** 023522

Cross-sectional TEM preparation of hybrid inorganic/organic materials systems by ultramicrotomy

H Kirmse[1], E Oehlschlegel[1], F Polzer[1], S Blumstengel[2], M Sparenberg[2] and F Henneberger[2]

Humboldt-Universität zu Berlin, Institut für Physik, [1]AG TEM, [2]AG Photonik, Newtonstraße 15, 12489 Berlin, Germany

Email: holm.kirmse@physik.hu-berlin.de

Abstract. Preparation of hybrid inorganic-organic systems (HIOS) for transmission electron microscopy (TEM) in cross sectional view is the key for understanding the interfacial structure. Strikingly different materials properties like hardness, cleavability and heat sensitivity limit the number of applicable preparation strategies. Successful preparation of a HIOS system combining ZnO and para-sexiphenyl (6P) is realized by ultramicrotomy. It is shown that the alignment of the cutting plane with respect to the (0001) cleavage plane of ZnO plays a decisive role for successful preparation of extended TEM lamellae and the preservation of the HIOS structure. In particular, for (0001) oriented ZnO substrates the optimum cut direction is parallel to the HIOS interface. In cross-sectional high-resolution TEM images (100) lattice planes of 6P are observed proving the appropriate preparation strategy.

1. Introduction

Both, inorganic semiconductors and conjugated molecules are subject of intense fundamental research. The individual systems, i.e. inorganic semiconductors or conjugated organic molecules, exhibit their own optoelectronic properties permitting numerous applications in everyday life. The combination of the two materials systems promises superior performance compared to the isolated system. The interface between both components plays a key role for the final performance. Access to the structure of the interface is provided by cross-sectional transmission electron microscopy (TEM). For TEM, appropriate samples have to be prepared. It was shown that the structure of the organic component embedded in a hybrid system can be preserved by conventional preparation with final broad beam Ar ion milling assisted by liquid nitrogen cooling [1]. The application of this strategy to our system ZnO/para-sexiphenyl (6P)/ZnO resulted in a structure shown in figure 1. Although atomic force microscopy (AFM) evidences that the 6P layer has been deposited in a well-ordered structure [2] no horizontally aligned (100) lattice planes of 6P are visible in the high-resolution TEM (HRTEM) image. The 6P region shows typical contrast of an amorphous material. Obviously, the crystalline structure of 6P was destroyed, presumably by energy transfer from the impinging Ar ions to the molecules. Hence, an alternative preparation strategy has to be identified excluding ion milling. Strategies utilizing chemical or electrochemical etching are excluded too since these methods presume chemical similarity of the individual layers which is not fulfilled for HIOS. For the preparation of soft matter, ultramicrotomy (UM) is a well-established method. Applications of UM to the preparation of hard matter were reported too [3], whereas preparation of combined hard and soft matter as functional material is still lacking. In this paper, the UM preparation of HIOS for cross-sectional TEM investigations is described. The results are discussed with special focus on both, the crystallographic orientation of the ZnO substrate and the alignment of the UM cutting plane relative to the cleavage plane of the substrate.

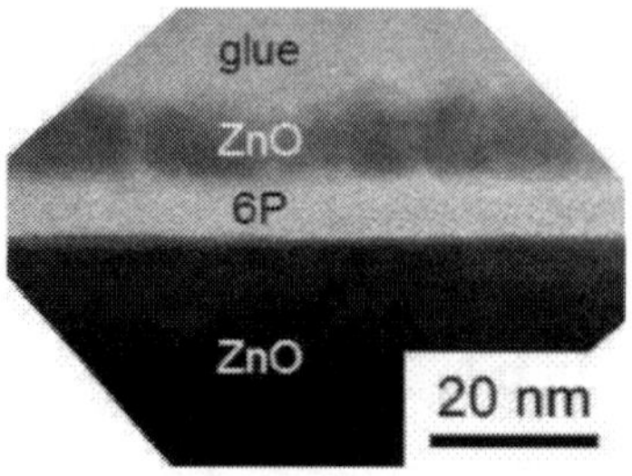

Figure 1. HRTEM image of the HIOS structure ZnO/6P/ZnO. Last step of TEM preparation was Ar ion milling combined with cooling by liquid nitrogen.

2. Samples and instrumentation

The HIOS structures were grown in a tandem molecular beam epitaxy (DCA450) equipped with separate growth chambers for 6P and ZnO. ZnO buffer layer and top layer are grown employing the standard and low-temperature epitaxial regime, respectively, described in detail elsewhere [4]. 6P growth is performed in the second chamber equipped with Knudsen type effusion cells. For growing the ZnO top layer, the substrate is kept at 100°C in order to maintain the structure of 6P. To control the growth direction of ZnO as well as the 6P film morphology, the hybrid stacks are fabricated on either $(01\bar{1}0)$ or (0001) oriented ZnO substrates. The set of samples grown for systematic studies of UM preparation is given in figure 2. The magnified insets give the projected atomic structure.

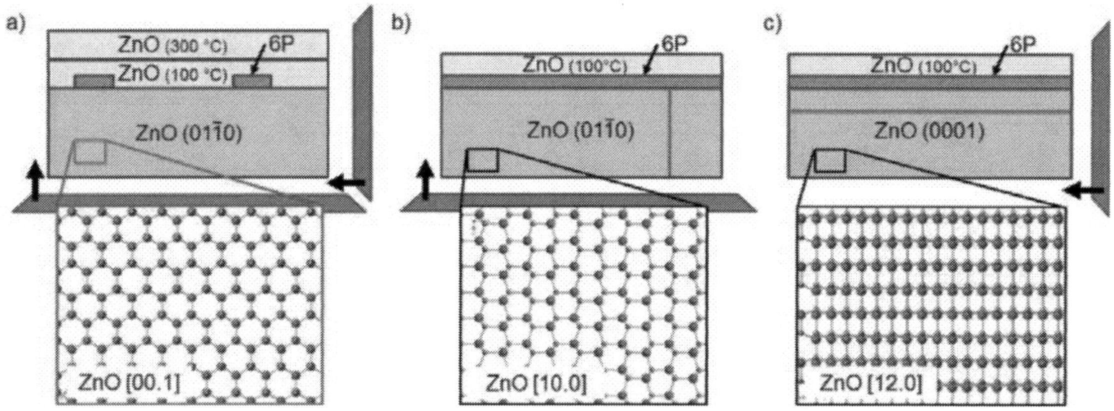

Figure 2. HIOS samples prepared by UM (a-c). The enlarged structural motives shown at the bottom represent the atomic arrangements with respect to the cutting direction (bold arrows) of the UM knife sketched in blue. In addition, the cleavage plane of ZnO is given in red.

For preparation of the samples for TEM a LEICA UltraCut7 room temperature ultramicrotome was utilized. The angle of the UM diamond knife was 35° and the cutting speed was set to 0.1 mm/s. Lamellae were transferred utilizing a magic loop tool from the water trough of the UM knife onto holey carbon film hydrophilized by glow discharge.

TEM imaging was performed in a JEOL JEM2200FS equipped with in-column energy filter. For enhancing the contribution of the organic component to the image contrast, a 10 eV energy slit was inserted selecting the zero-loss electrons and blocking those suffering electron losses of the carbon plasmon at about 25 eV. High-resolution TEM imaging was performed at an underfocus of 1000 nm.

3. Results and discussion

In order to systematically study the success of UM preparation, several alignments of the UM cutting plane, cutting direction, and cleavage plane of the ZnO substrate were investigated. In all cases face-to-face glued samples were used ensuring symmetric geometry of the sample for the cutting process.

In figure 3, the results of UM cutting *parallel* to the (0001) cleavage plane of ZnO are given (cf. figure 2a). When UM cut is performed *along* the inorganic/organic interface the HIOS is preserved only incompletely (cf. figure 3a). In particular, the ZnO columns grown atop 6P were partly removed

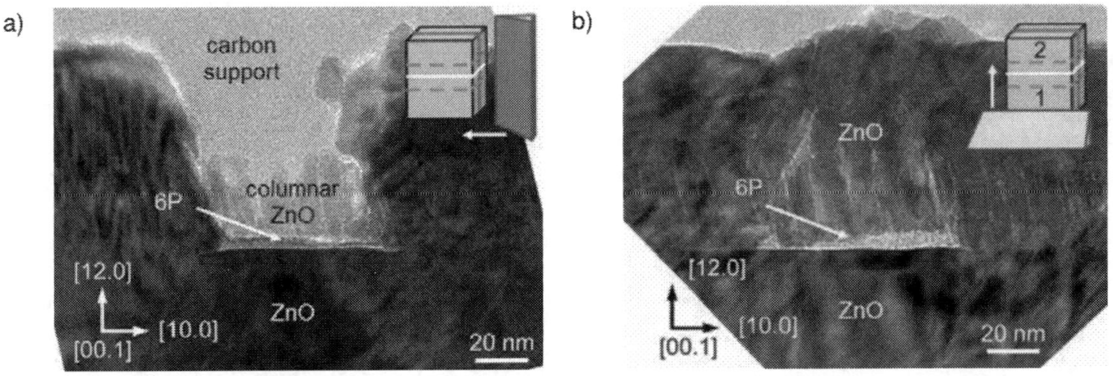

Figure 3. HIOS after UM preparation with cutting plane *parallel* to cleavage plane marked by red line. The direction of UM cut (see white arrows) was aligned a) *along* and b) *normal* to the HIOS interface.

due to shear forces acting normal to the column axis. Consequently, the apparent specimen thickness in this region is reduced. Moreover, the upper part of the ZnO layer grown atop 6P is completely missing. The inset of figure 3a shows schematically the face-to-face glued specimen and both the position of the UM knife and the cutting direction. The cleavage plane is marked in red.

In contrast to the alignment chosen for figure 3a, the HIOS is fully preserved when the cutting direction is aligned normal to the HIOS interface (figure 3b). There are two different scenarios for a face-to-face glued sample. When the UM cut progresses from the ZnO substrate towards the HIOS side of the sample the HIOS components are stretched and initial structure is not preserved (cf. area 1). On the opposite side of the sample (area 2), i.e. after crossing the glue line (see the white line in the inset of figure 3b), the HIOS components are kept attached to each other. As a result of the forces acting during UM cut, the lamellae are separated at the glue line into the initial sample parts. Moreover, the lamellae are fragmented into needles with their axis aligned parallel to the edge of the UM knife. Concerning preservation of the 6P structure one would expect to see (100) lattice planes aligned parallel to the [12.0] direction of ZnO. However, no indication for this is found. This can be attributed to the particular MBE growth procedure (cf. figure 2a). The growth temperature of 300 °C applied during the second step of ZnO overgrowth exceeded the critical temperature for amorphization of 6P.

Figure 4 shows results of a second UM preparation applied to the samples sketched in figure 2b) and c). In both cases the cleavage plane of ZnO was aligned *normal* to the UM cutting plane. It has to be pointed out that for both set-ups complete lamellae were gained and no fragmentation appeared. Figure 4a represents the result for UM cutting *normal* to the HIOS interface. Due to the nominal 6P layer thickness of 1 monolayer a visibility of (100) lattice planes cannot be expected. Nevertheless, the

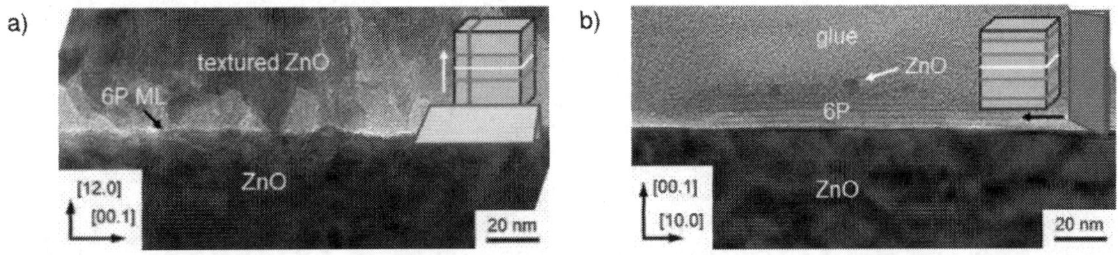

Figure 4. HIOS after UM preparation with cutting plane *normal* to cleavage plane (marked by red lines). The cutting direction was aligned *in-plane* to the cleavage plane (see white arrow in a) and black arrow in b)). Owing to the different substrate orientation the direction of UM cut was aligned a) *normal to* and b) *along* the HIOS interface.

correlation between ZnO substrate and the ZnO top layer was maintained. The presence of 6P is indicated by both, the ZnO top layer exhibiting textured structure as revealed by electron diffraction and the higher intensity at the HIOS interface in the TEM image due to weak interaction between organic material and transmitted electrons.

Full preservation of the crystalline structure of 6P is evident from figure 4b. Here, the UM cut was performed parallel to the HIOS interface. A 6P island of a height of about 20 nm is visible in contact with the (0001) oriented ZnO substrate. Lattice planes are clearly recognized within the 6P.

In order to understand the arrangement of 6P molecules at the HIOS interface an intensity profile is taken from figure 4b (see solid line in figure 5). The distance measured for the second, third, and fourth lattice plane amounts to about 2.5 ± 0.1 nm, which corresponds to the (100) lattice planes of 6P. The interpretation of the first lattice plane having a thickness of 1.6 nm is not straightforward. This becomes obvious when a second profile taken from an in-focus image of the same region is compared (see dotted line in figure 5). On the one hand the increase of contrast for large defocus is nicely reproduced. But on the other hand, a shift of the position of the lattice planes is observed. Since the HRTEM imaging process is strongly affected by defocussing, any direct interpretation of the interfacial structure becomes impossible. In order to assess the real interface structure, exit wave reconstruction has to be performed based on defocus series. The exit wave of the object can be obtained by back propagation of the image wave taking into account the contrast transfer function for the applied HRTEM imaging conditions.

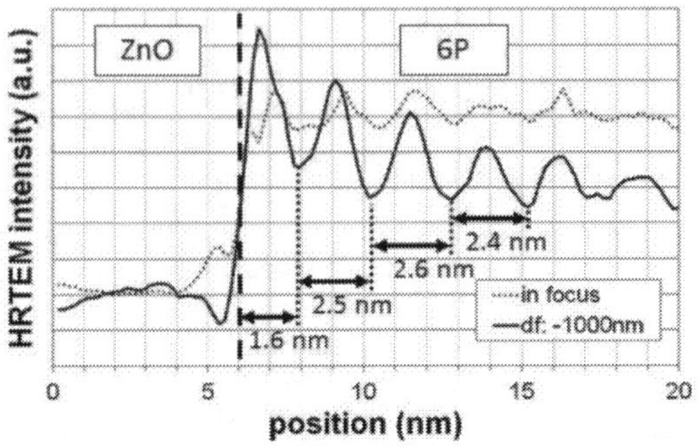

Figure 5. HRTEM image intensity profiles across the HIOS interface (dashed line) taken from figure 4b (solid) and from an image acquired at in-focus conditions (dotted).

Looking back to figure 4b, atop 6P nano-crystallites of the non-continuous ZnO cap layer are seen. The amorphous glue is still in contact with the layers indicating both, appropriate UM cutting procedure and preservation of the initial arrangement of the HIOS components.

4. Conclusions

UM is an appropriate method to prepare cross-sections of HIOS for TEM investigation. In comparison to ion milling based methods, UM is realized without thermal treatment, thus any modification of the organic component upon heating is ruled out. In order to achieve optimum TEM samples the alignment of both, UM cutting plane and cutting direction with respect to the cleavage plane of the substrate have to be considered. As long as the cutting direction is aligned in-plane to the cleavage plane of the ZnO substrate extended lamellae are gained. For samples with the HIOS interface oriented parallel to the (0001) plane of ZnO both, the structure of 6P and the contact to the substrate were preserved. 6P is deposited as a crystalline layer onto (0001) ZnO substrate with the (100) lattice planes parallel to the ZnO surface.

Acknowledgement

The work was financially supported by the German research foundation in the framework of the collaborative research centre SFB951 "HIOS".

References

[1] Dürr AC, Schreiber F, Kelscha M and Doscha H 2003 *Ultramicroscopy* **98** 51
[2] Blumstengel S, Glowatzki H, Sadofev S, Koch N, Kowarik S, Rabe JP and Henneberger F 2010 *Phys. Chem. Chem. Phys.* **12** 11642
[3] Glanvill S R 1995 *Micr. Res. Technique* **31** () 275
[4] Blumstengel S, Sadofev S, Kirmse H, Henneberger F 2011 *Appl. Phys. Lett.* **98** 031907

Journal of Physics: Conference Series **471** (2013) 012035 doi:10.1088/1742-6596/471/1/012035

The structure and electronic properties of copper iodide 1D nanocrystals within single walled carbon nanotubes

N A Kiselev[1], A S Kumskov[1], V G Zhigalina[1], N I Verbitskiy[2], L V Yashina[3], A L Chuvilin[4,5], A L Vasiliev[1,6], A A Eliseev[2]

[1] Institute of Crystallography RAS, 119333 Moscow, Russia
[2] Department of Materials Science, Moscow State University, 119992 Moscow, Russia
[3] Chemistry Department, Moscow State University, 119992 Moscow, Russia
[4] CIC nanoGUNE Consolider, E-20018 Donostia-San Sebastián, Spain
[5] IKERBASQUE, Basque Foundation for Science, 48011, Bilbao, Spain
[6] NRC "Kurchatov Institute", 123182, Moscow, Russia

E-mail: kiselev@ns.crys.ras.ru

Abstract. Copper iodide one-dimensional nanocrystals within single walled carbon nanotubes (1D CuI@SWCNTs), i.e. meta-nanotubes [1], were investigated by high resolution electron microscopy (HRTEM). In meta-nanotubes of diameter D_m = 1.3–1.4 nm produced by arc-discharge (AD) method close-packed hexagonal or deformed cubic 1D crystal anion sublattices were observed with cations in octahedral or tetrahedral positions. These two sublattices reversibly transform to one another. In catalysed chemical vapour deposition (CCVD) meta-nanotubes of diameters D_m = 1.5–2.0 nm cubic anion sublattices are formed. For diameters ≥2.0 nm three-dimensional (3D) crystallization is observed.

1. Experiment

1D CuI@SWCNTs were obtained using capillary filling of SWCNTs by CuI melt. HRTEM was performed in an FEI Titan 80-300 and a Titan 60-300 at 80 kV with aberration correction. Image simulations were performed with the multislice software of Gomez and Beltran [2]. The role of guest compounds on the electronic structure of meta-nanotubes was studied by X-ray photoelectron spectroscopy (XPS), near edge X-ray absorption fine structure (NEXAFS) and Raman spectroscopy. The nanocrystals were grown within nanotubes obtained by AD method [3] and by CCVD [4].

1.1 1D CuI@SWCNT (arc discharge, mean diameter D_m = 1.3–1.4 nm)

The 1D CuI crystals are characterized by a close-packed hexagonal (Figures 1 and 2) or a deformed cubic (Figure 3) anion sublattice. As was shown for the 1D CuI@SWCNT and for the 1D CuBr@SWCNT these two types of anion sublattices reversibly transform one to another inside the nanotube [3]. Copper cations can occupy octahedral positions or can be statistically distributed across two tetrahedral positions. The type of anion sublattice will be denoted in a shorthand form, e.g. hex for hexagonal, "o" for octahedral and "t" for tetrahedral cation positions, crystal model orientation 0° or 90° ((110) and (111) for cubic structure, correspondingly). In spite of a similar anion sublattice, the cation distribution introduces significant differences into the image. There are images that can be interpreted only with "o" cations (Figures 1, 3(a)) or only with "t" cations (Figure 2).

The image in Figure 3 (a) corresponds to the (hex (o), 0°) sublattice. In Figure 3 (b) a part of this image transfers into a chain of hexagons typical, from image simulation, of the (cubic (t), (111)) sublattice. For the cubic modification an essential compression along <111> transverse to the nanotube axis and elongation along the nanotube axis is observed because of the restricted inner channel diameter. The deformed cubic structure in the (110) and (111) projections is similar in appearance to a monoclinic one.

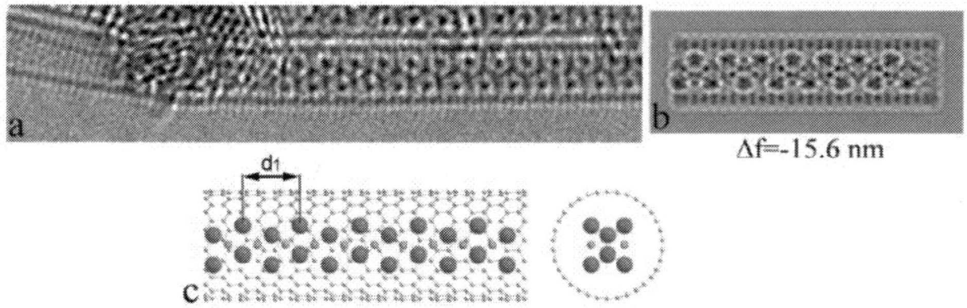

Figure 1. 1D CuI@SWCNT (17, 0) nanocomposite in hexagonal modification. (a) –(hex(o), 0°) image type and (b) – image simulation corresponding to "o" cation positions. In the left part of (a) a fragment of empty SWCNT (17, 0) is visible. (c) – the model for nanocomposite with copper cations located in "o" positions. d_1 = 0.72–0.73 nm.

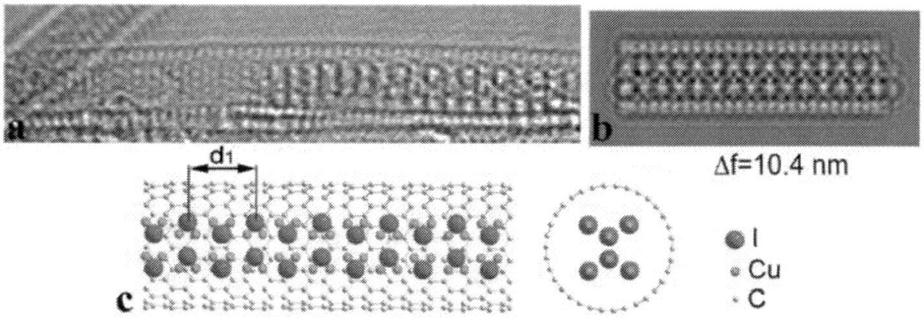

Figure 2. 1D CuI@SWCNT (17, 0) nanocomposite in hexagonal modification. (a) – (hex(t), 0°) image type; (b) – nanocomposite image simulation corresponding to "t" cation positions; (c) – the model.

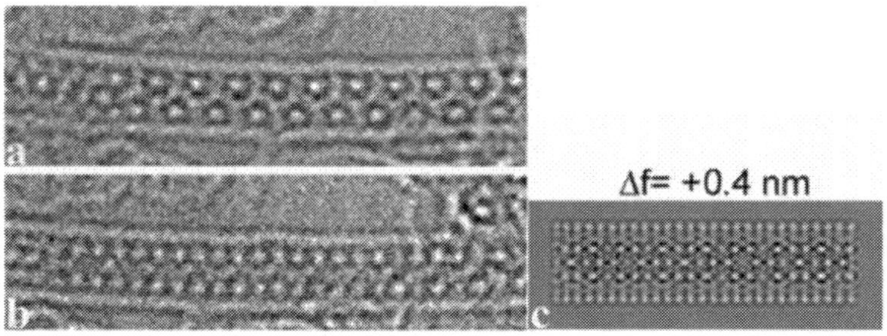

Figure 3. The 1D CuI@SWCNT(17,0): the (hex(o), 0°) sublattice in (a) is transformed into (cubic(t), (111)) (b); c – image simulation.

1.2 1D CuI@SWCNT (CCVD, D_m = 1.5–2.0 nm)

For the SWCNT (CCVD) with above mentioned diameter (D_m) only a cubic anion sublattice was observed. The first type of images is characterized by a hexagonal net of "subunits" (Figure 4a,c). The second type of images revealed an arrangement of "subunits" close to a square net (Figure 4a,b). According to the model and image simulation CuI crystallizes in the form of the 1D crystals with zinc blende or rock salt type unit cells connected by [001] edges and translated along <110>. For the first type of images the periodicity d_1 = 0.425 nm and the angle α = 62°. For the second type of images d_2 = 0.425 nm, β = 80°. Copper cations can reversibly transfer from tetrahedral to octahedral sites. According to the image simulation the 1D crystal cubic unit cell is 16.7% compressed along <110> axis in comparison with the 3D bulk structure.

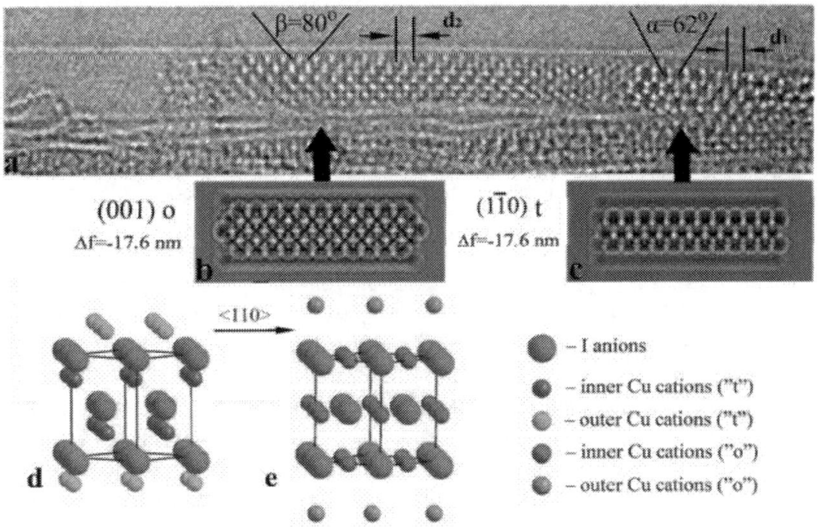

Figure 4. 1DCuI@SWCNT (CCVD, (13,13)): a – HRTEM of meta-nanotube; b – image simulation for the (001) projection with cations located in "o" positions; c – image simulation for the $(1\bar{1}0)$ projection with cations located in "t" positions; d – the computer model of the CuI crystal with "t" cations (ZnS type); e – the computer model of the CuI crystal with "o" cations (NaCl type). The <110> direction coincides with the nanotube axis.

1.3 3DCuI@SWCNT (CCVD, $D_m \geq 2.0$ nm)

3D CuI@SWCNTs were generated (Figure 5). The images are observed with motifs typical for the first or the second image type of the 1D CuI@SWCNT (D_m = 1.5–2.0 nm) images, but in comparison with the 1D crystal some additional rows of "subunits" are observed and 3D growth appeared. At the same time unit cells are oriented in the same manner as in the 1D CuI.

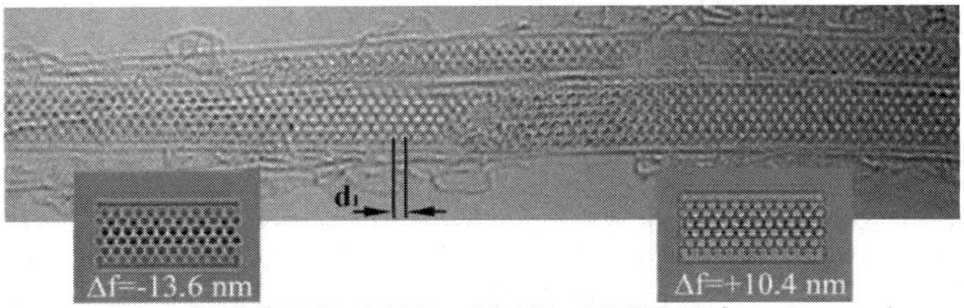

Figure 5. The 3D CuI@SWCNT (CCVD, (18,18)): HRTEM of meta-nanotube and image simulations for the $(1\bar{1}0)$ projection with cations located in "t" positions.

2. Electronic properties

The crystals of CuI@SWCNT exhibit an acceptor behaviour independent of nanotube diameter, as supported by XPS and Raman spectroscopy [4, 5]. In both samples large shifts of G mode (+8–15 cm^{-1}) towards higher frequencies were observed, indicating charge transfer from the nanotube wall to the guest crystals [6, 7]. Moreover, the effect is increasing with nanotube diameter, which was proved earlier to be a result of increasing number of CuI-SWCNT contacts per carbon atom of SWCNT. To clarify the nature of the charge transfer between CuI and SWCNT NEXAFS spectra of composites at the C 1*s* edges were obtained in the total electron yield mode by detecting a sample current. The main structures in the spectra reflect C 1*s* electron transitions to empty electron states of C $2p_z\pi$ and C $2p_{x,y}\sigma$ character [8]. The spectra of pristine SWCNTs and CuI@SWCNTs are practically identical except a shoulder-like peak at the low-energy side of the π-resonance (Figure 6).

The energy shift of this spectral feature ~1.2 eV below the π peak maximum exceeds significantly Fermi level downshift in composites [5], which indicates that it cannot originate from a direct SWCNT → CuI charge transfer, but is a result of orbital hybridization with a formation of new localized states. Taking into account symmetry of bonding and antibonding orbitals one can expect the new localized state to form significantly lower Fermi level of composite, causing depletion of valence electrons to this state and Fermi level downshift. This effect obviously increases the work function of the composite observed earlier [5, 9]. To clarify the origin of this new localised state a reference sample of CuI sputtered onto purified SWCNTs was prepared *in situ* in the preparation chamber of XPS spectrometer by thermal evaporation of dehydrated CuI powder (Aldrich) from an effusion cell in a vacuum of $\sim 3\times10^{-7}$ mbar. The NEXAFS spectrum of the reference sample illustrates no additional features near π-resonance, indicating no chemical interaction between SWCNT walls and CuI nanocrystals formed on the external surface (Figure 6). Thus, one can conclude that the localized state in CuI@SWCNTs occurs due to dimensionality and small size of the intercalated nanocrystals. Most probably dangling bonds and coordination insaturation of the CuI nanocrystal play a crucial role in CuI-SWCNT interaction, leading to coordination of nanocrystal by SWCNT carbon atoms through new bond formation.

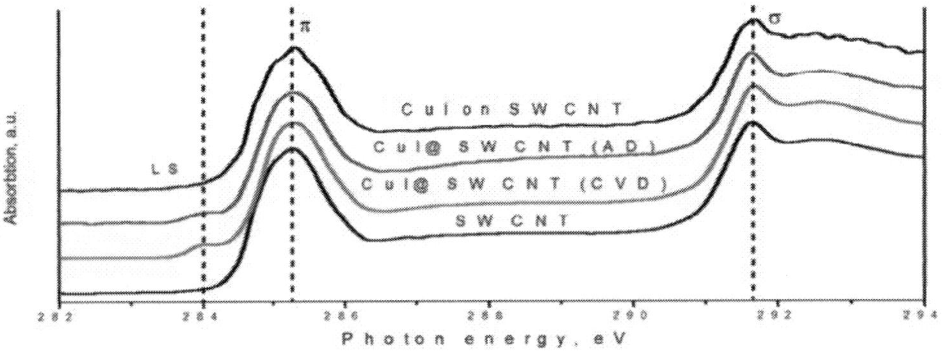

Figure 6. NEXAFS spectra of CuI@SWCNTs in comparison with pristine SWCNTs and a reference sample of CuI deposited onto SWCNTs.

References

[1] Sloan J, Monthioux M 2012 In *"Carbon Meta-nanotubes: Synthesis, Properties and Applications"*, ed. M Monthioux (Chichester: Wiley-Blackwell) p 225
[2] Gomez R A *et al* 2001 *Rev. Latin Am. Met. Mat.* **21** 46
[3] Kiselev N A *et al* 2012 *Journal of Microscopy* **246** 309
[4] Kumskov A S *et al* 2012 *Carbon* **50** (12) 4696
[5] Eliseev A A *et al* 2012 *Carbon* **50** (11) 4021
[6] Eliseev A A *et al* 2009 *Russian Chemical Reviews* **78** 833
[7] Corio P *et al* 2004 *Chemical Physics Letters* **383** 475
[8] Generalov A V *et al* 2010 *Fullerenes, Nanotubes and Carbon Nanostructures* **18** 574
[9] Eliseev A A *et al* 2010 *Carbon* **48** (10) 2708

Electron microscopy of the effect of heat treatment on thestructure of metal containing nanocomposites with silicon- carbon matrix

M Presniakov[1,2], A Popov[1], A Vasiliev[2,3]

[1]National Research University «Moscow Power Engineering Institute», Krasnokazarmennaya 14, Moscow, 111250, Russia
[2]National Research Centre «Kurchatov Institute», Akademika Kurchatova pl. 1, Moscow, 123182, Russia
3Shubnikov Institute of Crystallography, RAS, Leninskiy Prospekt 59, Moscow, 119333, Russia

E-mail: mpresniakov@gmail.com

Abstract. The influence of heat treatment in vacuum and in oxygen on the composition and distribution of tantalum containing nanocomposites in silicon-carbon matrix has been investigated. It is shown that the influence of heat treatment determines the interaction of oxygen with the atoms in the amorphous matrix as well as with nanocrystals of tantalum carbide. We propose a qualitative model of the processes occurring in these interactions.

1. Introduction

Amorphous silicon-carbon films (SCF) and amorphous films of metal nanocomposites within silicon-carbon matrix (MCN) are a class of modern materials with an extremely wide range of functional properties and potential applications [1]. The unique properties of the amorphous matrix, the ability to choose the composition and concentration of nanocrystals allow us to change, for example, the electrical conductivity by 18 orders of magnitude.

2. Materials and methods

Silicon-carbon films were prepared by plasma decomposition of a silicon-containing organic matter, polyphenylmethylsiloxane (PPMS)-$(CH_3)_3SiO(CH_3C_3H_5SiO)_3Si(CH_3)_3$ [2]. To obtain tantalum-based nanocomposites tantalum magnetron sputtering was conducted simultaneously with the growth of the SCF by plasma decomposition of PPMS.

The influence of thermal treatments in vacuum and oxygen atmosphere on the composition and distribution of the nano-sized material was investigated in films about 400 nm thick with an average 70 wt. % of tantalum deposited on a single crystal silicon substrate. Heat treatment of the films in vacuum or in an oxygen atmosphere was carried out at a temperature of 395 °C for 400 min [3]. The samples for transmission electron microscopy (TEM) were prepared by focused ion beam (FIB) with an FEI Helios NanoLab 600. Studies of membranes were made from films as-grown or heat-treated under the above modes, using a Titan 80-300 TEM/STEM (FEI, USA) with energy dispersive X-ray spectroscopy (EDXS).

3. Experimental

3.1 Scanning electron microscopy (SEM) of focused ion beam thinned membrane

The FIB made cross-section (Fig.1) of MCN films gave possibility to study the Ta distribution in the volume of the films and the influence of the anneal temperature.

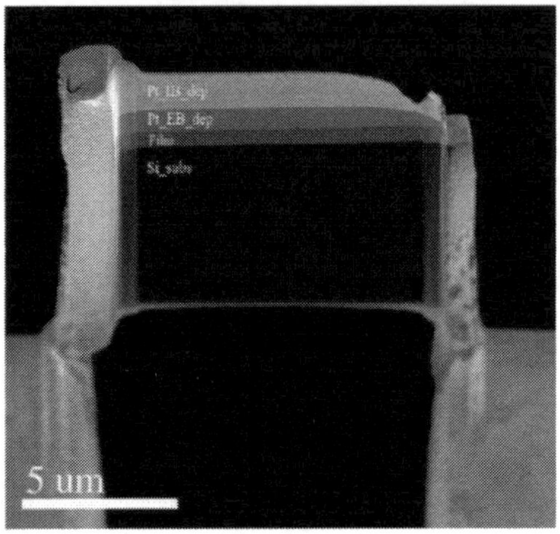

Figure 1. FIB made cross-sectional TEM sample (SEM image)

3.2 Transmission electron microscopy (TEM)

Transmission electron microscopy (Fig.2) shows that the introduction of tantalum leads to the formation of nanocrystals 2 to 6 nm in size over the entire depth of the film, where the lattice parameter, *d*, corresponds to that of TaC.

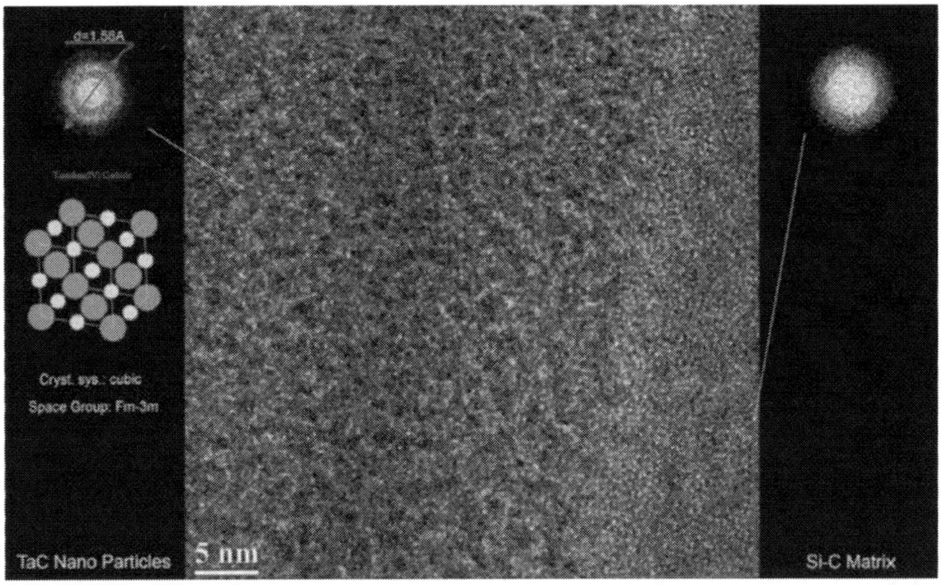

Figure 2. HR-TEM image of TaC nanoparticles (NP) in Si-C amorphous matrix (AM). Inset: FFT from NP and AM.

3.3 Scanning transmission electron microscopy (STEM)

Heat treatment in vacuum has practically no effect on the structure and composition of the films. Another situation is observed after film treatment in oxygen. An area that has a darker contrast in annular dark-field Z-contrast micrographs and does NOT contain any nanocrystals is formed near the free surface of the film. Moreover, the concentration of tantalum in this area is reduced substantially (Fig.3).

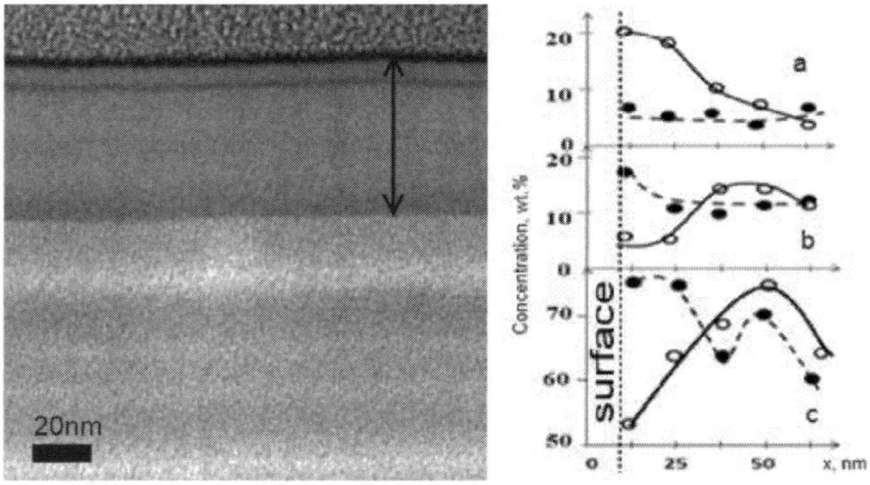

Figure 3. Annular dark-field STEM image of tantalum depletion in the cross-section of the sample after heat treatment in oxygen and plots of distribution of the concentration of oxygen (a), carbon (b) and tantalum (c) in the near surface state of the film after heat treatment in vacuum (dotted line) and in oxygen atmosphere (solid line)

4. Conclusions

Results of the study of the chemical composition and the thickness changes of the near-surface region of a tantalum implanted silicon carbide amorphous film after heat treatment in vacuum and oxygen are shown in Fig.3.

One can see from Fig.3a that the oxygen concentration in the surface region significantly increases as a result of heat treatment in oxygen atmosphere. In contrast, the carbon concentration (Fig.3b) in the surface region is decreased by about half. The change in the concentration of tantalum is more complex (Fig.3c): it is reduced (by about 20 wt. %) in the surface region, and there is a slight increase of Ta in the depth of the layer.

Based on the data a model of the processes occurring during the heat treatment in oxygen MNCs may be proposed. The effect of heat treatment on the structure of tantalum silicon-carbon nanocomposites consists of three stages (Fig. 4):

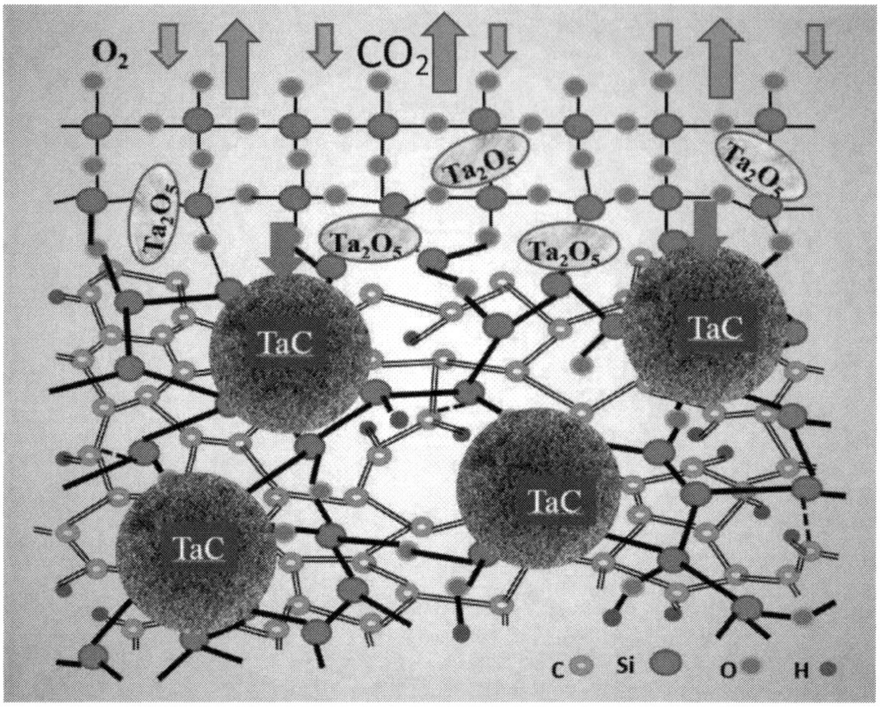

Figure 4. Influence of the heat treatment on oxygen on the MCSCN structure.

1st The interaction of oxygen with carbon atoms at the film surface forms volatile compounds of CO, CO_2 and volatilization of these carbon compounds from the surface region of the structural network.

2nd The interaction of oxygen with carbon and volatilization of formed compounds cannot explain the observed increase in the oxygen concentration in the surface layer. That is why the reaction of oxygen with SiO_x forming SiO_2 and reaction of oxygen with tantalum forming tantalum oxide may be expected. Moreover, since at the surface region there are no nanocrystals it is necessary to assume that the nascent tantalum oxide particles are amorphous. The latter assumption seems very reasonable since it is known that Ta_2O_5 films may persist in the amorphous state at temperatures up to 700 °C.

3rd The remaining tantalum carbide nanocrystals are displaced by the growth boundary of silicon oxide.

Acknowledgement
Authors are grateful to M. Shupegin for sample preparation. This work was supported by RFBR (project N 13-07-00173).

References

[1] Parhomenko Yu V, Malinkovich M D, Skryleva E A and Shupegin M L 2005 *Trans. Materials of Electronic Technology* **3** pp12-16

[2] Belogorokhov A I, Dodonov A M, Malinkovich M D, Parkhomenko J N , Smirnov A P and Shupegin M L 2007 *Sci. Materials of Electronic Technology* **1** pp 69-71

[3] Popov A I, Presnyakov M Y, Tabachkova N Yu, Shupegin M L and Eydelman K B 2012 *Proceedings of the VIII International Conference "Amorphous and microcrystalline semiconductors" (Saint Petersburg: Publishing House of the Polytechnic University)* **F-II** pp 126 - 127

Investigation of growth of thin layers of perovskite on native silicon dioxide by a combination of atomic force microscopy and transmission electron microscopy

A Taghi Khani and T Walther

Department of Electronic and Electrical Engineering, University of Sheffield, Mappin Street, Sheffield S1 3JD, UK

E-mail: a.taghikhani@sheffield.ac.uk, t.walther@sheffield.ac.uk

Abstract. Thin layers of $(Sr,Ba)TiO_3$ perovskite have been grown on native silicon dioxide by pulsed laser deposition at the Technical University of Darmstadt, Germany. Atomic force microscopy (AFM) has been used to investigate the surfaces of the native silicon oxide before and after over-growth by the perovskite in plan-view. Bright-field and dark-field scanning transmission electron microscopy (STEM) in a JEOL 2010F field-emission transmission electron microscope have been combined to investigate the layer stacks of $Si/SiO_2/(Ba,Sr)TiO_3$ in cross-section. The aim is to correlate surface roughnesses in plan-view geometry with interface roughness in cross-sectional geometry, with an emphasis on detecting percolation in the perovskite layers if they approach thicknesses of only a few unit cells.

1. Introduction

Many III-V compound semiconductors such as GaAs, InAs and InSb, have higher electron mobility than Si and also a direct band-gap which makes them more attractive for use in electronic and opto-electronic applications. However, silicon is mainly used in semiconductor devices due to its chemical stability, thermal properties, scalable processing of its oxide and cost. Hence III-V epitaxy on Si has been studied for years.

Because of the mismatch in lattice parameters between different semiconductors (other than AlAs/GaAs) there usually is a high density of dislocations at the interface. One solution to avoid this is to use an intermediate layer between silicon and III-V semiconductor. One choice for the intermediate layer would be a cubic perovskite-type oxide. All perovskites have lattice constants near 4Å which is close to $\sqrt{2}$ times the lattice constant of GaAs ($\approx$ 5.65Å); therefore GaAs could potentially be grown on the perovskite layer by 45° in-plane rotation which could give fewer dislocations compared to direct growth of GaAs on Si [1]. In this work we investigate the growth of thin layers of perovskite on native silicon dioxide before overgrowth by GaAs.

2. Experimental details

Four wafers, each consisting of a thin layer of perovskite grown on Si substrates via the native oxide by pulsed laser deposition, have been analyzed by atomic force microscopy (AFM), ellipsometry and scanning transmission electron microscopy (STEM). Two of the samples have thin layers of $SrTiO_3$

with nominal thicknesses of 2 and 8nm and the other two have layers of $BaTiO_3$ of same nominal thicknesses.

AFM images have been captured by a Veeco Nanoscope V instrument operated in tapping (non-contact) mode. Nanotec's WSxM 5.3 software [2] was used to analyse the AFM images. A VASE® J.A. Woollam Co., Inc. ellipsometer has been used to measure the thickness of the perovskite layers. To investigate the surface of thin perovskite layers and the interfaces between Si and the layers by electron microscopy, cross-sectional samples were prepared by gluing two pieces of wafers, one with nominally 2nm perovskite layer and one with nominally 8nm perovskite layer, face to face, gluing Si backing blocks to each side to make them 3mm thick, cutting the blocks by a diamond saw, grinding and polishing the cross-sectional samples and finally ion milling them in a Fischione 1050 argon ion mill at 4kV and 2mA until electron transparency. The electron microscope used in this study is a JEOL-2010F field-emission transmission electron microscope operated at 197kV, which has a scan unit and bright-field (BF) and annular dark-field (ADF) detectors for scanning transmission electron microscopy (STEM). The microscope is also equipped with a GATAN imaging filter with a charge coupled device (CCD) camera for image acquisition.

3. Results

Figure 1 shows the AFM images of areas of 5×5 μm^2 for each wafer with a thin layer of $BaTiO_3$ (a: 2nm, b: 8nm) or $SrTiO_3$ (c: 2nm, d: 8nm nominal thickness). The acquisition time was 1s for all images which consist of 512 × 512 pixels (yielding a sampling of 9.8nm/pixel).

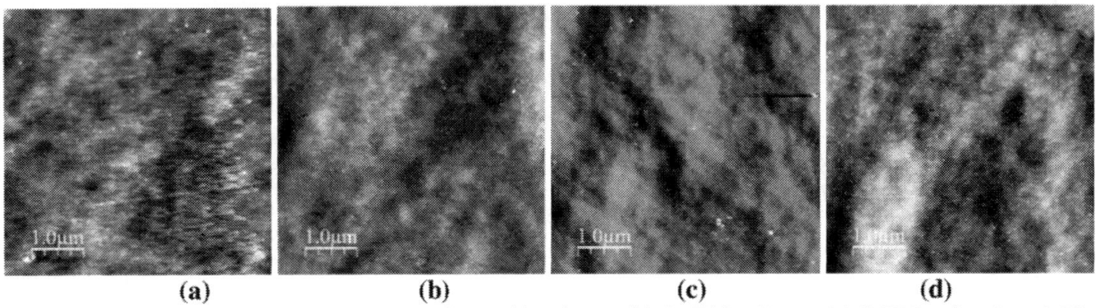

Figure 1. AFM images of 5μm × 5μm **(a)** $BaTiO_3$ (2nm) **(b)** $BaTiO_3$ (8nm) **(c)** $SrTiO_3$ (2nm) and **(d)** $SrTiO_3$ (8nm). The image contrast of black to white corresponds to a height of 10nm in all images.

Figures 2a and 2b are annular dark-field (ADF) and bright field (BF) STEM images of the cross-sectional sample of $BaTiO_3$ acquired with spot size of 0.3 nm and dwell time of 19.5μs. The beam convergence semi-angle is equal to 9.5mrad, the collection angles are equal to 0-10mrad for BF and 55-170mrad for ADF. The differences of contrast along the layer in bright-field confirm that the perovskite layer is not epitaxially grown but poly-crystalline. The field of view of 844nm is sampled onto 4096 × 4096 pixels at a sampling rate of 0.21nm/pixel.

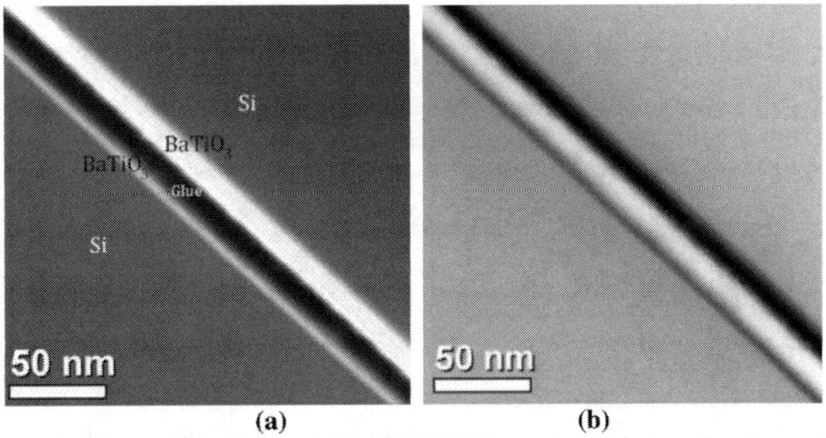

Figure 2. Central regions of **(a)** ADF and **(b)** BF-STEM images of cross-sectional sample of $BaTiO_3$ (2 and 8nm titanate layers appear bright in ADF and dark in BF and show surface roughness).

The corresponding BF and ADF-STEM images of cross-sectional samples of $SrTiO_3$ (2 and 8nm) are shown in figure 3. Again in bright-field, lateral fluctuations in contrast along the $SrTiO_3$ layers can be observed which indicates the layers are poly-crystalline. The field of view of 124nm was scanned with 2048 × 2048 pixels at a dwell time of 19.5μs, which gives a sampling equal to 0.06nm/pixel.

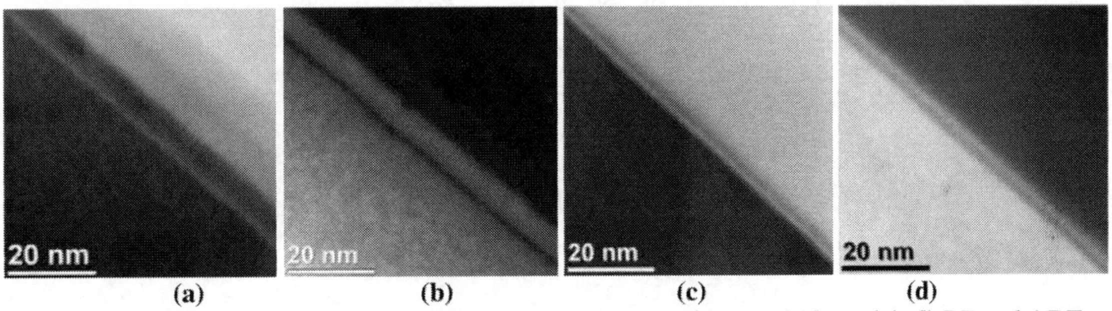

Figure 3. Central parts of **(a,b)** BF and ADF-STEM images of 8nm $SrTiO_3$ and **(c,d)** BF and ADF-STEM images of 2nm $SrTiO_3$.

To obtain the roughness from the STEM images, first the images have been converted to binary format using thresholding, and then a MATLAB code was applied to each image column to work out where the transition from 0 to 1 occurs, which locates the surface of the perovskite layer. Knowing these points for all image columns and calculating a straight line that best fits the data, the difference between the real data and the linear estimate describes the roughness of the surface of the layers. The standard deviation of this data is the root-mean-square (RMS) interface roughness in projection. Figure 4 shows the roughness data acquired from the BF-STEM image shown in Figure 2b for the 8nm BTO as an example.

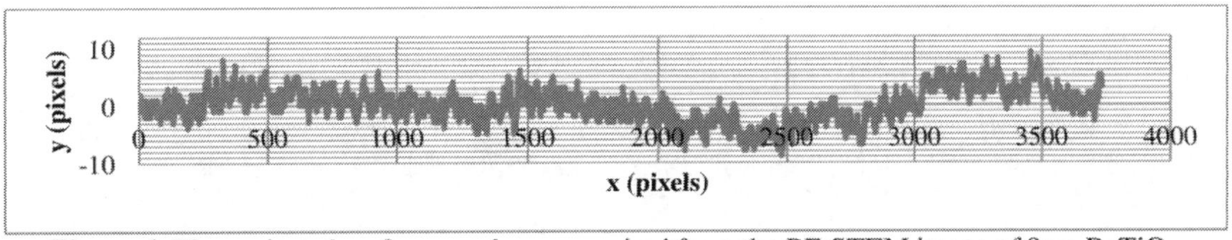

Figure 4. The projected surface roughness acquired from the BF-STEM image of 8nm $BaTiO_3$

Table 1 shows the RMS surface roughness of all four samples measured and calculated from both AFM and STEM images.

Table1. RMS roughness in nm, measured from the AFM images and the STEM images

	$BaTiO_3$(8nm)	$BaTiO_3$(2nm)	$SrTiO_3$(8nm)	$SrTiO_3$(2nm)
AFM	0.61	0.48	0.57	0.55
STEM	0.58	1.09	0.53	0.73

The surface roughness measured from AFM images and STEM images don't agree with each other for very thin layers. As STEM measures the interface roughness as projected along the sample thickness, small protrusions or indentations will be smeared out, therefore the roughness value from STEM images should be always smaller than the true values. Similarly, for very thin layers (2nm) the grains are probably narrower than the tip radius and also smaller than the sampling, and integration over several grains will also reduce the apparent surface roughness.

Using an ellipsometer the thicknesses of perovskite layers for all four wafers have been measured. Table 2 compares the thicknesses of the perovskite layers measured by ellipsometry and STEM.

Table2. Thickness of perovskite layer in nm, measured by ellipsometry and from the STEM images

	$BaTiO_3$(8nm)	$BaTiO_3$(2nm)	$SrTiO_3$(8nm)	$SrTiO_3$(2nm)
STEM	10.32	2.68	5.92	1.82
ellipsometry	9.22	2.31	5.59	2.39

The thicknesses of the perovskite layers measured by ellipsometry roughly agree with the measurements by STEM. Thicknesses of both $BaTiO_3$ layers and the nominally 8nm $SrTiO_3$ layer are ~10% larger than ellipsometry suggests, which may be due to uncertainties in the complex dielectric function of the perovskite.

4. Conclusion

The thicknesses and surface roughnesses of thin layers of poly-crystalline (Sr, Ba)TiO_3 grown on silicon substrate via native oxide have been investigated using atomic force microscopy, ellipsometry and scanning transmission electron microscopy. The roughnesses and thickness measured by different methods have been compared. From our data it appears likely that the thinnest perovskite layers of ~2nm thickness have an RMS roughness of at least 0.7-1.1 nm, which is ⅓ - ½ of the layer thickness and indicates that the layers will probably not be completely continuous and might have pin-holes

Acknowledgement
The authors would like to thank Prof. L. Alff and Dr. J. Kurian from TU Darmstadt for pulsed laser deposition of the thin perovskite layers. Thanks are also due to Dr. K. Kennedy and Mr. R. Frith from the National Centre for III-V Technologies of the University of Sheffield for their help with AFM and ellipsometry data acquisition.

References

[1] Liang Y, Kulik J, Eschrich TC, Droopad R, Yu Z and Maniar P 2004 *Appl. Phys. Lett.* **85,** 1217
[2] Horcas I, Fernandez R, Gomez-Rodriguez JM, Colchero J, Gomez-Herrero J and Baro AM 2007 *Rev. Sci. Instrum.* **78**, 013705

Journal of Physics: Conference Series **471** (2013) 012038 doi:10.1088/1742-6596/471/1/012038

Origin of a-plane (Al,Ga)N formation on patterned c-plane AlN/sapphire templates

A Mogilatenko[1,2], H Kirmse[2], S Hagedorn[1], E Richter[1], U Zeimer[1], M Weyers[1], G Tränkle[1]

[1]Ferdinand-Braun-Institut, Leibniz-Institut für Höchstfrequenztechnik, Gustav-Kirchhoff-Str. 4, 12489 Berlin, Germany
[2]Humboldt University of Berlin, Newtonstr. 15, 12489 Berlin, Germany

anna.mogilatenko@fbh-berlin.de

Abstract. a-plane (Al,Ga)N layers can be grown on patterned c-plane AlN/sapphire templates with a ridge direction along $[1\bar{1}00]Al_2O_3$. Scanning nanobeam diffraction reveals that the formation of a-plane layers can be explained by nucleation of c-plane (Al,Ga)N with $[11\bar{2}0]$(Al,Ga)N ‖ $[0001]Al_2O_3$ at the ridge sidewalls. Faster growth of the top $(11\bar{2}0)$(Al,Ga)N facet in the vertical direction leads to the overgrowth of c-plane (Al,Ga)N nucleated on the horizontal ridge and trench surfaces. Phase separation into binary GaN and AlN takes place during the first growth stages. However, this fades out and does not influence the composition of the final thick a-plane (Al,Ga)N layer.

1. Introduction

Most of today's III-nitride based device heterostructures are grown along the c-axis direction of the wurtzite structure (so called c-plane layers). However, optoelectronic devices obtained on the basis of c-plane nitrides suffer from spontaneous and piezoelectric polarisation fields which appear in the noncentrosymmetric wurtzite structure along the polar c-axis. To overcome this problem growth along non-polar or semipolar directions was proposed [1]. However, this growth usually requires application of a number of different substrates and/or substrate orientations. For example, growth of non-polar a-plane GaN and AlN layers is generally achieved on r-plane sapphire [2]. Recently, we have shown that predominant growth of thick a-plane Al-rich (Al,Ga)N layers is possible by epitaxial lateral overgrowth (ELOG) of patterned c-plane sapphire with ridges arranged along $[1\bar{1}00]Al_2O_3$ [3]. This ridge direction in ELOG of III-nitrides often leads to formation of strongly faceted non-planar surfaces as well as to problems with layer coalescence. This led to the common opinion that this ridge geometry is inappropriate for the fabrication of planar III-nitride layers of high quality. In this study we focus on the orientation and compositional homogeneity of (Al,Ga)N layers during initial growth stage on c-plane AlN/sapphire templates patterned along this $[1\bar{1}00]Al_2O_3$ direction.

2. Experimental details

In our experiment a 15 µm thick $Al_xGa_{1-x}N$ layer with a nominal composition of $x = 0.95$ was grown using hydride vapour phase epitaxy (HVPE) and ELOG on a patterned c-plane template consisting of a 600 nm thick AlN buffer layer on sapphire (see [3] for details). The ridges in the c-plane AlN/sapphire template were oriented along the $[1\bar{1}00]Al_2O_3$ direction, which results in a-plane (i.e.

(11$\bar{2}$0)Al_2O_3) trench sidewalls. Secondary electron microscopy (SEM) and scanning transmission electron microscopy (STEM) including high-angle annular dark-field (HAADF) imaging, energy dispersive X-ray spectroscopy (EDXS) and scanning nanobeam diffraction (SNBD) have been applied to understand the layer structure. The ASTAR/DigiSTAR system of NanoMEGAS has been used for mapping of crystallite orientation and phase in STEM. The crystallographic orientation maps were obtained by comparison of experimental electron diffraction patterns to a set of patterns calculated for different phases.

3. Experimental results

Figure 1a shows a cross-sectional SEM image of an a-plane (Al,Ga)N layer grown in our experiment. The a-plane orientation was confirmed by electron backscatter diffraction as published elsewhere [3]. During the growth there is simultaneous (Al,Ga)N nucleation on the horizontal c-plane surfaces, i.e. on the AlN/sapphire ridges (region 1) and in the trenches (region 2), as well as on the vertical ridge sidewalls (regions 3, 4). In order to understand the formation of a-plane layers on patterned c-plane AlN/sapphire templates we have analysed the layer structure during the first growth stages. Figure 1b shows an exemplary HAADF STEM image of a [1$\bar{1}$00]-oriented Al_2O_3 ridge with AlN template on top, which was overgrown by Al-rich (Al,Ga)N using HVPE. The HAADF STEM micrograph shows a non-uniform intensity distribution within the (Al,Ga)N layer with bright regions of triangular shape appearing at the ridge sidewalls. EDXS mapping suggests a preferential Ga accumulation at these positions (figure 1c). As result, there is no Ga incorporation into the (Al,Ga)N layer at the beginning of the growth, and the ridges are covered with binary AlN. The formation of ternary (Al,Ga)N is delayed most probably until the ridge sidewalls are completely overgrown by AlN. Notice that EDXS suggests the formation of binary AlN on top of the AlN template as well as in the trenches. However, there is an inhomogeneous intensity in the HAADF image with regions growing on the horizontal surfaces (regions 1, 2) appearing darker than regions growing on ridge sidewalls (regions 3, 4). We attribute these differences in intensity to different channelling conditions, which can appear in sample regions with the same composition, but different orientation.

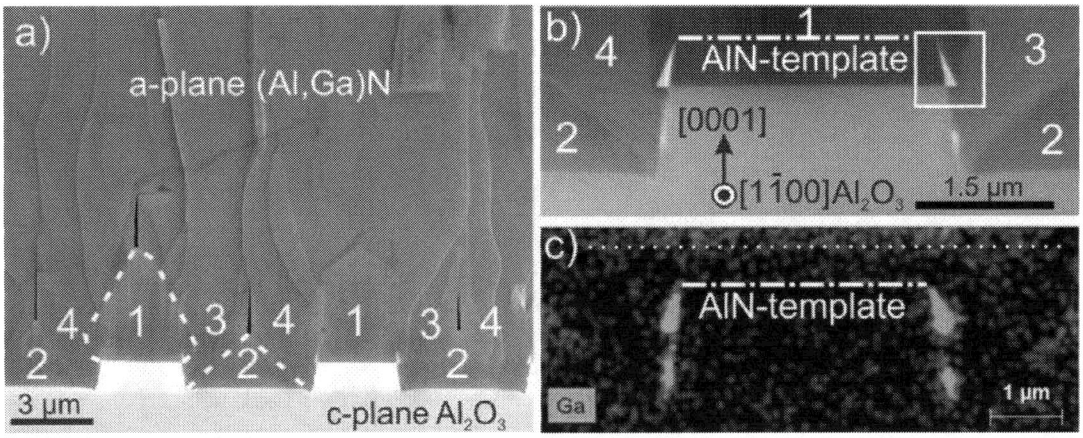

Figure 1. a) Cross-sectional SEM micrograph of (Al,Ga)N layer grown on patterned c-plane sapphire with ridge direction along [1$\bar{1}$00]Al_2O_3; b) HAADF STEM of an overgrown AlN/Al_2O_3 ridge; c) EDXS mapping of Ga-K_α peak intensity in the region shown in b). The white dash-dotted line indicates the interface between the AlN template and the (Al,Ga)N layer grown on top, whereas the white dotted line indicates the delayed onset of Ga incorporation into the growing (Al,Ga)N layer. The white rectangle in in b) marks the area for further SNBD analysis.

SNBD has been applied to prove this suggestion. Figures 2b and c show exemplary SNBD maps of a part of a [1$\bar{1}$00]-oriented Al_2O_3 ridge with AlN template on top (see the rectangle marked in figure

1b). The analysed sample region is shown in a virtual bright-field image in figure 2a, which was reconstructed by depicting the intensity of the none-diffracted beam for each pixel of the scan. According to the orientation maps the AlN template shows the classical orientation on c-plane sapphire: $(0001)[11\bar{2}0]$AlN ∥ $(0001)[1\bar{1}00]Al_2O_3$. During the HVPE layer deposition (Al,Ga)N grows with this orientation relation on top of the AlN template (region 1 in figure 1a) and in the trenches (region 2 in figure 1a). Simultaneously, competing growth takes place on the Al_2O_3 ridge sidewalls. The sidewall surface corresponds to the sapphire a-plane. As indicated by the colour zone selectors in the bottom line of figure 2 the layer orientation in this region corresponds to $(0001)[11\bar{2}0]$AlN ∥ $(11\bar{2}0)[0001]Al_2O_3$. This layer orientation is well known from growth experiments on planar a-plane sapphire [4]. Consequently, the formation of an a-plane (Al,Ga)N layer in our experiment can be explained by nucleation of c-plane (Al,Ga)N at the ridge sidewalls of the patterned sapphire substrate (figure 2e).

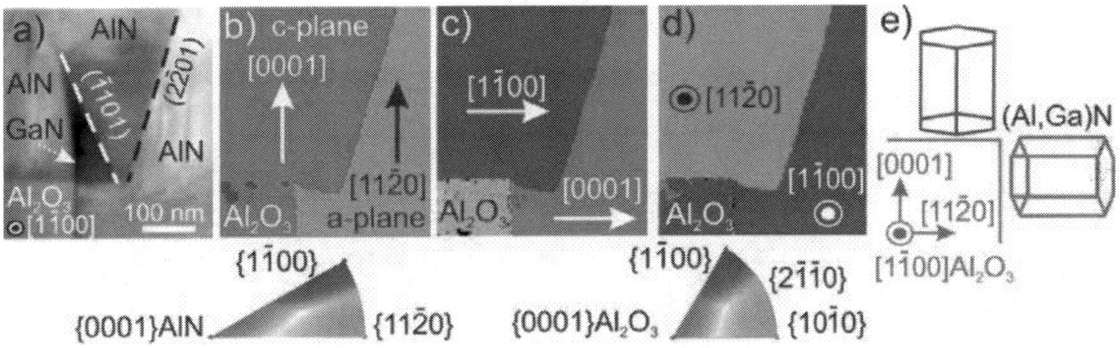

Figure 2. Bright field image (a) reconstructed from a set of electron diffraction patterns obtained by SNBD and crystallographic orientation maps (b), (c) and (d) for three different directions. The images are obtained at the sample region marked by a rectangle in figure 1b. Arrows mark the normal of the identified lattice planes, which can be extracted from colour zone selectors; e) sketch showing the wurtzite unit cell arrangement on a ridge and its sidewall.

Analysis of individual diffraction patterns shows that Ga-rich regions observed by EDXS correspond to wurtzite structure. Consequently, a phase separation into binary GaN and AlN compounds appears at the beginning of the growth. This phase separation can be explained by a preferential agglomeration of Ga atoms on non-polar surfaces. Notice that the a-plane material originates only from the a-plane sapphire sidewalls, whereas the m-plane sidewalls of the AlN template supply nucleation sites for c-plane GaN. These GaN regions contain a high density of basal plane stacking faults (BSFs), which can be explained by a high lattice mismatch between the (0002) lattice plane distances in the AlN template and GaN growing on its m-plane side facet. The presence of BSFs explains the appearance of vertical streaks in $[11\bar{2}0]$ electron diffraction patterns (DPs) of GaN (figure 3b), whereas no streaks were observed in DPs of the AlN template (figure 3c).

Analysis of crystal faceting can supply important information on the evolution of the growth fronts. As visible in figure 2a, the GaN region formed on the m-plane sidewall of the AlN template develops facets which are close to the $\{\bar{1}101\}$ facets generally observed during growth of c-plane III-nitride layers in ELOG experiments with this pattern geometry. However, GaN is subsequently overgrown by c-plane AlN growing on the top of the AlN template and propagating into the lateral direction with formation of $\{2\bar{2}01\}$ facets. The observed crystallite shape suggests that the overgrowth of c-plane GaN by c-plane AlN starts at the upright point of the triangular shaped GaN grain and propagates not only laterally, but also to the bottom (down to the base of the triangular-shaped GaN grain). The lateral propagation of the $\{2\bar{2}01\}$ facets in c-plane AlN might be slower compared to the vertical growth of a-plane (Al,Ga)N nucleating at ridge sidewalls, which would explain the fast overgrowth of c-plane (Al,Ga)N growing at the horizontal surfaces (regions 1 and 2 in figure 1a) and the coalescence of the a-plane (Al,Ga)N regions.

Using SNBD analysis it is possible to determine crystallite orientation and distinguish between different phases. We have tried to distinguish between AlN and GaN using the ASTAR software of NanoMEGAS. The phase identification was carried out by cross-correlation of each experimental electron diffraction pattern with calculated patterns for AlN and GaN. The obtained phase map is shown in figure 3a. As visible, the regions (i) and (ii) were correctly assigned to AlN and GaN, whereas phase identification in region (iii) failed (compare figures 2a and 3a). In particular, AlN nucleating at the ridge sidewall and viewed in the $[1\bar{1}00]$ zone axis was mistakenly assigned to the GaN phase. A detailed analysis of DPs suggests that this technique allows distinction between AlN and GaN viewed in the $[11\bar{2}0]$ direction (figures 3b,c). However, any accurate phase identification fails for the $[1\bar{1}00]$ zone axis. The lower accuracy of the phase recognition observed for the $[1\bar{1}00]$ zone axis can be explained by a lower number of reflections present in these electron diffraction patterns (compare figures 3c and d).

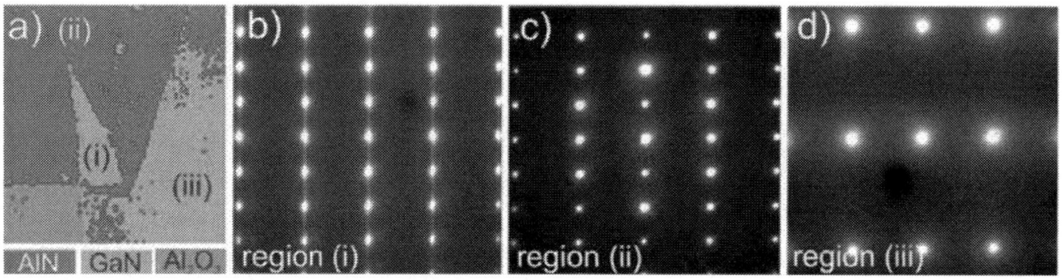

Figure 3. Calculated phase map (a) of the sample region shown in figure 2 and experimental diffraction patterns from (b) c-plane GaN (region (i)) viewed in $[11\bar{2}0]$GaN; (c) c-plane AlN (region (ii)) viewed in $[11\bar{2}0]$AlN; (d) a-plane AlN (region (iii)) viewed in $[1\bar{1}00]$AlN.

4. Conclusions

We have analysed composition and crystallographic orientation of (Al,Ga)N layers grown on c-plane AlN/sapphire substrates patterned in the $[1\bar{1}00]Al_2O_3$ direction. Nucleation of c-plane (Al,Ga)N on vertical a-plane ridge sidewalls of sapphire results in a-plane (Al,Ga)N orientation with respect to the horizontal growth front. Simultaneously, there is a formation of c-plane crystallites on the horizontal surfaces, i.e. on the top of the AlN/sapphire ridges and in the trenches. We suggest that a slow growth rate of $\{2\bar{2}01\}$ facets in c-plane (Al,Ga)N prevents coalescence of the c-plane (Al,Ga)N regions. In contrast, fast propagation of a-plane crystallites in the vertical $[11\bar{2}0]$ (Al,Ga)N direction is suggested to be responsible for the rapid overgrowth of the c-plane regions and formation of continuous a-plane layers. Phase separation into binary GaN and AlN is observed during the first growth stages, which leads to a delay in formation of ternary (Al,Ga)N.

Acknowledgements

This work is financially supported by the Deutsche Forschungsgemeinschaft (Project No. RI1224/1-1).

References

[1] Schwarz U, Kneissl M 2007 *Phys. Stat. Sol (RRL)* **1** A44

[2] Paskova T, Darakchieva V, Paskov P P, Birch J, Valcheva E, Persson P O A, Arnaudov B, Tungasmitta S, Monemar B 2005 *J. Cryst. Growth* **281** 55

[3] Mogilatenko A, Hagedorn S, Richter E, Zeimer U, Goran D, Weyers M, Tränkle G 2013 *J. Appl. Phys.* **113** 093505

[4] Tajima J, Togashi R, Murakami H, Kumagai Y, Takada K, Koukitu A 2011 *Phys. Stat. Sol. (c)* **8** 2028

Journal of Physics: Conference Series **471** (2013) 012039 doi:10.1088/1742-6596/471/1/012039

GaN-based radial heterostructure nanowires grown by MBE and ALD

L Lari[1,2], I M Ross[2], T Walther[2], K Black[3], C Cheze[4], L Geelhaar[4], H Riechert[4] and P R Chalker[3]

[1] Department of Physics, University of York, Heslington, YO10 5DD, York, UK
[2] Kroto Centre for High-Resolution Imaging and Analysis, Dept. Electronic & Electrical Engineering, University of Sheffield, Sheffield S1 3JD, UK
[3] Department of Engineering, University of Liverpool, Liverpool, L69 3GH, UK
[4] Paul-Drude-Institut für Festkörperelektronik, 10117 Berlin, Germany

E-mail: leonardo.lari@york.ac.uk

Abstract. A combination of molecular beam epitaxy (MBE) and atomic layer deposition (ALD) was adopted to fabricate GaN-based core/shell NW structures. ALD was used to deposit a HfO_2 shell of onto the MBE grown GaN NWs. Electron transparent samples were prepared by focussed ion beam methods and characterized using state-of-the-art analytical transmission and scanning transmission electron microscopy. The polycrystalline coating was found to be uniform along the whole length of the NWs. Photoluminescence and Raman spectroscopy analysis confirms that the HfO_2 ALD coating does not add any structural defect when deposited on the NWs.

1. Introduction

Vertically aligned radial nanowire (NW) heterostructures, consisting of a gallium nitride (GaN) core and an outer shell made of a material with a larger band gap, are good candidates as building blocks for UV optoelectronic and electronic devices [1]. A full control and understanding of their growth mechanism is required to tailor the geometry and the properties of these structures.

MBE growth of dislocation-free plain GaN NWs has already successfully been proven on Si and sapphire substrates [2, 3]. Although AlGaN would be the natural material of choice for the deposition of GaN-based heterostructures, the growth of high quality, free standing, axial and core/shell GaN/AlGaN NWs heterostructures using MBE, can be challenging [4, 5]. In the core/shell case, this difficulty is associated to the NWs growth mechanism, which requires Ga-rich growth conditions to obtain radial growth [2, 3]. This leads to an enlargement of the NW core/shell structures at the growth tips, resulting in irregular shell thickness and eventual coalescence of the NW as reported in [5].

Here we adopted an alternative route to manufacture GaN-based core/shell NW structures, using atomic layer deposition (ALD) of HfO_2 as large band gap material for the shell deposition on to MBE-grown GaN NWs. The microstructure and uniformity of the HfO_2 coatings on were studied using state-of-the-art analytical (scanning) transmission electron microscopy (S)TEM techniques in addition to photoluminescence (PL) and Raman analysis.

---

[1] To whom any correspondence should be addressed.
[4] Formerly at Qimonda, 81730 Munich, and NaMLab, 01099 Dresden, Germany.

2. Methods

GN NWs were grown by plasma-assisted MBE on a Si (100) substrate under N-rich conditions. The growth followed the recipe described in [6] with a N/Ga flux ratio of 5:1 for 75 minutes and a substrate temperature of 730°C. The as-grown sample was transferred to an Aixtron AIX 200FE AVD reactor, fitted with a modified liquid injection system [7]. 250 cycles of ALD were undertaken using $[(MeCp)_2HfMe(OMe)]$ dissolved in toluene (0.05M) as the Hf precursor and H_2O as oxidizing agent. Inter-cycle purges were obtained by introducing research grade purity (6N) argon as the carrier gas at the inlet of the reactor, the flux being regulated by a mass flow controller.

A sample for TEM investigations was then prepared using a dual beam microscope FEI Quanta 3D 200 equipped with a C gas injector and an Omniprobe lift-out system. A $2 \times 1 \times 2\ \mu m^3$ carbon mask was deposited on the sample before rough milling of a TEM lamella by focused Ga^+ ion beam. The lamella was then in-situ lifted-out and transferred to a Cu Omniprobe grid, followed by low current milling, until electron transparent (down to a thickness of 100nm).

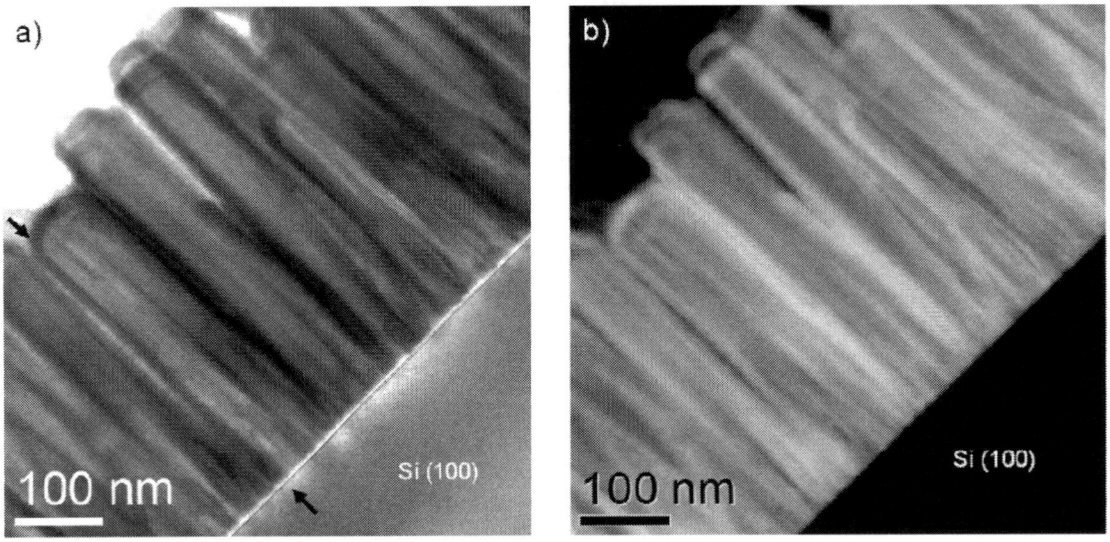

Figure 1. BF-STEM (a) and ADF-STEM (b) image showing the free standing NWs coated by a layer of Hf. Differences in contrast between core and coating are due to the larger density of HfO_2 with respect to GaN. The uniformity of coating is visible where the NWs are separated (see black arrows)

TEM analysis was performed using an aberration corrected JEOL 2200FS FEG-(S)TEM operating at 200kV and a JEOL 2010F FEG-TEM microscope operating at 197kV equipped with a scanning unit and an Oxford Instruments Si:Li Pentafet EDX detector with ultrathin window and Link ISIS 300 acquisition software. Raman spectra were acquired using a Jobin-Yvon LabRam HR consisting of a confocal microscope, coupled to a single grating spectrometer equipped with a notch filter and a CCD camera detector. All of the spectra were recorded in backscattering geometry. PL spectra of the NWs were recorded employing the same LabRam HR instrument. PL data was acquired at room temperature using the 325nm line of a He–Cd laser as the excitation source.

3. Results and discussion

Figure 1(a) and 1(b) show NWs after the ALD coating process and are imaged in their full length. Due to the difference in average atomic number between GaN and HfO_2, the HfO_2 coating appears darker than the GaN core in the Bright-Field (BF)-STEM image of Figure 1(a), whereas the contrast is switched in the Annular Dark Field (ADF)-STEM image of Figure 1(b). The NWs are uniformly coated from top to bottom by a layer of HfO_2. This is best visualized where NWs are not superimposing each other in the field of view, as indicated by the black arrows in Figure 1(a).

Although the GaN core diameters range from 20nm to ~80 nm, the HfO_2 coating thicknesses are independent of the size of the NW cores and the orientation of the NW surfaces (including the top one) which is ~15nm.

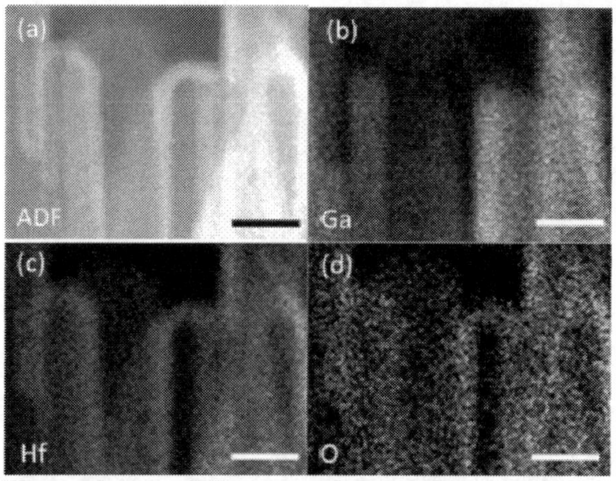

Figure 2. (a) ADF image of GaN/ HfO_2 Core/Shell nanowires structures. (b) gallium map showing the GaN cores. (c) and (d) hafnium and oxygen map highlighting the shell surrounding the core. Scale bars correspond to 100nm

EDX maps as shown in Figure 2 were acquired to confirm the elemental distribution in the sample. Maps were acquired from the area shown by the ADF image in Figure 2(a). EDX energy windows (200 eV wide) were centered on Ga-Kα (9.251 keV), Hf-Lα (7.898 keV) and O-Kα (0.525 keV). The GaN cores are clearly identified in Figure 2b) while Hf and O distributions follow exactly the coating around the cores [Figure 2(c) and 2(d)].

High-Resolution TEM images, as shown in Figure 3(a), acquired at the interface of the NWs with the Si substrate show, as expected, the presence of an amorphous Si_xN interlayer, typical of this mode of NW growth [3]. The polycrystalline coating reaches the NW base. The cubic silicon substrate is imaged along the [110] zone axis which is parallel to the [11-20] zone axis of the GaN wurtzite structure, as shown in the fast Fourier transform (FFT) image inset in Figure 3 (a).

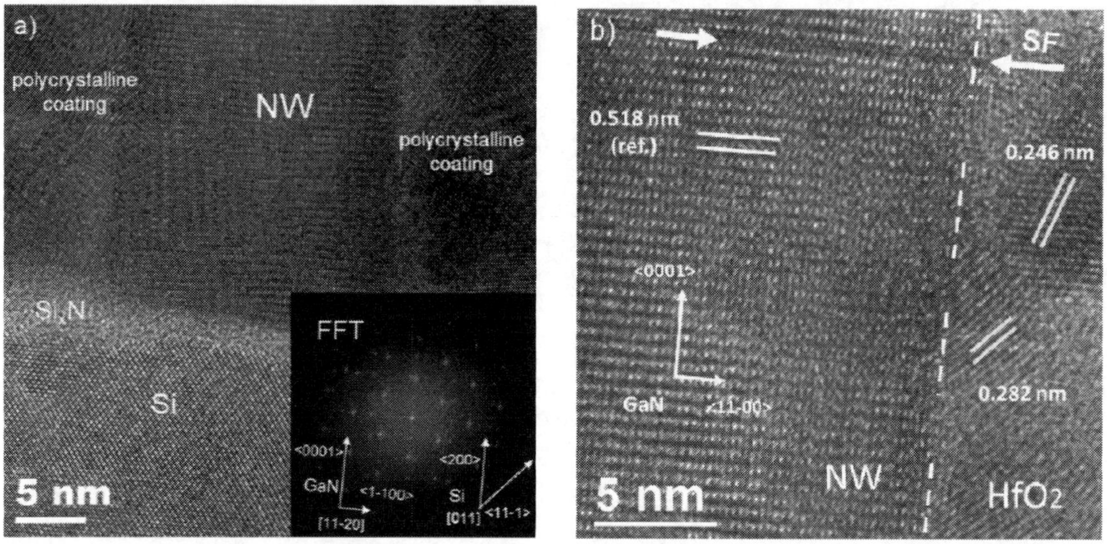

Figure 3. Aberration-corrected HR-TEM images of (a) NW to substrate interface and (b) NW side wall, where dashed lines indicate the interface with the coating. Inset in (a) is the FFT from the image showing the epitaxial relation between the NW and the substrate. A white arrow in (b) indicates a basal stacking fault (SF) present in the NW.

The as-grown GaN NWs are free-standing structures with the growth direction <0001> perpendicular within 5 degrees to the substrate [3]. The HfO_2 polycrystalline coating can be seen at the NW sides right down to the NW/substrate interface. The sharpness of the NW/coating interface is highlighted by dashed lines in Figure 3(b) where the value of 0.518nm [1] for the GaN (0001) lattice plane interdistance was taken as a reference for the internal calibration of the image. This allowed identification of the main lattice fringes within the polycrystalline HfO_2. The set of lattice fringes, shown in Figure 3(b), with distances measured as (0.246±0.003)nm and (0.282±0.003)nm can be attributed respectively to (1,0,-2) planes (0.2489 nm [8]) and (1,1,1) planes (0.2825 nm [8]) of the monoclinic (P 1 21/c 1; zirconia-like) HfO_2 structure. As highlighted by the white arrow in Figure 3(b) an intrinsic stacking fault is present along the NWs. This is a NW growth defect visible only along <11-20> directions [9].

In Figure 4(a), PL spectra show pronounced band edge emission peaks at 3.4eV both before and after the coating treatment. Similarly, UV Raman spectra shows enhanced longitudinal optical (LO) and 2LO modes in both cases (Figure 4(b)). Optical spectroscopy results indicate that the coating treatment did not induce any pronounced changes in the spectra, suggesting that the optical properties were not deteriorated.

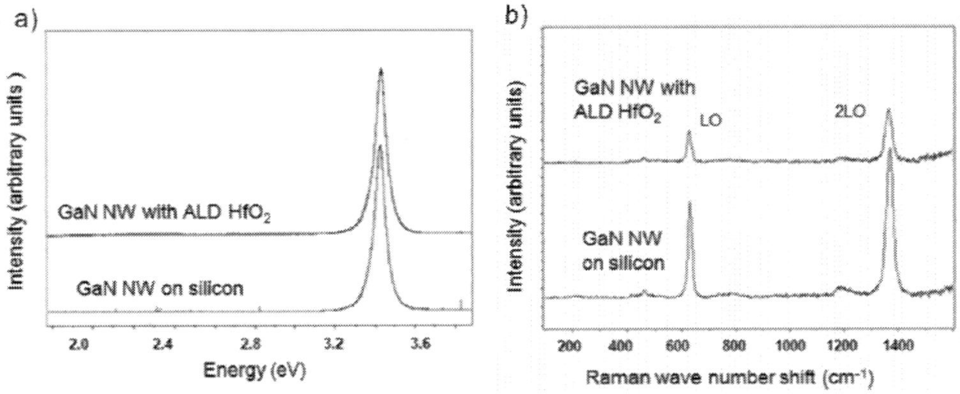

Figure 4. PL (a) and Raman Spectroscopy spectra (b) acquired at room temperature before and after the HfO_2 deposition by ALD.

4. Conclusions

In this work we have reported the successful growth of GaN/HfO_2 core/shell NW heterostructures by combining catalyst-free MBE growth of GaN NWs on Si (100) substrates with ALD of HfO_2. The uniformity of the HfO_2 coating along the whole length of the NWs was demonstrated as well as the sharpness of the NW/coating interface. Optical properties were found to remain unchanged by the subsequent coating. These results highlight the reliability and the potential versatility of this combination of deposition techniques for the growth of future optoelectronic devices components

Acknowledgements

Financial support by EPSRC grant EP/F02374X/1, EU PARSEM contract MRTN-CT-2004-005583 and IST project NODE 015783 is gratefully acknowledged.

References

[1] Ruterana P, Albrecht M and Neugebauer J (eds) 2003 *Nitride Semiconductors: Handbook on Materials and Devices* (Berlin: Wiley-VCH)
[2] Geelhaar L *et al.* 2007 *Appl. Phys. Lett.* **91** 093113
[3] Geelhaar L *et al.* 2011 *IEEE Journal of Selected Topics in Quantum Electronics* **17** 878
[4] Lari L *et al.* 2011 *Journal of Crystal Growth* **327** 27
[5] Lari L *et al.* 2010 *J. Phys.: Conf. Ser.* **209** 012011
[6] Chèze C *et al.* 2010 *Nano Research* **3** 528
[7] Potter RJ *et al.* 2005 *Chemical Vapor Deposition* **11** 159
[8] Ruh R and Corfield PWR 1970 *Journal of the American Ceramic Society* **53** 126
[9] Lari L *et al.* 2008 Phys. Stat. Sol. A **205** 2589

A luminescence study of doping effects in InP-based radial nanowire structures

D Lindgren[1], M Heurlin[1], K Kawaguchi[1,2], M T Borgström[1], M-E Pistol[1], B Monemar[1], L Samuelson[1] and A Gustafsson[1]

[1]Lund University, Solid State Physics and the Nanometer Structure Consortium, Box 118, S-221 00, Lund, Sweden
[2]Fujitsu Laboratories Ltd., Morinosato-Wakamiya 10-1, Atsugi, 243-0197 Japan

david.lindgren@ftf.lth.se

Abstract. We have used micro-photo- and cathodo-luminescence at low temperatures to study the effects of sulphur doping in InP and radial InP/InAs/InP structured nanowires. Samples with pure wurtzite crystal structure, with modulated wurtzite/zincblende crystal structure and with different radial InAs growth times were investigated. We observed a doping concentration gradient along the nanowires, the location of segments of different crystal structure and thickness fluctuations on the monolayer scale of the InAs layer.

1. Introduction

Nanowires (NWs) have been predicted to play a significant role in future electro-optical devices [1]. In order to realize devices, many parameters must be controlled simultaneously. One crucial parameter is the doping, which can have an effect on the crystal structure of the NWs. For instance, if InP NWs are doped with sulphur during growth, an otherwise mixed zincblende (ZB)/wurtzite (WZ) crystal structure can be turned into pure WZ [2]. When the doping is interrupted for specific time periods, the length and location of segments with the mixed crystal structure can be controlled along the NW [3]. These segments are referred to as stacking fault (SF) segments. The surface of the side facets of WZ NWs tend to be smooth, whereas for ZB they are corrugated. Any subsequent radial layer growth will consequently be influenced by the surface and therefore by the crystal structure.

In this work we show optical properties of InP NWs with pure WZ and with a modulated crystal structure, as well as InP/InAs/InP core/shell/cap NWs. For the latter, samples with radial InAs layers grown on both pure WZ and modulated crystal structure InP NW cores have been studied. The studies were done by micro-photoluminescence (μPL) spectroscopy and cathodoluminescence (CL) imaging.

2. Experimental details

The NW samples were grown using metal-organic vapour phase epitaxy (MOVPE). Two types of InP NW cores were prepared, one with pure WZ and one with a modulated crystal structure. The cores were about 2 microns long and about 60 nm in diameter. The modulated structures contained a number of roughly 100 nm long SF segments. For all InP NW core samples a second batch of samples with subsequent radial InP underlayer, InAs shell and InP cap growth were prepared. A series of samples using different InAs growth times on each type of NW core were also made. A more detailed description of the growth can be found elsewhere [3].

Optical measurements were performed on both NWs standing on their growth substrate (as-grown) and single NWs broken off and re-deposited on a gold-coated Si-substrate. The latter contains alignment marks, which facilitates CL and PL correlation measurements. The as-grown samples were studied both in top and side view. In side view the sample was cleaved through the middle along the NW length and mounted with the cleaved surface facing the laser or electron beam. This geometry is important especially for CL, as it allows studies on the full length of the NWs, the substrate, and the interface between them at the same time. The PL measurements were carried out at 4 K using the 532 nm line from a frequency doubled Nd-YAG laser as the excitation source using an excitation density of about 0.1 W/cm^2 and the spectral resolution below 1 meV. The CL measurements were carried out in an SEM operating at 5 kV and a probe current of 10-50 pA, with the sample at 7 K.

3. Results and discussion

3.1. PL of InP NW cores with and without SFs

The doping to achieve the pure WZ structure is often so high that it leads to degenerate doping, which broadens the emission above the 1.49 eV bandgap energy of the WZ InP. This is known as a Burstein-Moss shift [4]. The emission is also broadened to lower energies below the bandgap. Similar observations have previously been reported on a nearly identical structure by Wallentin et al. [5]. In our case we observe emission from single NWs ranging from below 1.4 eV up to about 2 eV, shown in blue in Figure 1.

The band alignment of InP WZ/ZB heterostructures is of type II, with the WZ conduction band (CB) edge above the CB of ZB. The extensions of the individual phases in the SF segments are thin enough to confine electrons in the ZB phase and holes in the WZ phase. This results in spatially indirect transitions, i.e. type II transitions. Such emission can be observed in the single NW spectrum shown in red, Figure 1, at 1.42-1.48 eV. For thin ZB sections in a predominantly WZ InP NW, this range correspond to extensions of 2-8 monolayers (MLs) according to Jancu et al. [6]. In addition, similar emission in the range up to 2 eV due to the doping is also present in this sample.

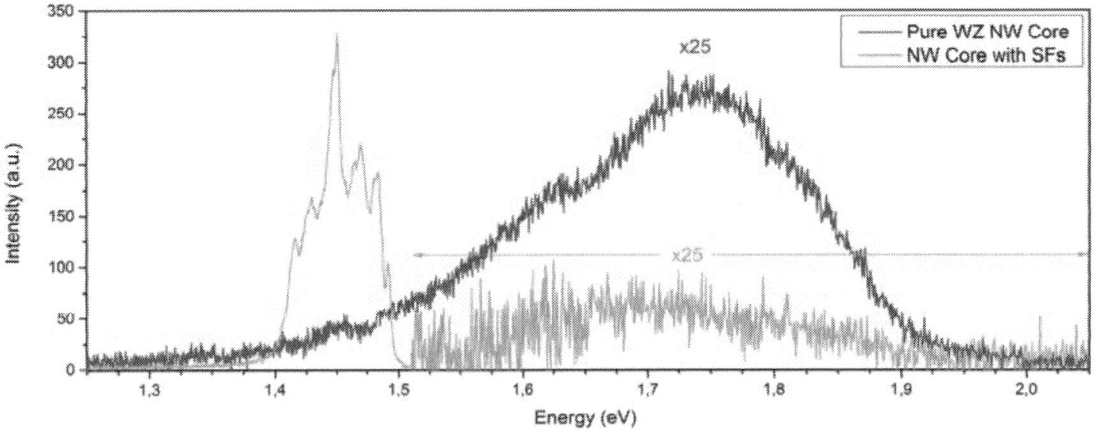

Figure 1. μPL spectra of a single InP NW with pure WZ crystal structure (blue) and with multiple SF segments (magenta). For comparison the spectrum of the pure WZ NW (and of the NW with SFs at energies above 1.51 eV) has been magnified.

3.2. PL of InP/InAs/InP Core/Shell/Cap with and without SFs

Radial growth of thin layers of InAs on pure WZ NW cores of InP leads to the formation of smooth quantum wells (QW). By varying the growth time of these thin layers, it is possible to shift the emission energy of the QWs. The QWs are well defined and the thickness varies in steps of single MLs in the radial direction. As the QWs are thin, the difference in quantization introduced by one ML is significant, and the thickness variations in single NWs lead to several emission peaks. In a series of NW growth runs with different growth times for the radial layer, we identified the emission peaks from QW thicknesses ranging from one to twelve MLs, measured on ensembles of NWs on as-grown samples [7]. Figure 2a shows an example of μPL spectra from single NWs with three different InAs growth times, 10 (black), 5 (blue) and 2 s (magenta).

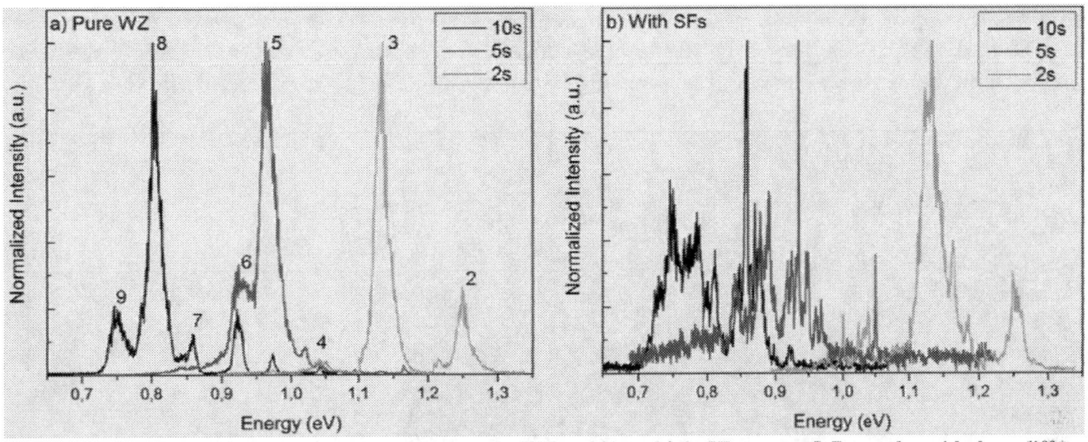

Figure 2. Single NW μPL spectra of InAs between a) pure WZ and b) multiple SF segment InP samples with three different InAs growth times. In a) the number of radial InAs MLs are indicated at the corresponding peak.

When the doping of the core is interrupted in short segments, we observe that the smooth QW breaks up into island formation at the SF segments. These islands form quantum dots (QDs) of various sizes and shapes and their emission is red-shifted, mainly due to the increased radial thickness. In Figure 2b, μPL spectra from a series of single NWs with multiple SF segments equally distributed along the NW and with three different InAs growth times are shown.

3.2.1. Identifying the location of the QDs. Using monochromatic CL imaging, the location of the QDs can be identified. Figure 3a shows an SEM image of an as-grown SF sample in side-view. The corresponding monochromatic CL images recorded at 0.95 and 0.87 eV are shown in (b) and (c), respectively. 0.95 eV corresponds to five MLs, the dominating thickness of the QWs in between the SF segments. This is supported by the PL spectrum in Figure 2a. By varying the detection energy (not shown here), different segments light up. At 0.87 eV the origin of the luminescence is less localized, but it is mainly found at the SF segments.

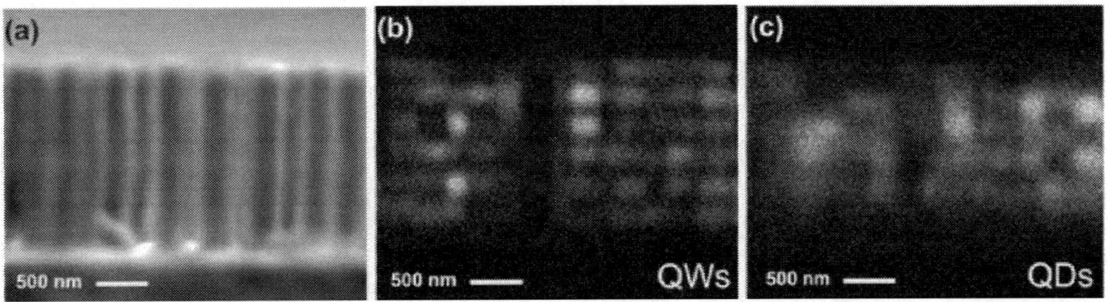

Figure 3. a) SEM image of as-grown samples with five SF segments and a 5 s InAs growth time seen in side-view. CL image recorded at b) 0.95 eV and c) 0.87 eV.

3.2.2. Identifying the location of the SF segments. The location of the SF segments has already been identified indirectly as the regions in between the QWs, Figure 3b. However, there is a second way to do this with CL imaging. Figure 4b shows a CL image recorded at 1.55 eV of an as-grown sample with one SF segment near the top of the NWs. This energy is above the bandgap of WZ InP, and related to the doping, as shown in Figure 1. The SF segments show up as dark regions in CL images recorded at energies above 1.49 eV. In addition, the main peak in Figure 4c is from the QW on the WZ parts of the NW. This peak corresponds to a one ML thick QW.

A feature that can be observed in the CL image is a doping concentration gradient going from high to low from the top to the bottom of the NW. This can be seen for all energies above 1.49 eV, as the CL images look similar, with the brightest luminescence at the top of the NWs. Interestingly we have previously seen the opposite behaviour when using S as dopant in ZB InP/InP Core/Shell NWs [8].

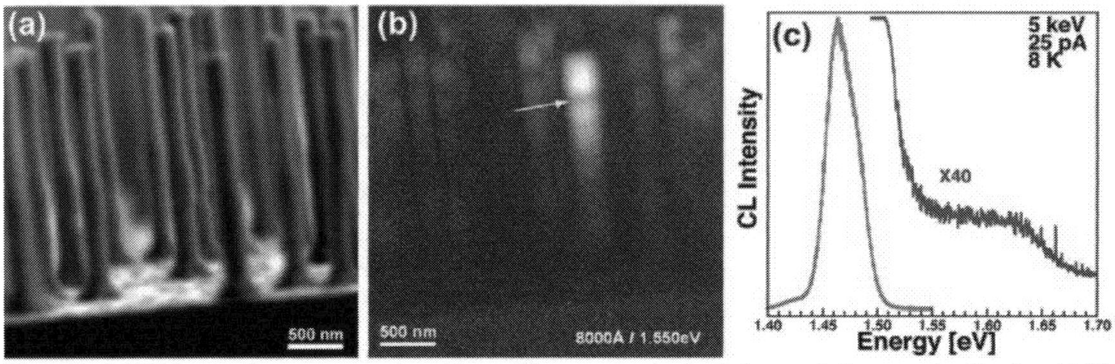

Figure 4. a) SEM image of as-grown samples with one SF segment and a 2 s InAs growth time seen in side-view. b) CL image recorded at 1.55 eV and c) CL spectra of the same region seen in a).

4. Summary

We have reported on μPL and CL measurements on InP based NW samples with pure WZ and modulated WZ/ZB crystal structure. The origin of the luminescence as well as the location of quantized structures and SF segments have been shown by CL imaging.

Acknowledgements

This work was performed within the Nanometer Structure Consortium at Lund University (nmC@LU), supported by the Swedish Research Council, by the Swedish Foundation for Strategic Research, by the Knut and Alice Wallenberg Foundation, by the Top-level Research Initiative "Nanotechnology and Energy Efficiency", and by the EU program AMON-RA (214812). This project is based on a project that was funded by E·ON AG as part of the E·ON International Research Initiative as well as by VINNOVA.

References

[1] Li Y, Qian F, Xiang J and Lieber C M 2006 *Mater. Today* **9** 18-27
[2] van Weert M H M, Helman A, van den Einden W, Algra R E, Verheijen M A, Borgström M T, Immink G, Kelly J J, Kouwenhoven L P and Bakkers E P A M 2009 *J. Am. Chem. Soc.* **131** 4578-9
[3] Kawaguchi K, Heurlin M, Lindgren D, Borgström M T, Ek M and Samuelson L 2011 *Appl. Phys. Lett.* **99** 131915
[4] Bugajski M and Lewandowski W 1985 *J. Appl. Phys.* **57** 521-30
[5] Wallentin J, Mergenthaler K, Ek M, Wallenberg L R, Samuelson L, Deppert K, Pistol M E and Borgström M T 2011 *Nano Lett.* **11** 2286-90
[6] Jancu J M, Gauthron K, Largeau L, Patriarche G, Harmand J C and Voisin P 2010 *Appl. Phys. Lett.* **97** 041910
[7] Lindgren D, Kawaguchi K, Heurlin M, Borgström M T, Pistol M-E, Samuelson L and Gustafsson A 2013 *Nanotechnology* **24** 225203
[8] Storm K, Halvardsson F, Heurlin M, Lindgren D, Gustafsson A, Wu P M, Monemar B and Samuelson L 2012 *Nat. Nanotechnol.* **7** 718-22

The measurement of electrostatic potentials in core/shell GaN nanowires using off-axis electron holography

S Yazdi[1], T Kasama[1], R Ciechonski[2], O Kryliouk[3] and JB Wagner[1]

[1]Center for Electron Nanoscopy, Technical University of Denmark, Lyngby, Denmark

[2]GLO-AB, Ideon Science Park, Lund, Sweden

[3]GLO-USA, Sunnyvale, CA, United States of America

sadegh.yazdi@cen.dtu.dk

Abstract. Core-shell GaN nanowires are expected to be building blocks of future light emitting devices. Here we apply off-axis electron holography to map the electrostatic potential distributions in such nanowires. To access the cross-section of selected individual nanowires, focused ion beam (FIB) milling is used. Furthermore, to assess the influence of FIB damage, the dopant potential measured from an intact NW is compared with a FIB prepared one. It is shown that in addition to the built-in potential between the p-type shell and unintentionally n-type under-layer there is a potential barrier between the core and under-layer which are both unintentionally n-type doped.

1. Introduction

Light-emitting diodes (LEDs) are predicted to be able to reduce global electricity consumption for lighting by over 50% [1]. Since GaN-based material systems span the entire visible spectrum, GaN is becoming the material of choice in LEDs. Nanowires (NWs) are promising candidates for building blocks of GaN LEDs due to their defect-free crystal structure, large surface to volume ratio, and exposed non-polar *m*-plane, $\{1\bar{1}00\}$, along which the active region of devices can be grown [2]. However, the lack of ability to control the distribution of dopants along the nanowires has hindered the progress of nanowire-based devices [3]. Finding a precise doping process is very challenging in core-shell structured nanowires. For the development of new doping methods, comprehensive feedback is required from a dopant characterisation technique. Unfortunately the nano-scale geometry of nanowires hampers conventional characterisation techniques such as four point probes (4PP) and Hall-effect measurements. Here, we use off-axis electron holography in the transmission electron microscope (TEM) to measure the electrostatic potential built-in across the core/shell of GaN nanowires.

In the absence of dynamical diffraction and magnetic field, the phase shift that the electron beam experiences as it passes through a TEM specimen is expressed by:

$$\varphi(x,y) = C_E \int_0^t V(x,y,z)dz, \qquad \text{(Eq. 1)}$$

where C_E is a constant depending on the acceleration voltage of microscope, t is the specimen thickness and $V(x, y, z)$ represents the electrostatic potential distribution in the specimen. This phase shift can be measured by off-axis electron holography. If the specimen thickness and the potential distribution along the beam direction, z, are known, $V(x, y)$ can be extracted from the phase image.

Journal of Physics: Conference Series **471** (2013) 012041 doi:10.1088/1742-6596/471/1/012041

2. Experimental details

The GaN NWs examined in this study were grown by catalyst-free metal organic chemical vapour deposition (MOCVD) on a sapphire substrate on which a GaN buffer layer was deposited. A schematic cross-section of the NW sample is shown in Fig. 1(a). The position of each nanowire was controlled using nano-imprint lithography [4]. Unintentionally n-type doped (hereafter referred to as "u-type") cores, shown in Fig. 1a, were nucleated and grown to a desirable length. Then, u-type under-layers (ULs) were grown around the core. By doping the shells with Mg (2.8% Mg/Ga ratio), pn junctions were formed between the u-type under-layers and p-type shells. A top-view scanning electron microscope (SEM) image of as-grown NWs is shown in Fig. 1(b). Two types of TEM specimen were prepared from the NWs using a dual beam (FIB/SEM) microscope equipped with a micromanipulator. The first type is a cross-sectional specimen prepared along the *a*-plane, $\{11\bar{2}0\}$, of the NWs using the *in-situ* lift-out technique and the second one is an intact single NW lifted up with the micromanipulator and placed on a TEM grid by electrostatic forces without using the FIB. Off-axis electron holograms were acquired at 300 kV in an FEI Titan 80-300 TEM, equipped with a rotatable Möllenstedt biprism.

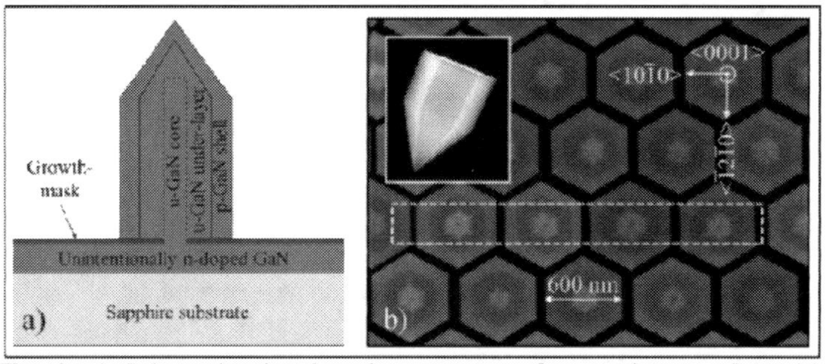

Fig. 1. a) Schematic showing the structure of nanowires used in this study. b) SEM micrographs acquired from as-grown nanowires along <0001> direction and <10$\bar{1}$0> direction (inset). The dashed box in this image shows the position from where the FIB specimen was prepared.

3. Results and discussion

3.1. FIB prepared specimen

A typical secondary electron image acquired with 2 keV electron beam from the cross-sectional surface of NWs after low voltage (2 kV) FIB polishing is shown in Fig. 2. The dopant contrast with higher secondary electron intensity on the p-type shell regions than on the u-type core and UL regions appeared only when the specimen surface was polished with less than 5 kV Ga ion beam. The dopant contrast in SEM was used as a criterion for minimising the FIB damage. Among available FIB voltages, from 30 kV down to 500 V in our dual beam microscope, 2 kV beam was chosen for the final specimen preparation step since the maximum dopant contrast was achieved after polishing the specimen surfaces with this beam voltage.

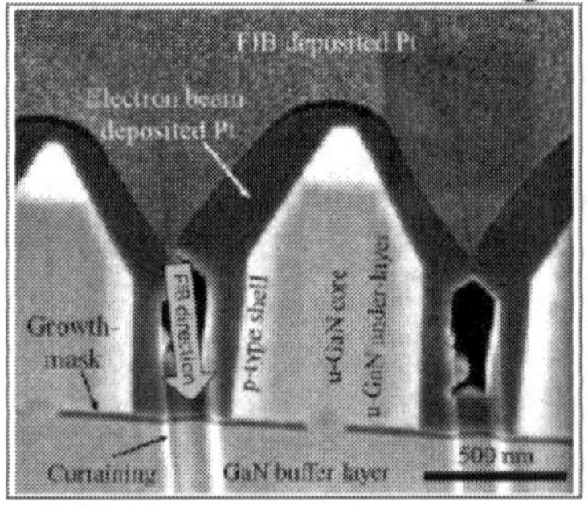

Fig. 2. A typical SEM image from a FIB prepared cross-sectional surface of NWs after 2kV FIB polishing. The SEM dopant contrast can be seen in this image. The FIB milling direction, shown by an arrow, was intentionally chosen not to be parallel to the NWs so that any possible thickness variations due to the curtaining effect would be distinguishable from electrostatic potential variations. It can be seen that the curtaining direction in the GaN buffer layer is different from the nanowire growth direction.

A reconstructed phase image from a FIB prepared cross-sectional NW is shown in Fig. 3(a). By tilting the specimen by ~5° away from the zone axis along the Kikuchi band center, the diffraction contrast was minimized while the UL-shell junction was kept edge-on. The crystalline thickness of the specimen was measured to be 300±10 nm using convergent electron beam diffraction. The measured crystalline thickness was in agreement, within the experimental error, with the thickness map obtained from the reconstructed amplitude image. The inelastic mean free path of 300 kV electrons in GaN was measured to be ~100 nm directly from a single GaN NW which has a known geometry and thickness. This value was used in calculating the thickness map from the amplitude image.

The entire 300 nm crystalline thickness was used to convert the reconstructed phase image into the potential map, without considering any "dead layer", a crystalline layer which is electrically damaged by FIB. The electrostatic potential profile across the NW, from the dashed box in Fig. 3(a), is shown in Fig. 3(b). The measured built-in potential across the UL-shell junction was only 0.3 V, significantly smaller than about 3V expected from the intended dopant concentrations used during the growth. Also surprisingly it was found that in addition to this potential barrier between the p-type shell and u-type UL, there is almost the same potential barrier between the u-type core and u-type UL. Further systematic studies are in progress to understand this additional potential barrier.

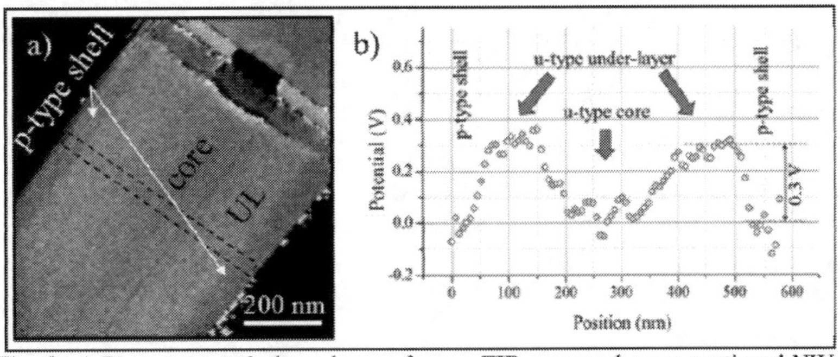

Fig. 3. a) Reconstructed phase image from a FIB prepared cross sectional NW.
b) Potential profile from the region marked with the dashed box in (a).

3.2. *Intact nanowire*

To understand how significant the effect of FIB damage is on built-in potential measurements, electron holography on single intact NWs was carried out. As these NWs have a core-shell structure, the electrostatic potential variation along the electron beam direction needs to be taken into account in addition to the specimen thickness when converting the phase image into the potential map. For a core-shell NW, such as one shown in Fig. 4.a, Eq. 1 can be simplified as:

$$\varphi(x) = C_E V_{shell} t_{shell}(x) + C_E V_{core} t_{core}(x), \qquad \text{(Eq. 2)}$$

where V_{core} and V_{shell} refer to the electrostatic potential of the core and shell respectively. $t_{core}(x)$ and $t_{shell}(x)$ represent the core thickness and shell thickness at position x respectively. By calculating these thicknesses, using the geometry shown in Fig. 4(a), and substituting them into Eq. 2 the following expression is obtained for the phase shift of the electron beam:

$$\varphi(x) = C_E \begin{cases} (a + 2x\tan 30^\circ)V_{shell}\ ; & 0 \le x < \frac{\sqrt{3}}{2}(a-b) \\ 2(a-b)V_{shell} + \left(b + 2\left(x - (a-b)\frac{\sqrt{3}}{2}\right)\tan 30^\circ\right)V_{core}\ ; & (a-b)\frac{\sqrt{3}}{2} \le x < a\frac{\sqrt{3}}{2} \\ 2(a-b)V_{shell} + \left(b - 2\left(x - (a+b)\frac{\sqrt{3}}{2}\right)\tan 30^\circ\right)V_{core}; & a\frac{\sqrt{3}}{2} \le x < (a+b)\frac{\sqrt{3}}{2} \\ \left(a + 2\left(a\sqrt{3} - x\right)\tan 30^\circ\right)V_{shell}\ ; & (a+b)\frac{\sqrt{3}}{2} \le x < a\sqrt{3} \end{cases} \qquad \text{(Eq. 3)}$$

where a and b are the side lengths of the hexagonal shell and core respectively. According to Eq. 3, a phase step of $\Delta\varphi = C_E b V_{bi}$ close to the core-shell interface at $x = \frac{\sqrt{3}}{2}(a-b)$ and $x = \frac{\sqrt{3}}{2}(a+b)$ should be measurable.

The amplitude and phase images from an intact NW are shown in Fig. 4(b) and (c) respectively. By removing the phase slope, due to the thickness variation, in region B in Fig. 4(c), the phase step across the pn junction between the p-type shell and u-type UL appeared in the phase image, as can be seen in Fig 4(f). The potential profile in Fig. 4(g) is obtained by dividing the phase profile from the dashed box in Fig. 4(f) by $C_E b$ ($V_{bi} = \Delta\varphi / C_E b$). The measured built-in potential across the shell-UL junction is only 0.4 V. This means that in our FIB prepared cross-sectional specimen the effect of FIB damage on the potential measurements is not severe. Therefore, the smallness of the measured built-in potential cannot be explained solely by FIB damage. Other possible sources of error in electron holography measurements such as charging and electron-hole pair generation as a result of electron irradiation also need to be assessed. It is also likely that in these NWs the dopant atoms have not been distributed properly during the growth or they have not been activated completely. Work is in progress to investigate these possibilities.

The built-in potential close to the tip of the NW in region A (Fig. 4(c)) was measured to be 0.55 V. The flattened phase image in this region, obtained by removing the thickness variation, and its corresponding potential profile are shown in Fig. 4(d) and (e) respectively. Since this potential is along the *c*-axis, it is speculated that the internal polarisation field affects the value of this potential.

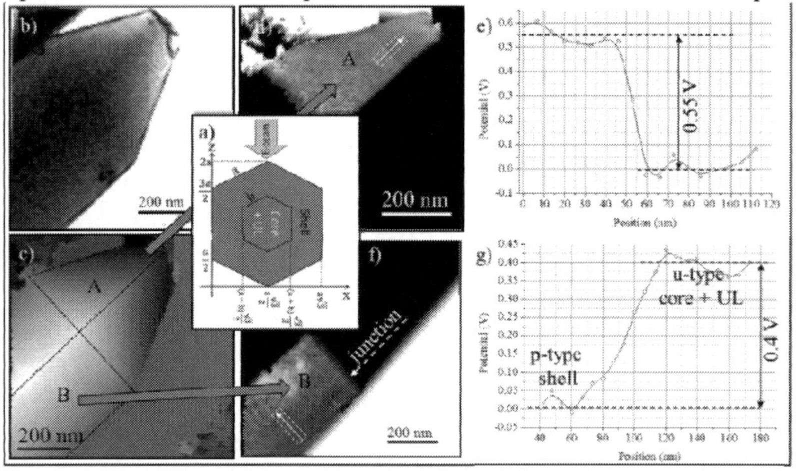

Fig. 4. a) The NW geometry and its orientation relative to the electron beam. b) Reconstructed amplitude and c) phase image from an intact NW. The phase image after removing the phase slope d) in A region and f) in B region. These regions are marked in (c). e) and g) show the potential profiles from the regions marked with dashed white boxes in (d) and (f) respectively.

4. Conclusion

Off-axis electron holography has been applied to a FIB prepared cross sectional GaN NW specimen for measuring electrostatic potential distributions. The built-in potential across the UL-shell junction was found to be 0.3 V in a FIB prepared specimen and 0.4 V in an intact NW. Therefore FIB damage cannot be solely responsible for the smallness of measured potentials. It was also found that there is a potential barrier between the core and UL. Further studies are in progress to expand our understanding of potential distributions in GaN NWs. This can lead us to fabricate more efficient GaN based LEDs.

References

[1] Tsao JY 2004 *Circuits and Devices Magazine IEEE* **20** 28
[2] Storm K *et al.* 2012 *Nature Nanotechnology* **7** 718
[3] Wallentin J *et al.* 2011 *J. Mater. Res.* **26** 2142
[4] Mårtensson T *et al.* 2004 *Nano Lett.* **4** 699
[5] The authors acknowledge funding from 'Nanowires for solid state lighting'. NMP.2011.2.2-3, Project ID:280773

Journal of Physics: Conference Series **471** (2013) 012042 doi:10.1088/1742-6596/471/1/012042

Towards a better understanding of trench defects in InGaN/GaN quantum wells

F C-P Massabuau, L Trinh-Xuan, D Lodié, S -L Sahonta, S Rhode, E J Thrush, F Oehler, M J Kappers, C J Humphreys and R A Oliver

Department of Materials Science and Metallurgy, University of Cambridge, Pembroke Street, Cambridge CB2 3QZ, United Kingdom

E-mail: fm350@cam.ac.uk

Abstract. Trench defects are a commonly occurring feature in InGaN/GaN quantum well (QW) structures. This defect appears at the surface of a structure as a trench enclosing a region of material with peculiar emission properties. Transmission electron microscopy was used to characterise the sub-surface structure of such defect. It consists of a basal-plane stacking fault (BSF) located in the QW stack and bound by a vertical stacking mismatch boundary (SMB) which runs towards the surface and which opens up into pits, which merge to form a trench. Atomic force microscopy and cathodoluminescence were performed on the same individual defects in order to directly correlate the morphology with the emission properties. A strong correlation has been established between the thickness of the trench and the redshift and intensity of the emission of the enclosed region suggesting that bright trench defects emitting at a longer wavelength nucleate early during the growth. Data also suggest that the SMB may act as a non-radiative recombination centre.

1. Introduction

InGaN/GaN quantum well (QW) structures are a key component of high efficiency opto-electronic devices [1]. However, such structures are characterised by a high density of defects such as threading dislocation [2], V-pits [3] and trench defects (trenches at the surface of the structure enclosing a region of material) [4], which might affect their optical properties. Although threading dislocations and V-pits have attracted an intense research effort, trench defects are still poorly understood.

This study investigates the structure of trench defects by transmission electron microscopy (TEM) and tries to link the morphology of a defect with its optical properties using a "multi-microscopy analysis" based on atomic force microscopy (AFM) and scanning electron microscopy with cathodoluminescence (SEM-CL).

2. Experimental methods

InGaN/GaN multiple QW structures were grown by metal-organic vapour pase epitaxy (MOVPE) in a Thomas Swan 6 x 2-inch close-coupled reactor. Trimethylgallium (TMG), trimethylindium (TMI) and ammonia (NH_3) were used as precursors with H_2 as the carrier gas for GaN growth and with N_2 as the carrier gas for InGaN growth. GaN pseudo-substrates grown on *c*-plane (0001) sapphire with a miscut of 0.25 ± 0.10 ° towards (11-20) were employed. Following the growth of approximately 2.2 nm of $In_{0.175}Ga_{0.825}N$ at 745°C, a protective capping layer of about 1 nm of GaN is grown at 745°C. Then the

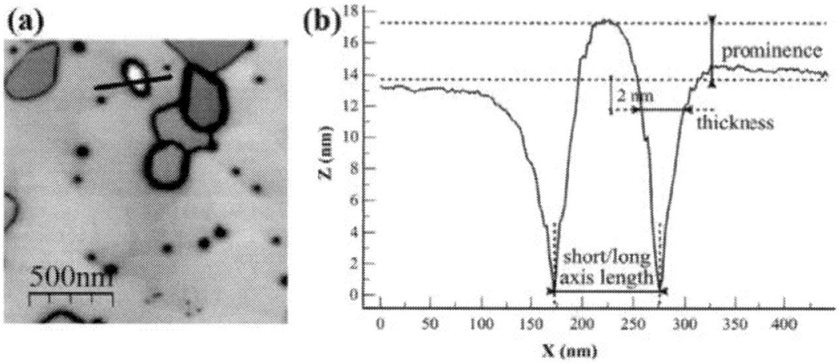

Figure 1. (a) AFM image of trench defects and (b) line profile associated with the black line in (a) illustrating the metrics recorded.

growth of GaN continues while the temperature is ramped-up to 860°C, the growth temperature is maintained at 860°C till 7.5 nm of GaN in total is grown. The thicknesses and composition of the structure were assessed by X-ray diffraction using Vickers et *al.*'s method [5].

AFM in tapping mode was performed to assess the morphology of the defects. Figure 1 shows a typical AFM image of trench defects (figure 1(a)) and illustrates the metrics which are used to characterise the defects' geometry (figure 1(b)): data about the prominence, thickness and area of about 200 defects were collected. The prominence corresponds to the difference in height between the enclosed region of the defect and the average height of the surrounding area. The thickness was measured as the thickness of the V-shaped ditch 2 nm below the level of the surrounding area. Finally, approximating a trench defect as an ellipse, the area was calculated using the measurements for the short and long axis length of the ellipse (measured at the bottom of the ditch on both sides of the defect).

SEM and CL were carried out on the same defects using a process explained elsewhere [6]. Emission spectra were recorded along a line going from the surrounding material on one side of a trench defect to the surrounding material on the other side. Using a two-Gaussian peak model to fit the data as described in reference [6], the redshift and intensity ratio were recorded. The redshift corresponds to the difference between the peak emission wavelength of the material enclosed within the defect and the peak emission wavelength of the surrounding material. According to this definition, a positive value of the redshift means that the inner region of the trench defect emits at a higher wavelength than the surrounding material. Finally, the intensity ratio corresponds to the ratio between the integrated intensity emitted from the enclosed region of a defect and the integrated intensity from the surrounding region. Therefore, according to this definition, an intensity ratio above unity relates to a trench defect with an enclosed region emitting more intense light than its surroundings.

3. Results

3.1. Sub-surface observation

The structure of trench defects was investigated by TEM. It can be seen in figure 2(a) that a basal plane stacking fault (BSF) is located in the QW stack below the defect. The BSF is observed to be bound by a stacking mismatch boundary (SMB) [7] which runs vertically to the apex of the V-shaped ditch (figure 2(d)). In plan view, the SMB is observed to consist of many straight segments oriented at 60° to each other (figure 2(b)). Thickness reduction fringes associated with a V-pit are observed in figure 2(c), therefore considering the facets of a V-pit are towards {1-101} [8], it can be inferred that the SMB is oriented parallel to <1-100>. Moreover, hexagonal fringes similar to that of a V-pit are observed around the SMB and their centre matches with the change in orientation of the SMB.

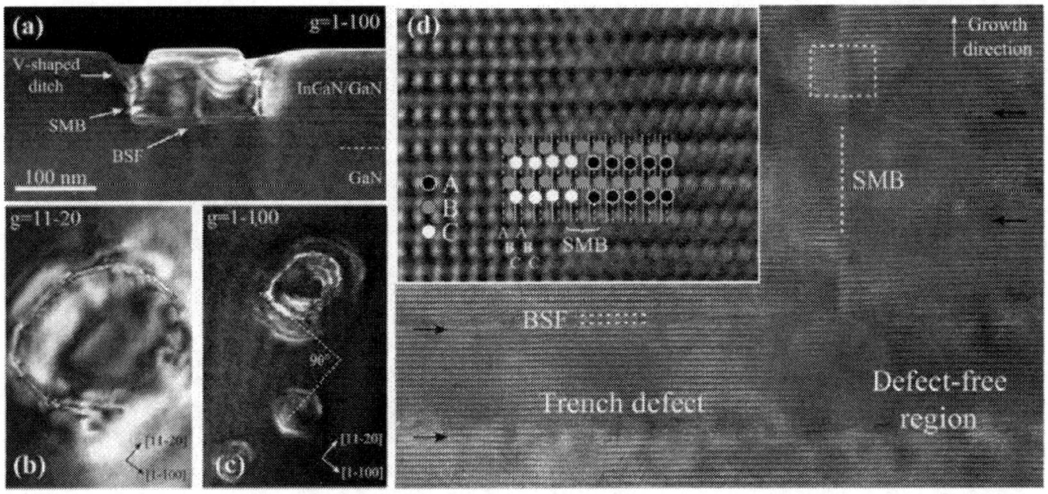

Figure 2. (a) Dark field cross-sectional TEM image of a trench defect using *g* = *1-100*. Dark field plan-view TEM image of a same trench defect using ***g*** = *11-20* (b) and ***g*** = *1-100* (c). (d) High-resolution TEM image of the SMB. The position of the InGaN QWs is indicated with black arrows. The inset represents a Wiener filtered magnified region of the broken line box.

Hence, trench defects are in fact made of a BSF located in the QW stack and terminated by a SMB. The vertical SMB is oriented parallel to <1-100> and a hexagonal pit is generated whenever a change of orientation of the SMB occurs, and the coalescence of these pits results in the trench feature which is observed at the sample surface.

3.2. Correlation between morphology and optical properties
A "multi-microscopy analysis" has been performed in order to link the morphology characterised by AFM to the emission properties of exactly the same defects observed by CL (inset of figure 3(a)). It was observed that while the surface area of the enclosed material, measured by AFM, was completely unrelated to the emission, a slight correlation could be established between the prominence of the inner region and the redshift [6]. This trend can be explained by the increased QW thickness which is associated with the growth of additional material in the enclosed region, thus emitting light at a longer wavelength. Finally, a strong correlation has been established between the thickness of the V-shaped ditch and both redshift and intensity ratio (figure 3). Considering the sub-surface structure of the defect, the trench thickness relates to the depth at which the SMB opens up into a pit. Therefore a thick trench necessarily results from an SMB located deep into the QW stack and therefore to a BSF that nucleated at the early stages of the QW growth (configuration 1 in figure 3(a)).

4. Discussion

Although it remains unclear why the enclosed region within a trench defect can emit light at a higher intensity than its surrounding material, figure 3(b) can be exploited to unveil information about the SMB. It has already been described that a wide trench relates to defects with a BSF located at the bottom of the QW stack (configuration 1 in figure 3(b)). But for thin trenches – i.e. defects where the SMB generates pits close to the surface, two extreme cases arise: either the SMB is short and thus the BSF is located close to the surface (configuration 2a in figure 3(b)), or the SMB is long and therefore the BSF is located at the bottom of the stack (configuration 2b in figure 3(b)). By comparing a defect in configuration 1 (for which the intensity is high) and a defect in configuration 2b (for which the intensity is much lower), it can be inferred that the presence of the SMB running through a significant part of the QW stack induces a loss of intensity. Therefore the SMB may act as a non-radiative recombination centre.

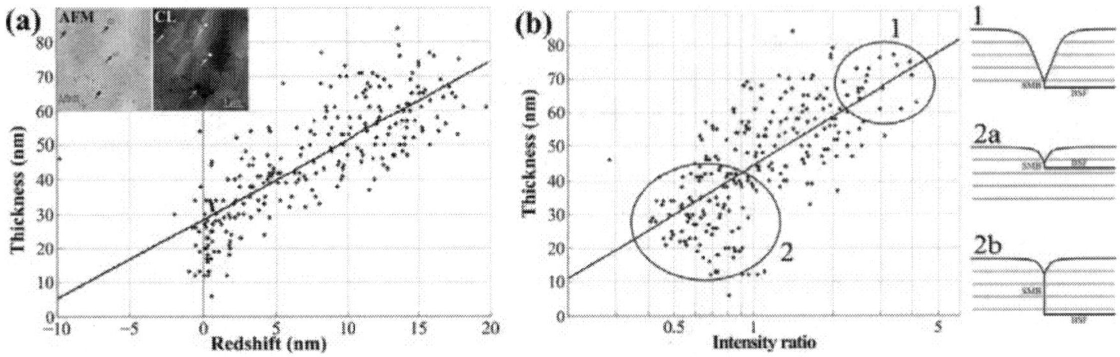

Figure 3. (a) Correlation between the thickness of trench defects and their redshift. The inset illustrates how AFM and CL were performed on the same features. (b) Correlation between the thickness of trench defects and the intensity ratio.

5. Conclusion

The structure of trench defects has been investigated by TEM. It consists of a BSF bound by a vertical <1-100>-oriented SMB which opens up as hexagonal pits whenever it changes orientation.

A "multi-microscopy analysis" involving AFM and CL has been performed in order to correlate the morphology and emission of these defects. It results that in order to emit a more intense light at a longer wavelength, defects need to have their BSF located deep into the QW stack. It is also suggested that the SMB may act as a non-radiative recombination centre.

References

[1] Nakamura S, Mukai T, and Senoh M 1994 *Appl. Phys. Lett.* **64** 1687
[2] Bennett S E 2010 *Materials Science and Technology* **26** 1017
[3] Wu X H, Elsass C R, Abare A, Mack M, Keller S, Petroff P M, DenBaars S P, Speck J S and Rosner S J 1998 *Appl. Phys. Lett.* **72** 692
[4] Bruckbauer J, Edwards P R, Wang T, and Martin R W 2011 *Appl. Phys. Lett.* **98** 141908
[5] Vickers M E, Kappers M J, Smeeton T M, Thrush E J, Barnard J S and Humphreys C J 2003 *J. Appl. Phys.* **94** 1565
[6] Massabuau F C –P, Trinh-Xuan L, Lodié D, Thrush E J, Zhu D, Oehler F, Zhu T Z, Kappers M J, Humphreys C J and Oliver R A 2013 *J. Appl. Phys.* **113** 073505
[7] Cho H K, Lee J Y, Kim C S, Yang G M, Sharma N and Humphreys C 2001 *J. Cryst. Growth* **231** 466
[8] Kim I -H, Park H -S, Park Y -J and Kim T 1998 *Appl. Phys. Lett.* **73** 1634

Microstructural characterisation of Bi_2Se_3 thin films

N V Tarakina[1], S Schreyeck[1], T Borzenko[1], S Grauer[1], C Schumacher[1], G Karczewski[1,2], C Gould[1], K Brunner[1], H Buhmann[1] and L W Molenkamp[1]

[1]Experimentelle Physik III, Physikalisches Institut and Wilhelm Conrad Röntgen-Research Centre for Complex Material Systems, Universität Würzburg, Am Hubland, D-97074 Würzburg, Germany
[2]Institute of Physics, Polish Academy of Sciences, Al. Lotnikov 32/46, 02-668 Warsaw, Poland

E-mail: nadezda.tarakina@physik.uni-wuerzburg.de

Abstract. The microstructure of Bi_2Se_3 thin films grown by molecular beam epitaxy on Si(111), InP(111)B and Fe-doped InP(111)B substrates has been studied in detail using scanning transmission electron microscopy. Films grown on Si(111) and InP(111)B substrates show the formation of twin domains: rotation twins (with the grain boundary perpendicular to the substrate) and lamellar twins (with the grain boundary parallel to the substrate). The presence of twins was confirmed by atomic-force microscopy (AFM) and X-ray diffraction (XRD). At the interface between Bi_2Se_3 film and Si(111) or InP(111)B substrates poorly crystallized layers of about 1 nm and 1.8 nm thickness, respectively, followed by well-crystallized Bi_2Se_3 layers, were found. The use of a Fe-doped InP (111) substrate with a rough surface enables the suppression of twin formation.

1. Introduction

Bismuth selenide, Bi_2Se_3, has been studied for many years as a promising material for thermoelectric [1-4], photoelectrochemical [5] and optical recording devices [6]. Together with Bi_2Te_3 and Sb_2Te_3 it was predicted to be a 3D topological insulator (TI) with semiconductor behaviour in bulk and protected topological states at the surface [7-8]. Topological surface states make Bi_2Se_3 thin films very attractive for applications in spintronics and quantum computing. Hence, the growth of high-quality thin films from this material is a first step towards exploiting its unique properties.

Molecular beam epitaxial growth of Bi_2Se_3 has been realized so far on different substrates: Si(111) [9-12], $SrTiO_3$(111) [13, 14], Al_2O_3(110) [15], GaAs(111)B [16], CdS(0001) [17], InP(111)B [18], double-layer graphene grown on SiC(0001) [19]. Many variable parameters, which have to be taken into account during the growth, together with crystallographic features of the thin film and the substrate can noticeably influence the quality of the film. Bi_2Se_3 crystallizes in a rhombohedral unit cell, space group $R\text{-}3m$, with unit cell parameters in hexagonal setting of a = 4.143(5) Å and c = 28.636(20) Å [20]. The crystal structure of Bi_2Se_3 can be described as a cubic close-packing of Se and Bi atoms. Three selenium and two bismuth layers alternate along the c-direction forming a so-called quintuple layer. One unit cell of Bi_2Se_3 consists of 3 quintuple layers connected to each other via van der Waals bonding. Bi_2Se_3 thin films connect to the substrate via van der Waals bonding as well, which enables growing Bi_2Se_3 on substrates with big lattice mismatches between substrate and film.

The main goal of this work is to study the microstructure of the Bi_2Se_3 thin films grown by molecular beam epitaxy (MBE) on three different substrates, which should help to better understand the growth mechanism of Bi_2Se_3, and to further control the films' quality.

2. Results and Discussion

The Bi_2Se_3 thin films were grown on three different substrates (Si(111), InP(111)B, Fe-doped InP(111)B) by MBE in an ultra-high-vacuum MBE chamber with high purity Bi (99.9999%) and Se (99.9999%) source fluxes provided by standard effusion cells. Independent of the choice of substrate, all studied films were grown at 300 °C, which, as was found previously [18], is an optimal temperature to grow Bi_2Se_3 by MBE. In the case of Si(111), prior to the growth, Bi was deposited on a Si(111)-(7×7) surface at 200 °C and then the substrate was heated up to 400 °C to obtain a Si(111)-($\sqrt{3}\times\sqrt{3}$)-Bi surface. InP(111)B and Fe-doped InP(111)B wafers were etched with hydrofluoric(HF) acid before growth to remove oxide at the substrates surface. After growth, the samples were cooled down in a Se atmosphere to avoid the formation of Se vacancies. Details of the growth are given in Refs. [18, 21].

Selected-area electron diffraction (SAED) patterns of Bi_2Se_3/Si, Bi_2Se_3/InP and Bi_2Se_3/InP(Fe) samples taken along $[100]^*_{Bi2Se3}$ show a distribution of sharp spots, confirming the epitaxial growth of Bi_2Se_3 on these substrates with the following crystallographic relations: normal to the growth plane $[001]_{Bi2Se3}||[111]_{substrate}$, in-plane $[100]_{Bi2Se3}||[1\text{-}10]_{substrate}$ or $[110]_{Bi2Se3}||[1\text{-}10]_{substrate}$. Bi_2Se_3 reflections were indexed in a trigonal unit cell with parameters a = 4.2 Å and c = 28.7 Å. For the films grown on Si and InP substrates, SAED patterns display additional spots coming from twinned domains which can be attributed to the formation of both rotation twin domains (with twin boundary perpendicular to the substrate) and lamellar twins (with twin boundary parallel to the substrate). In the case of growth of Bi_2Se_3 on Si(111), in some of the recorded SAED patterns the $00l$ rows of spots are split into two components; the magnitude of the splitting increases with increasing distance from the direct beam. This splitting indicates a tilt (of about 1°) of the rotation twins with respect to each other in the bc plane. Surface AFM measurements and X-ray diffraction in-plane rotational phi-scans obtained from these samples coincide with SAED data, confirming the presence of twins all over the film (Figure 1). No reflections from twin domains have been observed on the SAED patterns from the Bi_2Se_3 film grown on a Fe-doped InP substrate.

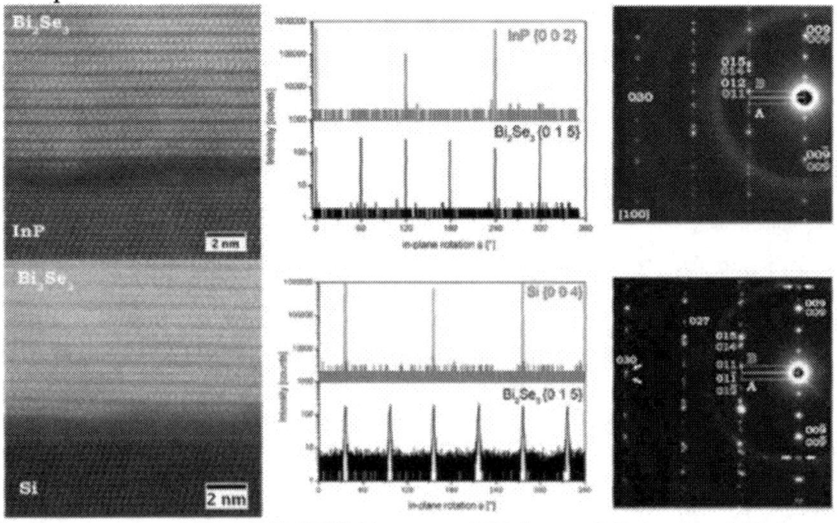

Figure 1. Cross-sectional HAADF-STEM images of Bi_2Se_3 on InP (top) and Si (bottom) with corresponding SAED patterns and X-ray diffraction in-plane rotation phi-scans.

The films grown on Si and InP substrates have a poorly crystallized layer at the interface about 1nm and 1.8nm thick, respectively, which is followed by well-crystallized Bi_2Se_3 layers. By poorly crystallized layer we mean that the interface is not uniformly crystallized over the whole film; at some regions we observe an amorphous layer and at some regions perfect quintuple layers directly at the interface. It is worth to mention that in the case of Bi_2Se_3/Si(111), in some regions the layers are not exactly parallel to the substrate (Figure 1), which coincides well with electron diffraction data (tilt of twin domains of about 1° relative to each other in the *bc* plane).

To study the formation of twins in detail we recorded high-angle annular dark-field scanning transmission electron microscopy (HAADF-STEM) images of Bi_2Se_3/InP at different regions with twin domains (Figure 2). The rotation twin formation in Bi_2Se_3 can be explained as follows: after the first atomic layer of Se is formed, a Bi layer can grow on top of it in two equally possible cubic close-packed atomic configurations. When two such equivalent but different regions meet a twin boundary is formed perpendicular to the growth plane. The observed rotation twins do not display sharp boundaries. Lamellar twins form a twinned domain boundary parallel to the growth plane. Three nonequivalent atomic positions in the Bi_2Se_3 crystal structure allow three possibilities to form a lamellar twin: at the middle Se, edge Se and Bi positions within a quintuple layer. The experimentally observed lamellar twins show the formation of twins only at the edge Se layer in the quintuple layer, which connects to the following layer only by van der Waals interactions (Figure 2). The same type of planar twins has been observed experimentally and predicted by *ab initio* calculations for bulk Bi_2Te_3 [22]. We believe that the formation of such a twin is probably most plausible from an energetic point of view for the case of Bi_2Se_3 as well.

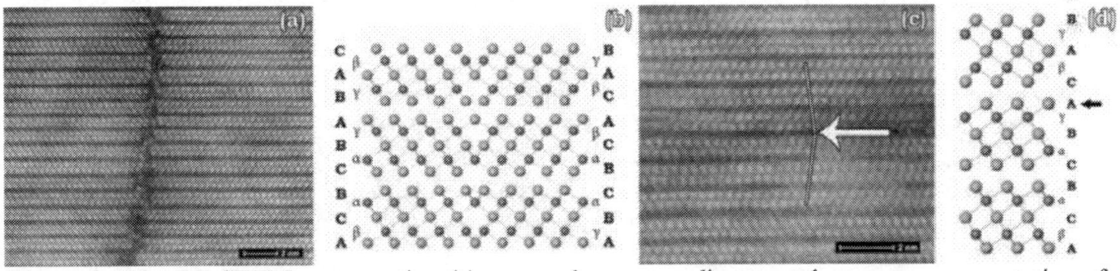

Figure 2. HAADF-STEM cross-sectional image and corresponding crystal structure representation of (a), (b) rotational and (c), (d) lamellar twins in Bi_2Se_3 grown on InP(111)B.

In the case of Bi_2Se_3 grown on Si(111), twinned grains are slightly twisted in the *ab* plane, because the left and the right twins which appear in the same HAADF-STEM image cannot be equally resolved at the same time; a slight tilt of about 1° perpendicular to the grain boundary is required to bring the left twin perfectly oriented along the [100] direction.

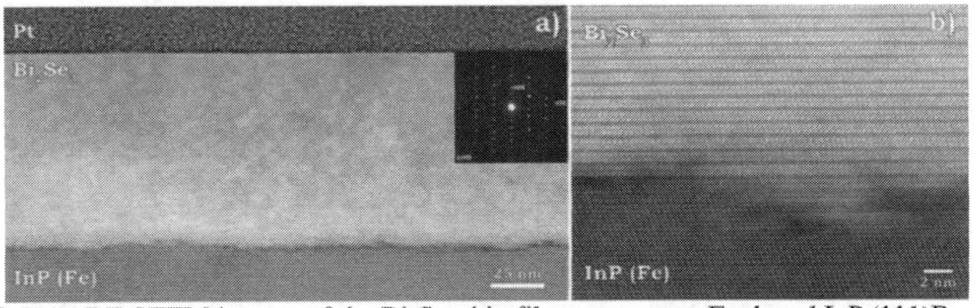

Figure 3. HAADF-STEM images of the Bi_2Se_3 thin film grown on a Fe-doped InP (111)B substrate. (a) Overview, (b) enlargement at the interface region.

In comparison with flat Si(111) and InP(111)B substrates, electron diffraction and X-ray diffraction patterns taken from the Bi_2Se_3 thin film grown on a Fe-doped InP (111)B substrate with a rough surface show no presence of twin domains. Bi_2Se_3 fills the cavities in the substrate, forming a perfect interface between the substrate and the first quintuple layer of the film (Figure 3). The absence of twins was also demonstrated by XRD and AFM experiments. It is worth to mention that this sample also showed a 89% reduction of the carrier density compared to the carrier density obtained for Bi_2Se_3 flakes exfoliated from flat InP(111)B.

3. Conclusions

The microstructure of Bi_2Se_3 thin films grown by molecular beam epitaxy on Si(111), InP(111)B and rough Fe-doped InP(111)B substrates were studied in detail. We observed the formation of twin domains (rotational and lamellar) and a poorly crystallized layer at the interface in the case of both studied flat substrates. The use of a rough Fe-doped InP(111)B substrate allows to suppress twin formation and results in a clear interface between the substrate and the first quintuple layer in the film.

Acknowledgements

This work has been funded by the EU ERC-AG Program (project 3-TOP) and the Alexander von Humboldt Foundation. N.V.T. acknowledges funding by the Bavarian Ministry of Sciences, Research and the Arts.

References

[1] Al Bayaz A, Giani A, Al Khalfioui M, Foucaran A, Pascal-Delannoy F and Boyer A 2003 *J. Cryst. Growth* **258** 135
[2] Sun Z, Liufu S and Chen L 2010 *Dalton Transact.* **39** 10883
[3] Kadel K, Kumari L, Li W Z, Huang J Y and Provencio P P 2011 *Nanoscale Res. Lett.* **6** 57–1
[4] Sun Z, Liufu S, Chen X and Chen L 2010 *CrystEngComm* **12** 2672
[5] Waters J, Crouch D, Raftery J and O'Brien P 2004 *Chem. Mater.* **16** 3289
[6] Watanabe K, Sato N and Miyaoko S 1983 *J. Appl. Phys.* **54** 1256
[7] Fu L, Kane C L and Mele E 2007 *J. Phys. Rev. Lett.* **98** 106803-1
[8] Zhang H, Liu C-X, Qi X-L, Dai X, Fang Z and Zhang S-C 2009 *Nat. Phys.* **5** 438
[9] Li H D, Wang Z Y, Kan X, Guo X, He H T, Wang Z, Wang J N, Wong T L, Wang N and Xie M H 2010 *New J. Phys.* **12** 103038-1
[10] He L, Xiu F, Wang Y, Fedorov A V, Huang G, Kou X, Lang M, Beyermann W P, Zou J and Wang K L 2011 *J. Appl. Phys.* **109** 103702-1
[11] Bansal N, Kim Y S, Edrey E, Brahlek M, Horibe Y, Iida K, Tanimura M, Li G-H, Feng T, Lee H-D, Gustafsson T, Andrei E and Oh S 2011 *Thin Solid Films* **520** 224
[12] Wang Z Y, Li H D, Guo X, Ho W K and Xie M H 2011 *J. Cryst. Growth* **334** 96
[13] Chen J, Qin H J, Yang F, Liu J, Guan T, Qu F M, Zhang G H, Shi J R, Xie X C, Yang C L, Wu K H, Li Y Q and Lu L 2010 *J. Phys. Rev. Lett.* **105** 176602-1
[14] Zhang G, Qui H, Chen J, He X, Lu L, Li Y and Wu K 2011 *Adv. Funct. Mater.* **21** 2351
[15] Tabor P, Keenan C, Urazdhin S and Lederman D 2011 *Appl. Phys. Lett.* **99** 013111-1
[16] Richardella A, Zhang D M, Lee J S, Koser A, Rench D W, Yeats A L, Buckley B B, Awschalom D D and Samarth N 2011 *Appl. Phys. Lett.* **97** 262104-1
[17] Kou X F, He L, Xiu F X, Lang M R, Liao Z M, Wang Y, Fedorov A V, Yu X X, Tang J S, Huang G, Jiang X W, Zhu J F, Zou J and Wang K L 2011 *Appl. Phys. Lett.* **98** 242102-1
[18] Schreyeck S, Tarakina N V, Karczewski G, Schumacher C, Borzenko T, Bruene K, Buhmann H, Gould C, Brunner K and Molenkamp L W 2013 *Appl. Phys. Lett.* **102** 041914
[19] Song C-L, Wang Y-L, Jiang Y-P, Zhang Y, Chang C-Z, Wang L, He K, Chen X, Jia J-F, Wang Y, Fang Z, Dai X, Xie C-X, Qi X-L, Zhang S-Z, Xue Q-K and Ma X 2010 *Appl. Phys. Lett.* **97** 143118–1
[20] Nakajima S 1963 *J. Phys. Chem. Solids* **24** 479
[21] Tarakina N V, Schreyeck S, Borzenko T, Schumacher C, Karczewski G, Brunner K, Gould C, Buhmann H and Molenkamp L W 2012 *Cryst. Growth Des.* **12** 191
[22] Medlin D L, Ramasse Q M, Spataru C D and Yang N Y C 2010 *J. Appl. Phys.* **108**, 043517-1

Journal of Physics: Conference Series **471** (2013) 012044 doi:10.1088/1742-6596/471/1/012044

The microstructure of Ge/Si layers grown at low temperature

V V Roddatis[1], A L Vasiliev[2,3] and M V Kovalchuk[2,3]

[1] CIC energiGUNE, Parque Tecnológico de Álava. Albert Einstein 48, 01510 Miñano, Álava, Spain.
[2]NRC «Kurchatov Institute», pl. akademika Kurchatova, 1, 123182, Moscow, Russia
[3]Shubnikov Institute of Crystallography RAS, Leninsky pr. 59, 117333, Moscow, Russia

a.vasiliev56@gmail.com

Abstract. Multilayer Si/Ge heterostructures with the thickness of Ge layers varying from 2 to 12 monolayers (ML) were formed by molecular beam epitaxy (MBE) on the (001) Si substrates at 300°C (Ge) and 450°C (Si). The study of the Si/Ge heterostructures was performed by transmission and Cs corrected scanning transmission electron microscopy (STEM). It was shown that the growth of Ge layers up to thickness of 5 ML occurs through the Frank - van der Merwe mechanism. For thicker Ge layers the growth mechanism of the Si-Ge heterostructure changes to Stranski - Krastanov with Si-Ge islands having the shape of inverted pyramids. The Si-Ge layer intermixing was discussed.

1. Introduction

Si/Ge heterostructures have come into focus of intensive research owing to the possibility of using these systems in the future quantum-optical applications. The growth mechanism of $Si_{1-x}Ge_x$ layers on Si substrates were considered in a number of surveys, for example Paul [1] and Brunner [2]. It is known that up to a critical thickness h_c the growth of a pseudomorphic two-dimensional layer proceeds through the Frank - van der Merwe (FM) mechanism. When the layer thickness exceeds h_c, stress relaxation occurs through the formation of surface relief of the $Si_{1-x}Ge_x$ layer and the appearance of 60^0 misfit dislocations (MDs) with a Burgers vector a/2<110>. In this case the growth of $Si_{1-x}Ge_x$ layer occurs via Stranski - Krastanov (SK) mechanism, leading to the formation of self-assembled $Si_{1-x}Ge_x$ islands in heterostructures with a variety of morphological types depending on the growth parameters [3,4]. The size of these islands formed on the surface of the growing Ge layer decreases with lowered growth temperature (T_g) due to the reduction of the surface thermal diffusion coefficient of the deposited atoms. $Si_{1-x}Ge_x$ islands adopted the shape of inverted pyramids were observed in multilayer structures with thick Ge layers (> 3 nm) grown at T_g=250-350°C by conventional transmission electron microscopy (TEM) [5-8] and analyzed with energy dispersive X-ray (EDX) microanalysis [6, 8] and other techniques. Crystalline Si-Ge islands with a size of the groove being more than 9 nm long grew into the underlying layers of Si. The authors [5-8] explained the formation of such islands by the intermixing of Si and Ge due to the diffusion caused by stresses in the layer. Most studies of intermixing process were performed for few monolayers of Ge on Si. To be used in applications Ge nanoislands should be embedded in a crystalline semiconductor matrix, yet the morphology and structure of these "buried" layers depend on the growth parameters of subsequent layers and may differ significantly from the bare layer of Ge. Therefore, we performed the study of the Ge/Si multilayer heterostructures with different Ge layer thickness, grown by MBE at low

Journal of Physics: Conference Series **471** (2013) 012044 doi:10.1088/1742-6596/471/1/012044

temperatures with the application of Cs-corrected STEM to investigate the growth mechanism of Ge and Si layers and intermixing processes at the Si/Ge interfaces.

2. Experimental

Multilayer Si/Ge nanostructures were grown by MBE on low-doped p-type (001) Si substrates. After the standard substrate surface cleaning a Si buffer layer with a thickness of 200 nm was grown at a temperature of 750°C by MBE. Then, heterostructures composed of five layers of Ge with nominal thicknesses d_{nom} =2, 3, 5, 7, 10 and 12 MLs (denoted further as Ge2ML…Ge12ML) separated by Si layers with the thickness of 10 nm were deposited by MBE (one ML thickness corresponds to $d(004)_{Si}$=0.136 nm). The deposition rate of Ge and Si was equal to 0.005 nm/s and 0.1 nm/s, respectively. The growth temperature of Ge was T_{Ge}=300 °C. The intermediate Si layers were grown by the two-stage process: the first layer of 2 nm was grown at T_{Si}=300 °C, while the subsequent layer of ~7 nm was grown at T_{Si}=450°C. All heterostructures were grown at the University of J. Kepler (Linz, Austria) by M.M. Rzaev (Lebedev Physical Institute, Moscow).

Cross sections of the heterostructures were prepared for transmission electron microscopy (TEM) study by the standard ion milling (Ar^+) procedure. The study was performed using a TITAN 80-300 TEM/STEM (FEI, USA) with a spherical aberration corrector (probe corrector) operating at an accelerating voltage of 300 kV. High-resolution (HR) dark-field images were recorded using high-angle annular dark-field detector (HAADF (Fischione, USA)). An energy-dispersive X-ray spectrometer - EDXS (EDAX, USA) was used to study the chemical composition of the heterostructures.

3. Results and discusion

HAADF STEM images of Ge2ML, Ge3ML and Ge5ML clearly demonstrate that the growth of the heterostructures occurs via the FM mechanism: the thickness of the Ge layers is uniform (Figure 1). Due to the considerable difference in atomic numbers, the Ge layers appeared much brighter than the Si layers. The Si spacer layers also follow a layer-by-layer growth mode. The Si/Ge interface roughness and the Ge layer thickness were 1 ML at the top and bottom of Ge/Si interfaces. In these samples the misfit dislocations (MD) are absent, the lattice mismatch between Si and Ge is compensated by the tetragonal distortions. The intensity profiles obtained from 2 ML, 3 ML and 5 ML samples contain a similar peculiarity (Figure 2). The right shoulder, which corresponds to the upper interfaces of Ge, is 0.6-0.7 times wider than the left one, indicating a large diffusiveness of the upper Ge interfaces. The intermixing process occurs within one ML at the lower interface.

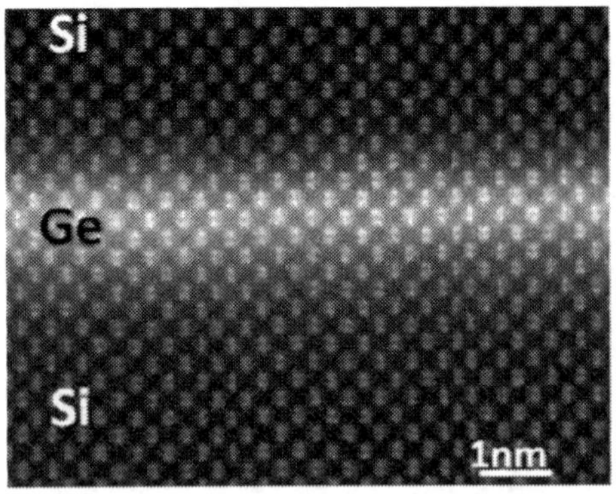

Figure 1. HRSTEM HAADF image of Ge layer in the Ge5ML heterostructure.

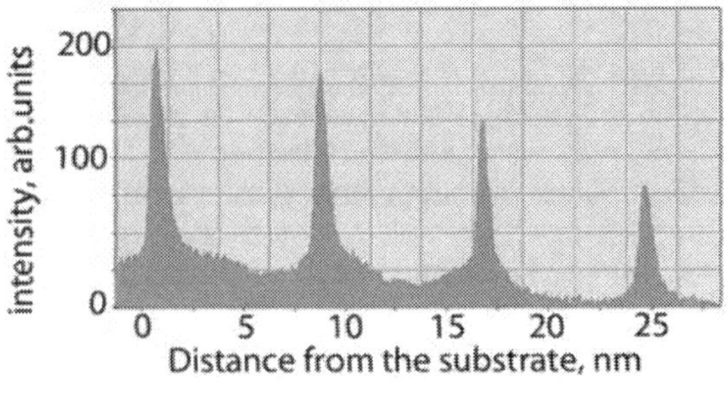

Figure 2. The intensity profile across the Ge3ML heterostructure.

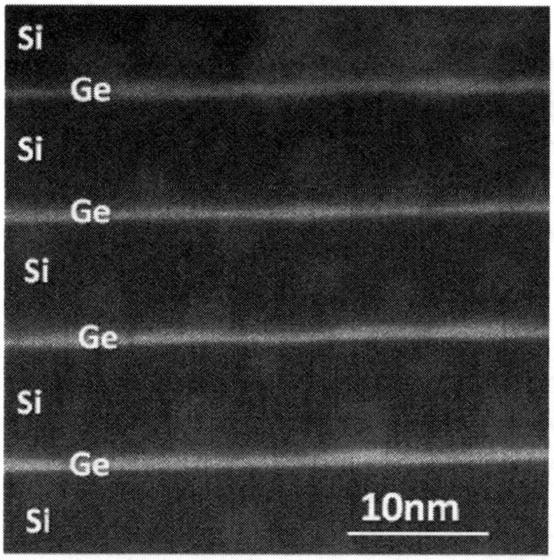

Figure 3. STEM HAADF image of Ge7ML heterostructure.

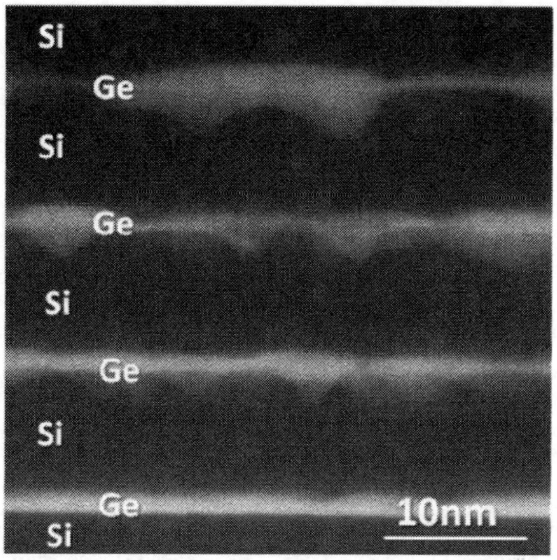

Figure 4. STEM HAADF image of Ge12ML heterostructure.

In the Ge7ML heterostructure the morphology of the Ge layer is different from the previous one (Figure 3). Thickness variations of the Ge layers became pronounced and both interfaces are "wavy". The thickness of the Ge layer determined from these images changed by a factor of 2. However, the defects associated with the Ge-Si lattice mismatch were not found, neither in the HAADF HR STEM nor in the bright-field (BF) HRTEM images. The stress could be released through the tetragonal distortions together with surface topography and Si/Ge intermixing of the layers. The morphology of Ge layers indicated that for that heterostructure the Ge layer growth mechanism could change to SK. However, the lattice of Ge remains tetragonally distorted and that promotes subsequent layer-by-layer growth of the Si spacer.

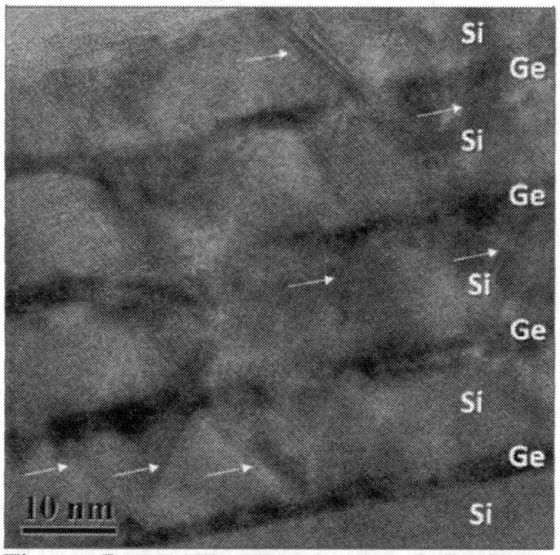

Figure 5. BF HR TEM image of Ge12ML heterostructure. The Ts and SFs are shown by arrows.

Once the thickness of the Ge layer reaches 10 ML (1.42 nm according to TEM data), the morphology of the heterostructures changes again: the Stranski-Krastanov mechanism of the film growth (Figure 4), where all the Si-Ge interfaces (except for the Ge/Si buffer) are not flat and inverted pyramids at the bottom of the Ge layers appear. The dimensions and density of inverted pyramids in the layers increases towards the surface of the heterostructures. If Ge/Si intermixing takes place, the morphology of all the interfaces in the heterostructure has to be similar, so not the intermixing was responsible for the formation of inverted pyramids. The shape of these "hut" islands is very close to the ones described in [9] for the Ge islands formed on the substrate with pits. The defects such as stacking faults (SFs), twins (Ts) and MDs, release the strains. These defects were studied in BF HRTEM mode (Figure 5).

The Ge/Si lattice misfit was accommodated by the formation of a 60^0 dislocations with Burgers vector ***a***/2<011>. The second mechanism of misfit accommodation was the formation of SFs and Ts accompanied by dislocations with the same Burgers vector projection (***b***=1/6<112>) in the Ge12ML heterostructures. BF HRTEM images demonstrate that Ts and SFs threaded one or two (Si+Ge) layers of the heterostructure, but in most cases these defects were terminated at the Si/Ge interface. The density of defects, estimated from the BF TEM images is~ $3 \cdot 10^{13} cm^{-2}$ in the Ge10ML heterostructure and becomes three times higher (up to $10^{14} cm^{-2}$) in the Ge12ML heterostructure. The Fast Fourier transforms (FFT) of an inverted pyramid obtained from the HAADF HR STEM image of the Si spacer layer and the inverted pyramid shows the shift to the larger inter-planar distance in Ge relative to the Si ones. This indicated the stress relaxation in the Ge layers and changes in growth mechanism from FM to SK. Sequentially, the growth mechanism of the Si spacer layers also switches to SK. The appearance of SFs and Ts enhanced the roughness of the Si spacer layers surfaces during their growth, because the growth rate of Si along <001> and <221> differs by a factor of 10 [10]. That was the reason for the appearance of faceting on Si spacer layers surface, which are similar to the artificial pits, formed on the Si surface. It was found in [11] that the Ge islands nucleated preferentially at the bottom of the pits. The roughness of Ge layers bottom interfaces was increasing from 1.26 nm to 2.16 nm for Ge10ML and Ge12ML heterostructures, respectively and that we associate with the increase of Si-Ge lattice mismatch due to the stress release for the thicker Ge layers and the increasing unevenness of the Si spacer top surface due to SFs and Ts. As it was discovered in [6], due to the formation of the misfit dislocations, samples with a larger Ge layer thickness tend to have less extent of Si-Ge intermixing. We observed an increase of inverted pyramid dimensions and their density in the 12ML sample relative to 10ML sample which was another indication that intermixing was unlikely as the mechanism of inverted pyramid formation.

4. Conclusion

A study of the structural properties of multilayer Si/Ge heterostructures grown by MBE at low temperature T_g =300°C was performed. For the Ge layer thicknesses up to 1 nm the formation of heterostructures occurs through the FM mechanism. The increase of the Ge layer thickness up to 1.1 nm leads to a change of the Ge layers morphology and growth mode to SK. Further increase of the Ge layer thickness causes the formation of 60° misfit dislocations, stacking faults and twins, and the change of the Si surface morphology and switching of the Si spacer growth mechanism to SK. These peculiarities of Si growth and not the Si/Ge intermixing results in the formation of the inverted Ge pyramids.

References

[1] Paul D J 2004 *Semicond. Sci. Technol.* **19** 75.
[2] Brunner K 2002 *Rep. Prog. Phys.* **65** 27.
[3] Voigtländer B 2001 *Surf.Sci. Rep.* **43** 127.
[4] Rastelli A, Stoffel M, Denker U, Merdzhanova T, and Schmidt O.G 2006 *Phys. Stat. Sol.* (a) **203** (14) 3506.
[5] Cheng H H, Chia C T, Markov V A, Guo X J, Chen C C, Peng Y H, and Kuan C H 2000 *Thin Solid Films* **369** 182.
[6] Soo Y L, Kioseoglou G, Huang S, Kim S, Kao Y H, Peng Y H, Kuan C H and Cheng H H 2000 *Appl.Phys.Lett.* **78** 3684.
[7] Lee J R, Lin S C, Lu C R, Lin J H, Chia C T, Chang H H 2008 *J. Phys .Chem. of Solids* **69**, 490.
[8] Cheng H H, Huang W P, Mashanov V I, and Sun G 2010 *J. Appl.Phys.***108**, 044314.
[9] Schülli T U, Vastola G, Richard M-I, Malachias A, Renaud G, Uhlík F, Montalenti F, Chen G, Miglio L, Schäffler F and Bauer G 2009 *Phys. Rev. Lett.* **102**, 025502.
[10] Csepregi L, Kennedy E F, Mayer J W, and Sigmon T W 1978 *J. Appl. Phys.* **49**, 3906.
[11] Chen G, Lichtenberger H, Bauer G, Jantsch W, Schaffler F 2006 *Phys.Rev B*.**74**, 035303.

Journal of Physics: Conference Series **471** (2013) 012045 doi:10.1088/1742-6596/471/1/012045

Measuring the work function of TiO_2 nanotubes using illuminated electrostatic force microscopy

Graham X R Smith[1], Rolf Crook[1] and Jay D Wadhawan[2]

[1]Energy Research Institute, Energy Building, Faculty of Engineering, University of Leeds, Woodhouse Lane, Leeds, LS2 9JT, UK
[2]Department of Chemistry, University of Hull, Cottingham Road, Kingston-Upon-Hull, HU6 7RX, UK

E-mail: pmgxrs@leeds.ac.uk

Abstract. The varying nature of TiO_2 nanotube work function as a function of illumination wavelength has been determined using illuminated electrostatic force microscopy. The dark work function was found to be 4.902 eV, with the largest change in work function due to illumination being at 300 nm, which was higher than the work function for bulk TiO_2 (4.899 eV). The change in work function due to illumination arises from the flattening of the energy bands at the surface due to charge migration.

1. Introduction

It is a well known fact that solar energy has the potential to provide all of the Earth's current power needs many times over, but the reason behind it not becoming a major industry as of yet is because of the difficulties in harnessing the power that is available. There are a huge variety of solar cells now being studied across the globe, including single-crystalline Si, dye-sensitized solar cells, quantum dot solar cells, polymer/organic cells, nanocrystal/nanostructure solar cells, multi-junction solar cells. Even within the latter of these streams of solar cell research there are several areas, including copper-indium-gallium-selenide (CIGS), gallium arsenide (GaAs) and gallium-indium-phosphide ($GaInP_2$) [1-4]. The research presented here has applications to nanostructure solar cells and multi-junction solar cells.

A problem associated with multi-junction solar cells is their loss of efficiency due to mis-aligned energy bands across the materials. As the electrons try to move through the device to produce the electrical current, they lose energy across the junctions between materials which results in them either falling back and recombining or only contributing a very small current, and thus the efficiency of the overall device is lowered. This research aims to look at how illuminated electrostatic force microscopy can be used as an aid for better material selection to help prevent this.

Titanium dioxide (TiO_2) is a relatively abundant, well-characterised material that has been used extensively for solar applications. It is commonly used as a photocatalytic material, both in experimental/research use for splitting water, and also as an end product for keeping windows or glass clean. By having TiO_2 in a nano-form, it allows us to build a theoretical multi-junction cell, as the nanotubes are formed on top of bulk TiO_2. These nanotubes are also currently being heavily studied for a great range of applications, but most importantly in relation to this research, they are one of the best materials at splitting water photocatalytically [2] – a process facilitated by a solar cell. It's also useful to consider how the work function of the nanotubes will differ from the bulk material, as well as whether they cause a shift in absorption of TiO_2 from the UV to the visible range of the electromagnetic spectrum.

2. Experimental

2.1. Sample Preparation
The titanium dioxide nanotubes were prepared by simultaneous anodisation and etching using a two-electrode electrochemical set-up, with electropolished titanium foil as the working electrode and a carbon rod as the counter/reference electrode, and non-aqueous electrolyte of 0.01% ammonium fluoride in glacial acetic acid. A bias of +40V was applied for 30 minutes. The sample was then washed with deionised water and dried under nitrogen. One side of the sample was then manually polished, and fixed to a manually polished sample puck using conductive carbon tape.

2.2. Illuminated Electrostatic Force Microscopy (iEFM)
The EFM measurements were taken on a Bruker MultiMode 8 microscope attached to a Nanoscope IV controller and an electronics extender module. The probes used for the measurements were Bruker SCM-PICs, with a force constant of 0.2 N m^{-2} and a resonant frequency of approximately 13 kHz. These probes have pyramidal tips that are 13 μm in height. They are made of silicon and coated with PtIr, having a work function of 5.18 eV. The range of applied bias for the EFM was from -4V to +4V.

The illumination came from a Bentham SSM150Xe xenon arc lamp with a typical output of 4 mW attached to a liquid light guide and a lens. This produced a light spot of approximately 1mm diameter on the sample. The wavelengths used in the experiments ranged from 200 – 700 nm.

3. Results and Discussion

3.1. SEM images

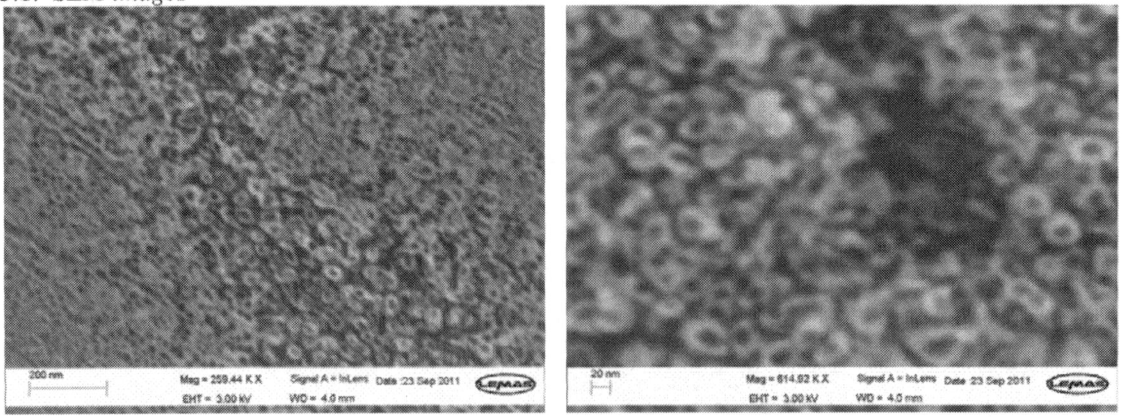

Figure 1 TiO_2 nanotubes seen at the surface of the sample. The figure on the right is a magnified version of the left figure. It gives a better indication of the size of the nanotubes, as they have an average diameter of 32nm.

The images above show that the nanotubes were formed on the surface of the TiO_2. Energy dispersive X-ray (EDX) analysis showed that the main components within the nanotubes were Ti (75%), O (20%) and F (5%). It is possible that N is also present, but the peak for N is hidden behind Ti in EDX. Both F- and N-doping are expected due to the ammonium fluoride in the synthesis.

3.2. Atomic Force Microscopy (AFM)

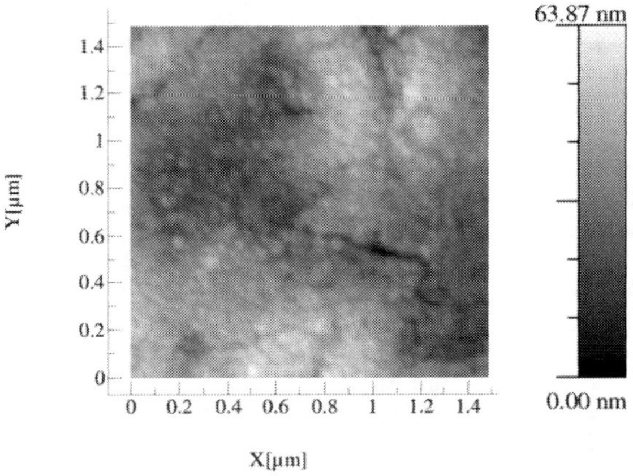

Figure 2 AFM image of TiO_2 nanotubes formed by simultaneous etching and anodisation of TiO_2 foil.

The nanotubes can be seen clearly in the AFM image in the form of round marks. They do not appear as rings due to tip convolution effects being greater than the diameter of the nanotubes (~32 nm).

3.3. iEFM Results

The work functions of the TiO_2 and the TiO_2 nanotubes were calculated by determining the contact potential difference between the tip and the sample. By knowing the work function of the tip (5.18 eV), the work function of the TiO_2 and the nanotubes can then be calculated using the equation:

$$\Delta V_{CPD} = \phi_t - \phi_s$$

where ΔV_{CPD} is the contact potential difference, and Φ_t and Φ_s are the work functions of the tip and the sample respectively. This was calculated to be 4.899 ± 0.02 eV for the bulk TiO_2.

The table below shows the calculated values for the TiO_2 nanotubes in the dark and under illumination:

Table 1 Work function of TiO_2 nanotubes under illumination at wavelengths between 200 and 700 nm

wavelength (nm)	work function (eV)
Dark	4.902
200	4.877
250	4.877
300	4.871
350	4.882
400	4.889
450	4.892
500	4.893
550	4.900
600	4.906
650	4.904
700	4.905

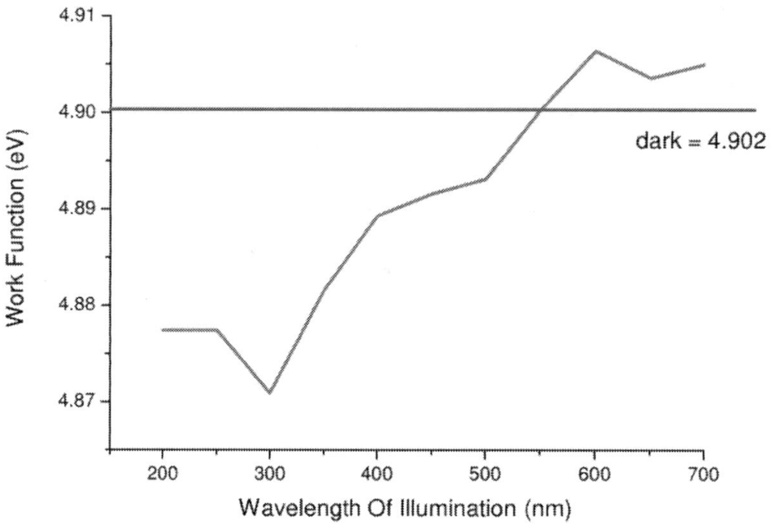

Figure 3 Plot of the work function of TiO_2 nanotubes varies due to illumination.

The greatest change in work function occurs under illumination at 300 nm which coincides with the bandgap of TiO_2 being 3.2 eV. The decreased work function under illumination is due to the flattening of the energy bands at the surface as the electrons migrate into the bulk. Decreases in the work function seen at wavelengths greater than the band gap are due to surface states available on the TiO_2 nanotube surface as a result of fluorine and nitrogen doping. These produce trapped oxygen vacancies where electrons can migrate across the band gap in two stages via lower energy before migrating into the bulk.

4. Conclusion

We have shown that the work function of TiO_2 nanotubes can be measured using electrostatic force microscopy to be 4.902 eV. We have also shown that this work function varies depending on the wavelength of light that is illuminating the sample, with the greatest change in work function seen for an illuminating wavelength of 300 nm. This matches up with the expected change for TiO_2, which typically has the greatest photoresponse in the UV region due to its bandgap being 3.2 eV. The changes seen at higher wavelengths are due to the complex nature of the surface states commonly found on TiO_2. The relevance of these results is that by understanding how the work function changes under light, it will be possible to select more appropriate materials for multi-junction solar cells, thereby reducing energy loss.

Acknowledgements

Thanks to the White Rose University Consortium for funding towards this research.

References

[1] Chae J and Kang M. 2011 *Journal of Power Sources* **196**, 4143
[2] Günes S, Neugebauer H and Sariciftci N S 2007 *Chem. Rev.* **107**, 1324
[3] Cartlidge E 2007 *Physics World*, **July**, 20
[4] Lee Y-J, Ruby D S, Peters D W, McKenzie B B and Hsu, J W P 2008 *Nano Letters* **8**, 1501
[5] Osterloh F E 2013 *Chemical Society Reviews* **42**, 2294

Journal of Physics: Conference Series **471** (2013) 012046 doi:10.1088/1742-6596/471/1/012046

Delayering of Microelectronic Devices Using an Adjustable Broad-Beam Ion Source

A C Robins, R R Cerchiara, P E Fischione, M F Boccabella, J M Matesa, L M Marsh and Z Zhang

E. A. Fischione Instruments, Inc., 9003 Corporate Circle, Export, PA 15632 USA

Email: rr_cerchiara@fischione.com

Abstract. Analysis of the integrated circuits of a microelectronic device depends on delayering. Focused ion beam (FIB) or broad ion beam (BIB) milling are effective complementary methods of delayering. FIB provides higher removal rates, but is limited in the effective area that can be revealed per unit time, while BIB provides lower removal rates, but has the advantage with respect to the size of the field of view produced. Microstructural features and the appearance of defects were identified and tracked for two model systems: Cu vias and Cu TSVs (through-silicon vias).

1. Introduction

Analysis of the integrated circuits of a microelectronic device depends on delayering. As the scale of the microstructure has decreased below the 32 nm node, the need for greater precision and planarity of the etched surface has increased. Options for rapid material removal include mechanical, plasma, and ion-based techniques. [1] Mechanical methods require the use of abrasives and lubricants that can produce residual contamination and damage.

Focused ion beam (FIB) or broad ion beam (BIB) milling are effective complementary methods of delayering. FIB provides higher removal rates, but is limited in the effective area (> 10 μm x 10 μm) that can be revealed per unit time. BIB provides lower removal rates, but has the advantage with respect to the size of the field of view (> 1 mm x 1 mm) produced.

2. Materials and methods

2.1. Materials

Dies containing Cu vias and Cu TSVs (through-silicon vias) were produced by proprietary processes.

2.2. Methods

A recently developed ion source technology was applied to the delayering process. [2] Adjusting (expanding) the diameter of the Ar ion beam at a given accelerating voltage (kV) allows areas to be planarized approaching 10 mm x 10 mm. An ion beam diameter of ~ 2 mm was used for the current tests detailed below. Laser detection of sample height permitted sample transfers between the ion milling instrument and the field emission scanning electron microscope (FESEM) without loss of the precise milling geometry. One or more ion sources were tilted to a low incident angle with respect to the plane of the given sample. A time series of FESEM images was acquired to document the delayering process. Microstructural features and the appearance of defects were identified and tracked for two model systems: Cu vias and Cu TSVs (through-silicon vias).

3. Results and discussion

3.1. Cu vias

Following mechanical preparation, a cross-section was planarized by ion milling with the following parameters: single ion source, 4 kV, 7° incident angle, 45% focus, and ± 70° rocking angle for 45 minutes. 1 μm of Cu overburden was observed (see figure 1). After mounting an identical sample in plan-view, this overlayer was removed using the following optimized parameters: 2 ion sources, 4 kV, 4° incident angle, 45% focus, and 360° rotation for 34 minutes. Further ion milling at 30 seconds or 1 minute iterations with imaging in a field emission scanning electron microscope (FESEM) revealed voids that had been apparent in the cross-section.

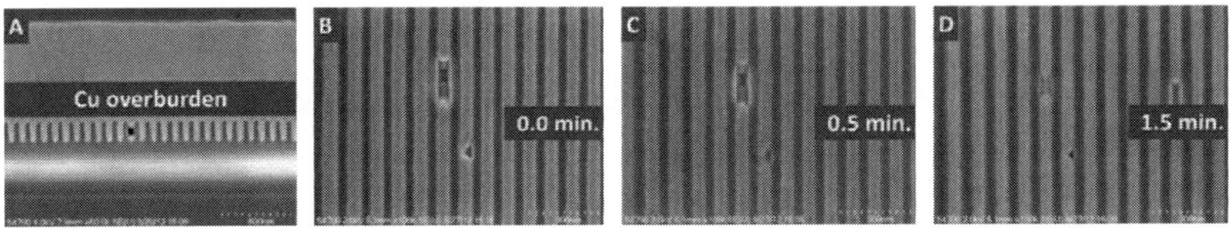

Figure 1. SEM images of Cu vias: Cross-sectional view before (A) and plan-views after (B, C, and D) removal of overburden.

3.2. Cu TSVs

Following mechanical preparation, a cross-section was planarized by ion milling with the following parameters: single ion source, 4 kV, 7° incident angle, 45% focus, and ± 55° rocking angle for 35 minutes. 2 μm of Cu overburden was observed (see figure 2). After mounting an identical sample in plan-view, the sample was delayered by ion milling with the following parameters: 2 ion sources, 4 kV, 5° incident angle, 45% focus, and 360° rotation. Changes in the microstructure of an individual TSV were tracked and the process terminated once the ion-milled surface was clearly below the terminus of an axial defect.

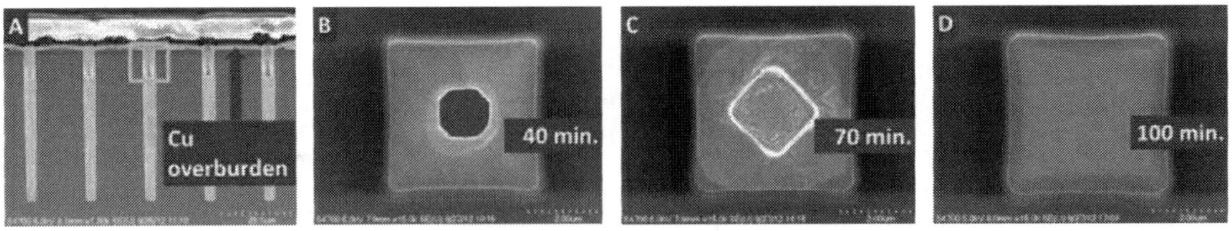

Figure 2. SEM images of Cu TSVs: Cross-sectional view before (A) and plan-views (B, C, and D) after removal of overburden.

3.3. Delayering plus FIB

After delayering, FIB can be used to create a site-specific cross-section of an individual defect. A typical scenario is presented in figure 3. An *in situ* FIB lift-out sample is obtained. Residual amorphization and implantation due to the interaction of the sample with Ga ions may occur during this process. Post-FIB clean up using the Model 1040 NanoMill® TEM specimen preparation system is the final step prior to imaging in a field emission transmission electron microscope (FETEM) [3].

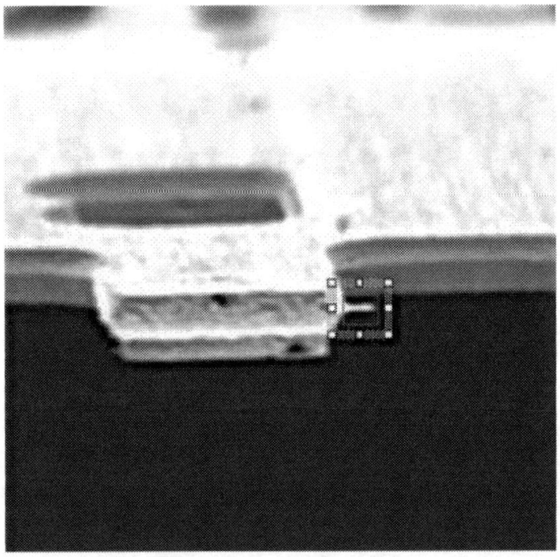

FIB *in situ* lift out.

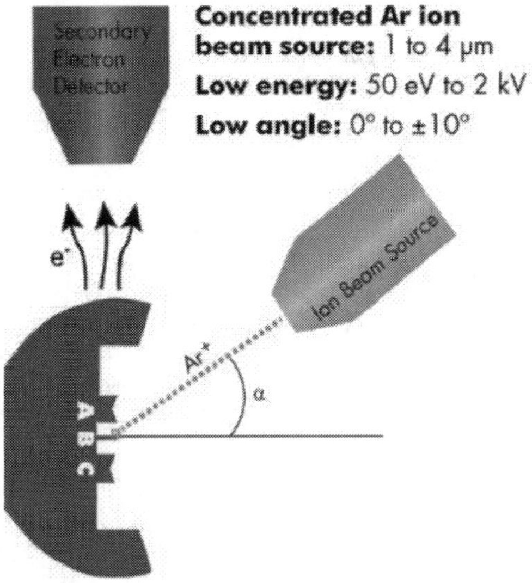

NanoMilling℠ process.

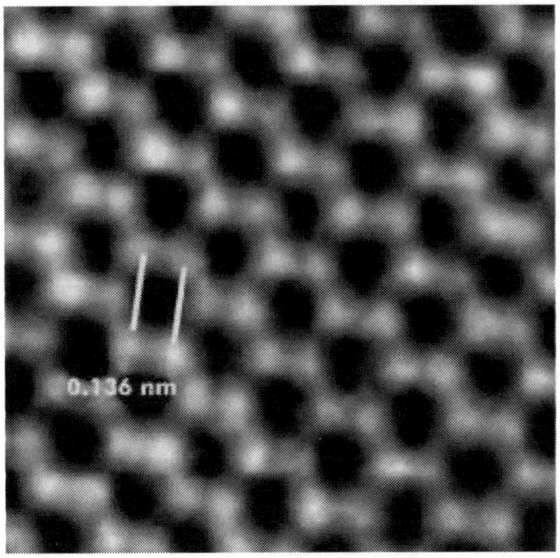

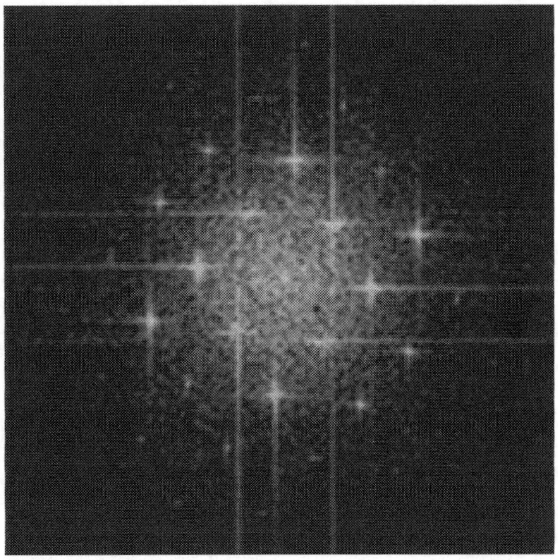

High Resolution TEM image and FFT of Si <110>.

Figure 3. FIB and post-FIB processing of a delayered microelectronic material.

4. Summary

An adjustable broad beam Ar ion source was used to delayer microelectronic materials. Overburden and defects were assessed in cross-section, then iterative ion milling and SEM imaging of Cu vias and TSVs produced corresponding microstructural information as a function of depth. FIB and post-FIB processing of an *in situ* lift-out sample can yield site-specific information concerning individual defects.

Acknowledgements
The high resolution TEM image and FFT of Si are courtesy of Angus Kirkland and Crispin Hetherington, University of Oxford, UK.

NanoMill is a registered trademark and NanoMilling is a service mark of E.A. Fischione Instruments, Inc.

References

[1] Torrance R and James D 2007 *Custom Integrated Circuits Conf 2007* (Washington, DC: IEEE) pp 429-36

[2] Cerchiara R R, Fischione P E, Lange J H, Woods P, Matesa J M, Boccabella M F and Robins A C 2011 *Microsc. Microanal.* **17**(S2) pp 390-1

[3] Cerchiara R R, Fischione P E, Liu J, Matesa J M, Robins A C, Fraser H L and Genc A 2011 *Microscopy Today* **19**(1) pp 31-3

Journal of Physics: Conference Series **471** (2013) 012047 doi:10.1088/1742-6596/471/1/012047

Study of site controlled quantum dot formation on focused ion beam patterned GaAs substrate

H Zhang, I M Ross and T Walther
Department of Electronic and Electrical Engineering, Mapping Building, University of Sheffield, Sheffield S1 3JD

E-mail: t.walther@sheffield.ac.uk, elp08hz@sheffield.ac.uk

Abstract. In this project, we combine Ga^+ focused ion beam (FIB) patterning and self-assembly of InGaAs quantum dots in order to produce regular quantum dot (QD) arrays. The GaAs substrate is patterned before transporting into molecular beam epitaxy growth chamber to do anneal and overgrowth. The thickness of deposited InAs is precisely controlled. Quantum dots are expected to nucleate at specific locations, where the ion beam has previously implanted gallium atoms. The properties of the quantum dots formed are related to the FIB beam parameters, which include accelerating voltage, probe current, dwell time and pitch. We studied and statistically analyzed the relationship, focusing on their diameter, height and distribution. Scanning electronic microscope (SEM) and atomic force microscope (AFM) images have been used to analyze the patterns and QDs before and after overgrowth. Micro-PL study was performed to test the QD opto-electronic property. Scanning transmission electron microscope (STEM) cross sectional analysis and X-ray mapping is performed.

1. Introduction

Quantum dots (QDs) are small artificial crystal structures, whose electronic properties depend strongly on the size. One of the dominating fabrication techniques is Stranski-Krastanov growth by molecular beam epitaxy (MBE). It is a top-down approach, which is a quick and easy way to grow QD on a large area. But the inherently random position of nucleation puts limits on down-sizing devices nowadays. If only a few quantum dots are included in a device, the location and uniformity of quantum dots will have a strong influence on the reproducibility of the electronic and opto-electronic performance across multiple devices. The bottom-up growth by patterning the substrate before overgrowth is an alternative approach which can produce regular QDs on patterned substrates with better position control and homogeneity [1].

We studied InGaAs quantum dots overgrowth on Ga^+ focused ion beam (FIB) patterned GaAs (001) substrate. QDs are expected to nucleate at specified locations structure by the ion beam. The aim is to study how patterns fabricated with different ion beam parameters could lead to different nucleation results.

2. Patterning and overgrowth

The patterns were fabricated in a JEOL 6500F scanning electronic microscope (SEM) equipped with an Orsay Physics Ga+ FIB column, which makes it a dual-beam SEM/FIB instrument. A Raith lithography system was used to design and fabricate the patterns. The patterning ion beam had different parameters including accelerating voltage, probe current, dwell time and pitch. After FIB patterning, in-situ secondary electron images (SEI) were captured at each individual patterned region. Before transferring the patterned substrate into the MBE chamber for quantum dot growth, atomic force microscope (AFM) images were also captured in tapping (non-contact) mode. During overgrowth, the sample was degassed at 400°C for 1 hour and then annealed under an arsenic flux at 580°C for 30 minutes, during which the surface was monitored using reflection high-energy electron diffraction (RHEED). A clear (2x4) reconstruction was observed after this process, which is indicative of a smooth and relatively oxide-free surface. Then a 20nm GaAs layer was grown at 580°C and a growth rate of 0.5ML/s. The thickness of this layer was deemed sufficient to reduce the effect of any surface ion damage and yet not to fill in the FIB-generated patterns. Finally, InAs quantum dots were grown at 520°C and a growth rate of 0.02ML/s by depositing around 1.5 monolayers of material. The amount of deposited material was judged to ensure that quantum dots nucleated mainly at patterned locations and not on the flat surface between or outside the patterned regions [2]. The finished sample was imaged again by SEM and AFM respectively. Two pairs of SEM and AFM images are shown below in figure 1 and figure 2. The images clearly indicate island structures were formed during overgrowth, at specified locations where ion beam had scanned before.

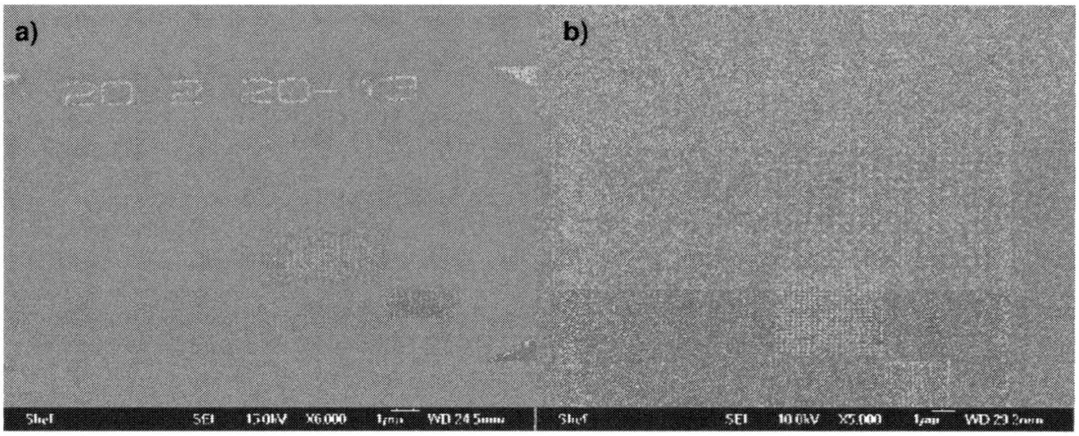

Figure 1. Secondary electron images captured before (a) and after (b) overgrowth. The patterning ion beam parameters are accelerating voltage of 20kV, probe current of 4pA and dwell time 20ms.

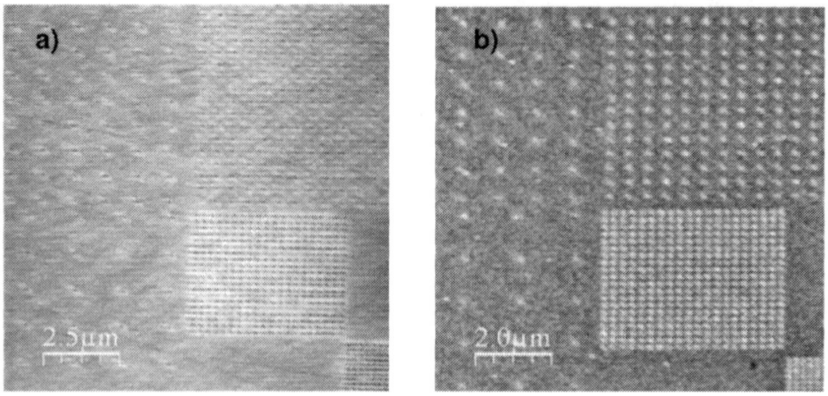

Figure 2. AFM images captured at the same region as fig. 1, before (a) and after (b) overgrowth.

Apart from the island structures shown above, two other types of structure formed, rings and holes. Figure 3 shows the relationship between the patterning ion beam parameters and the structures produced. The result shows that short dwell times and small aperture sizes (low doses) produced dots while long dwell times and big aperture sizes (high doses) produced rings or holes. The height and diameter of islands are measured by analysing AFM images. The measurements were collected from three patterns formed at 20keV ion energy and 5pA probe current with three different dwell times (1ms, 5ms and 20ms), as well as from one pattern formed at 20keV ion energy but with the probe current increased to 40pA at a dwell time of 1ms. From the left part of figure 4 can be seen that the island heights are almost constant, covering a narrow range from 12nm to 25nm, with values slightly increasing with pitch. The diameter scales with the square root of the pitch, as can be seen from the right part of figure 4, and is also increasing with the dwell time and probe current, which is not so obvious from the diagram. The result confirms the possibility of controlling the nucleated quantum dot size. However, it is clearly difficult to create islands under 100nm in diameter and below 10nm in height, which will be a prerequisite for strong quantum confinement effects in optical spectra.

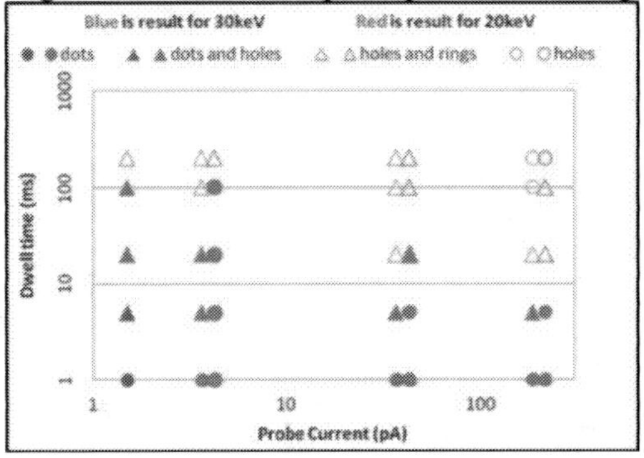

Figure 3. Different ion beam parameters and the resulting structures formed over the patterns. The ion dose is increasing from bottom left to top right.

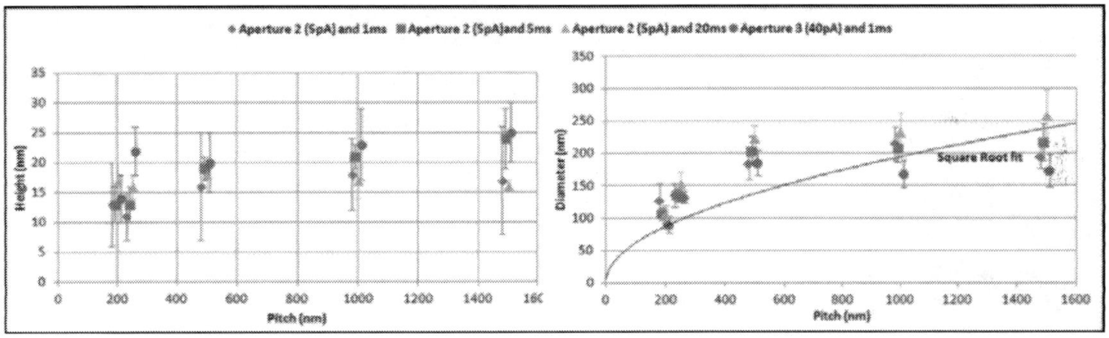

Figure 4. Height and diameter of islands vs. pitch

3. Micro photoluminescence and TEM cross sectional analysis

Micro photoluminescence (μ-PL) studies were performed at room temperature. The exciting laser (green) had a diameter of ~5μm, a wavelength of 532nm and 1-2mW power. μ-PL measurements were performed scanning the laser beam along a straight line from positions outside the patterns, where QDs had randomly nucleated on the surface, across the entire patterned area with a step size of 5μm. The spectra are not as sharp as expected [3]. There are two rather broad peaks in the spectra at ~878nm (1.41eV) and ~1152nm (1.08eV). They are related to the underlying GaAs substrate and the InGaAs QD signal. From the latter peak wavelength we can estimate the compositional ratio of Ga:In atoms as 74:26 (i.e. $In_{0.26}Ga_{0.74}As$). The FIB patterned area shows higher signals for both peaks, where the signal from the GaAs substrate increases more rapidly. The enhanced optical contributions from both the pure GaAs and the InGaAs islands may be explained by indium atoms from the wetting layer formed upon epitaxy having been more efficiently integrated into the islands when the surface has previously been FIB patterned, confirming the formation of QDs.

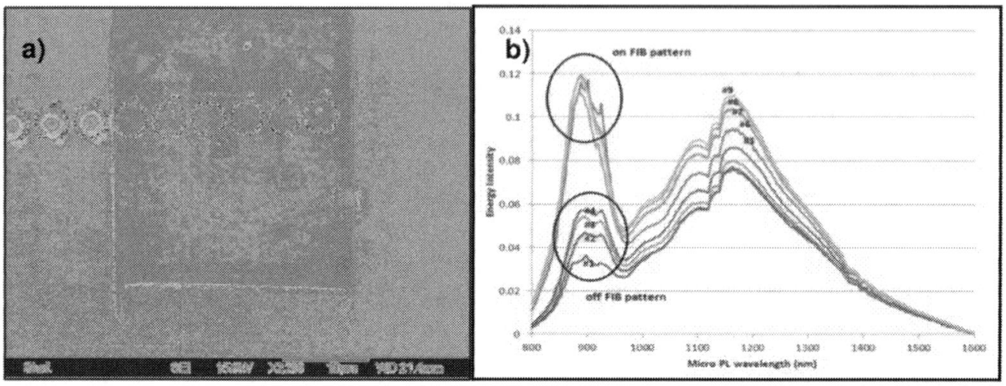

Figure 5. Set of experimental μ-PL spectra on and off the pattern region (a). The intensities at two wavelengths, 878nm (GaAs) and 1152nm (InGaAs), were extracted at each scan position (b).

STEM cross sectional analysis was performed on a FIB prepared membrane which has a row of island embedded. Annular dark-field and bright-field STEM imaging and X-ray maps of GaK, AsK and InL are shown in figure 6. All images are background subtracted with the maximum element counts listed below. We calculate the In composition from their relative intensity ratios. The X-ray maps indicate a peak indium concentration of $x \approx 0.16$ near the top of the islands.

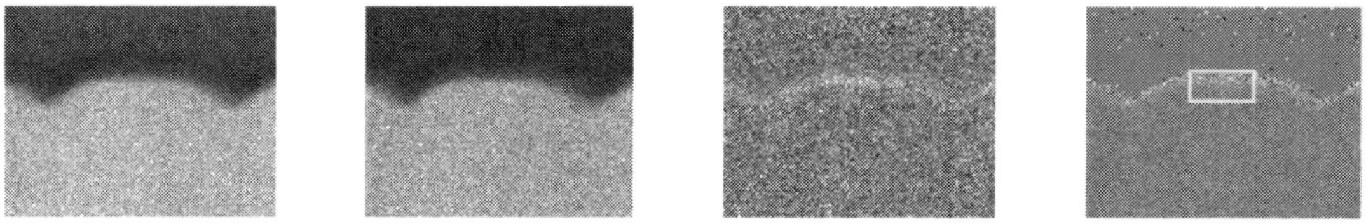

a) GaK (max=108) **b)** AsK (max=75) **c)** In L (max=24) **d)** In% map

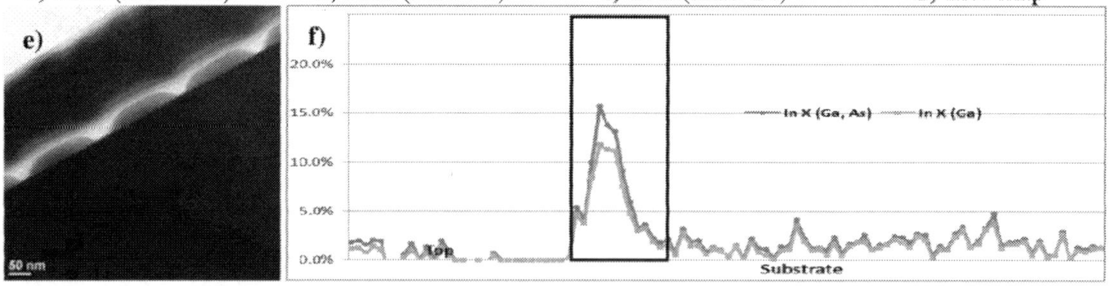

Figure 6. X-ray elemental maps of Ga (a), As (b) and In (c); (d) is an indium concentration map. (e) is an ADF image of the cross sectioned islands; (f) shows concentration profiles from the top of the island to the substrate, calculated from the marked area shown in (d).

References

[1] Gray JL, Hull R and Floro JA 2006 *J. Appl. Phys.* **100**(8) 084312
[2] Kiravittaya S, Heidemeyer H and Schmidt OG 2004 *Physica E* **23**(3-4) 253-9
[3] Srujan M, Ghosh K, Sengupta S and Chakrabarti S 2010 *J. Appl. Phys.* **107**(12) 123107
[4] Zhang H, Ross IM, Hopkinson M, Zhang S and Walther T 2012 *Proc. 12th Int. Conf. on Nanotechnology (IEEE-Nano)*, Birmingham, UK, eds. K Jiang and RE Palmer, abstract #8576, p. 716-718

Journal of Physics: Conference Series **471** (2013) 012048 doi:10.1088/1742-6596/471/1/012048

Correlation of Microstructure and Transport Properties of Multilayered Graphene Spin Valves on SiO_2/Si

L Lari[1,2,3], G Hughes[4], K Muramoto[5], A Hirohata[6,7], M Shiraishi[5] and V K Lazarov[2,3]

[2]Department of Physics, University of York, Heslington, YO10 5DD, York, UK
[3]The JEOL Nanocentre, University of York, Heslington, YO10 5BR, York, UK
[4]Department of Materials, University of Oxford, Parks Road, Oxford, OX1 3PH, UK
[5]Graduate School of Engineering Science, Osaka University, Osaka, Japan
[6]Department of Electronics, University of York, Heslington, YO10 5DD, York, UK
[7]PRESTO, Japan Science and Technology Agency, Kawaguchi, Japan

E-mail: leonardo.lari@york.ac.uk

Abstract. Multi-layered graphene based spin valves were deposited on SiO_2/Si substrates and the Co ferromagnetic and non-magnetic electrodes were patterned using electron-beam lithography. Aberration-corrected (scanning) transmission electron microscopy imaging and energy dispersive X-ray spectroscopy were employed to study the interfacial structure of the device components from thin cross-sectional lamellae obtained by focused ion beam microscopy. The robustness of the spin injection with a constant spin polarization under positive and negative bias voltage was attributed to the interface smoothness which leads to suppression of spin scattering at the interface between the ferromagnet and the multi-layered graphene.

1. Introduction

Multi-Layered Graphene (MLG) on SiO_2/Si substrates is a promising building block for spintronic devices [1]. Recently, MLG based spin valves with unprecedented robustness of spin polarisation have been demonstrated under positive and negative bias voltage [2, 3]. The room temperature magneto-resistance (MR) of these devices is constant with respect to the applied bias voltage, indicating that the spin injection is independent of the applied bias. This is a very desirable property that is lacking in metallic systems spin valves deposited on semiconductors and has been attributed to the high quality of the interface between graphene and ferromagnets. However, the quality of the interface was assessed by indirect methods only, namely through the comparison of bias-voltage dependence versus injection current in "local" and "non-local" schemes [2] but no micro-structural investigations were performed on the devices.

In this work we investigate the microstructure of these high performance MLG spin device structures by using aberration-corrected (scanning) transmission electron microscopy ((S)TEM) imaging and energy dispersive X-ray spectroscopy. This approach, by a direct correlation between the MLG spin valves' magneto-transport properties and their structure, allows verification of the model of spin scattering suppression at the interface as described in [2].

[1] To whom any correspondence should be addressed.

2. Methods

2.1. Device deposition and magneto-transport studies

The devices were fabricated on SiO_2/Si substrates. MLG flakes were obtained by the peeling method using adhesive tape from highly oriented pyrolytic graphite (HOPG; NT-MDT Co.) and polyamide-oriented highly oriented graphite (Kaneka Co. super graphite) [4]. A MLG flake was deposited on the substrate before electron-beam lithography patterning of ferromagnetic (Co) and non-magnetic (Au/Cr) electrodes (Figure 1a). The MR effect in these devices was studied in the "non-local" geometry as described in [2,3].

2.2. Electron microscopy methods

A thin cross-sectional lamella was obtained by focused ion beam (FIB) microscopy from areas across four-terminal MLG spin device structures (across the black line as shown in Figure 1a). Electron microscopy analysis was done using a JEOL JEM-2200FS (scanning) transmission electron microscope (S)TEM equipped with CEOS probe and image aberration correctors. Analytical attachments include a high angle annular dark field (HAADF) detector and a Thermo scientific, Noran 7 energy dispersive X-ray spectroscopy system. Imaging conditions for HAADF STEM analysis were as following: convergence semi-angle of 24 mrad, HAADF inner and outer collection semi-angles were 110 mrad and 180 mrad respectively.

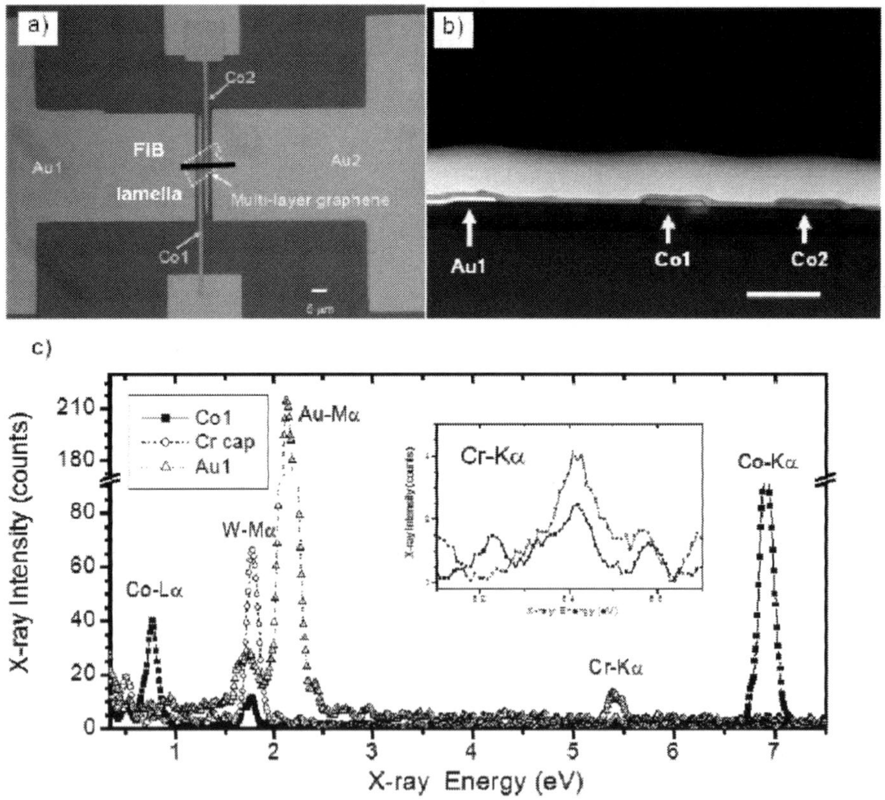

Figure 1. a) Light microscopy image of a MLG spin valve before deposition of Cr cap layer. The black line indicates the lamella removed by FIB; b) device cross-section STEM image showing an Au/Cr contact and Co electrodes (scale bar corresponds to 1 micron); c) EDX point analyses from Co1 electrode (filled squares), Cr cap (open circles), and Au1 electrode (open triangles). Inset the Cr-Kα peaks for Co electrode and the cap layer

3. Results and discussion

From transport measurements on these devices, the spin polarization of the injected spin current was found to be constant up to +2.7V and -0.6V in positive and negative bias applications [2]. This remarkable property was attributed to the high degree of smoothness of the interface of the ferromagnets with the graphene multilayer [2].

The HAADF image in Figure 1b) shows an overview of the first Au/Cr electrode *Au1*, and the two Co ferromagnetic electrodes labelled as *Co1* and *Co2*. EDX point analyses (Figure 1c) confirm the composition of the device components: the spectrum from Au1 electrode (open triangles) shows a marked Au-Mα peak at 2.12 keV as well as the Cr-Kα peak at 5.41 keV, the spectrum from Cr cap layer (open circles) shows the Cr-Kα peak. Finally a strong Co-Lα peak at 0.68 keV is shown in spectrum from the Co1 electrode (filled squares). Due to the presence of the FIB protective layer, the W-Mα peak at 1.77 keV is present in all the three spectra. An overview of the layered structure at the terminal Co/MLG is shown in Figures 2a) and 2b). A 285nm silicon dioxide layer can be seen on top of the Si doped substrate. The thickness of the MLG is 50 nm, corresponding to about 150 monolayers of graphite. On top of the MLG a 0.7 micron wide and 60 nm thick Co electrode is shown with a good uniformity and parallel surfaces. Above the electrode, a 60nm thick Cr capping layer is shown, the material further on top corresponds to tungsten, residue of the protection mask deposited in the FIB in order to preserve the structure during the ion milling.

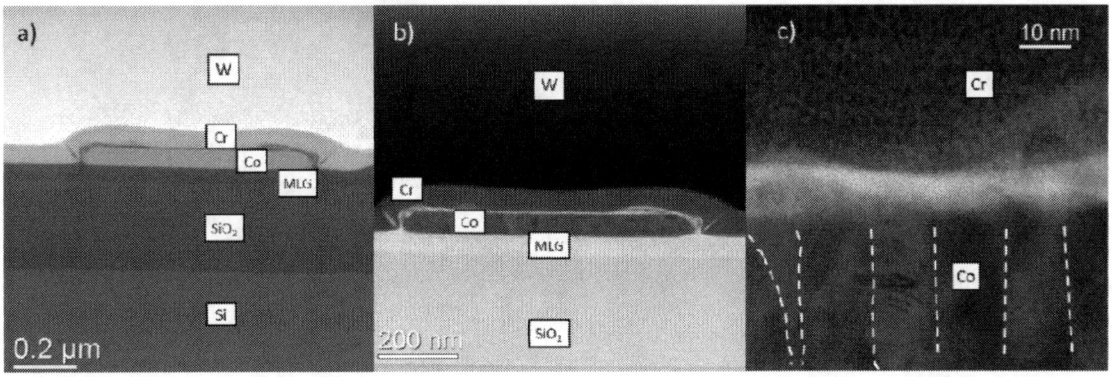

Figure 2. a) High angle annular dark and b) bright field STEM images of the multilayered structure of the device at the Co/MLG terminal; c) interface Cr capping layer to Co ferromagnet. Dotted lines highlight the multigranular structure of the electrode.

The most important area of the sample for the spin current injection consists of the two ferromagnets, the Co electrodes [2,3]. We found that the Co electrodes consist of coalesced grains with an average width of about 10 nm, as shown in Figure 2c). Metallic Co is known to have two major crystallographic structures: hexagonal close-packed (hpc) structure (P63/mmc) with lattice parameters a=0.2170, c=0.2045 and face centred cubic (fcc) structure (F m-3m) with lattice parameter a=0.3544 [5].

Fast Fourier Transform (FFT) analysis was applied to HRSTEM images from several grains finding areas along [001] zone axis with lattice spacing of (0.177±0.002)nm corresponding to the (002) planes of the cobalt cubic structure at the top surface of the Co. However, the presence of the hcp structure of the cobalt cannot be completely ruled out. In fact, in Figure 3a), two adjacent grains labelled G1 and G2 are imaged at the interface with the MLG. The stacking planes of the MLG are parallel to lattice planes visible in the off-axis grain G1. Their distance was measured to be (0.205±0.002)nm which is compatible with both the (0002) planes of the hcp structure (0.2045 nm) or the (1-11) planes of the fcc structure (0.2046 nm). Grain G2 shows a large variation of lattice arrangement at the nanometre scale due to the presence a high density of crystal defects in addition to surface amorphisation due to the FIB sample preparation. The FFT from the G2 area shows a cubic symmetry of the Bragg spot pattern, but the presence of multiple spots make it difficult to ascertain the crystal structure of the grain. However, a set of planes of Bragg reflections (highlighted by white circles in the FFT in Figure 3c) are clearly aligned with respect to the MLG stacking planes showing a good quality of the interface.

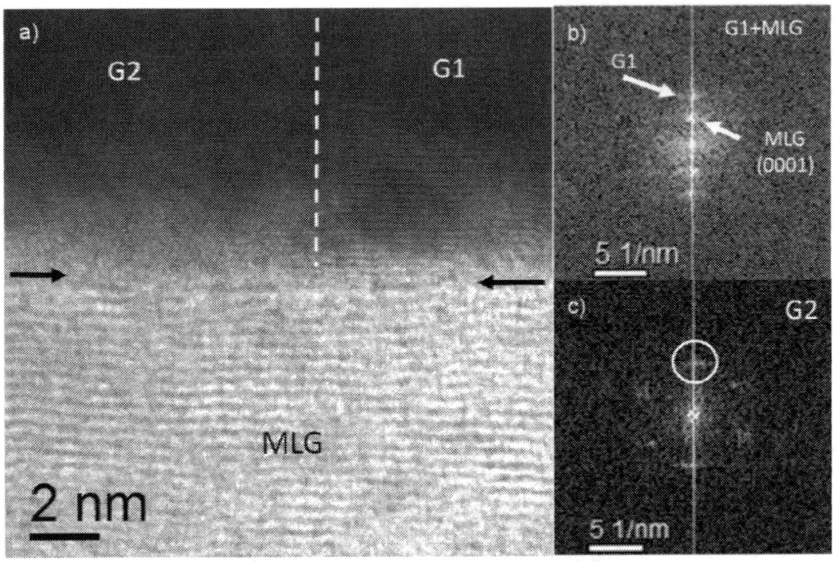

Figure 3. a) BF-STEM image of the interface Co-MLG showing two grains in the Co with different epitaxial relationships with respect to the MLG; b) FFT of interface G1 and G2 showing the stacking of Co planes parallel to the MLG (0001). c) FFT of Grain 2 white circles highlight the Co planes aligned parallel to the MLG stacking planes.

4. Conclusions

Aberration corrected (S)TEM analytical and imaging techniques were applied to study the microstructure of MLG-based spin valves devices, showing high robustness of spin polarisation under positive and negative bias. The smoothness of the interface between the multilayered graphene and the ferromagnetic (Co) electrode was confirmed by bright field STEM imaging. Our results support the theory that the robustness of the spin injection can be attributed to the suppression of spin scattering at the interface between the ferromagnet and the graphene.

References

[1] Ohishi M, Shiraishi M, Nouchi R, Nozaki T, Shinjo T, and Suzuki Y 2007 *Jpn. J. Appl. Phys.* **46** L605

[2] Shiraishi M, Ohishi M, Nouchi R, Mitoma N, Nozaki N, Shinjo T, and Suzuki Y 2009 *Adv. Funct. Mater.* **19** 3711

[3] Muramoto K, Shiraishi M, Mitoma N, Nozaki T, Shinjo T, and Suzuki Y 2009 *Appl. Phys. Express* **2** 123004

[4] Murakami M, Nishiki N, Nakamura K, Enhara J, Okada K, Kouzaki T, Watanabe K, Hoshi T and Yoshimura S 1992 *Carbon* **30** 255

[5] Owen E A and Madoc Jones D 1954 *Proceedings of the Physical Society* (London) **67** p456

The microstructure of Si surface layers after He^+ and Ar^+ plasma immersion ion implantation

J M Chesnokov[1], A L Vasiliev[1,2], V F Lukichev[3] and K V Rudenko[3]

[1] National Research Centre "Kurchatov Institute", Moscow, Russia
[2] Institute of Crystallography Russian Academy of Sciences, Moscow, Russia
[3] Institute of Physics and Technology Russian Academy of Sciences, Moscow, Russia

E-mail: a.vasiliev56@gmail.com

Abstract. Processes accompanying the amorphization of Si surface layer after Ar^+ and He^+ low energy plasma immersion ion implantation were studied by transmission electron microscopy and microanalysis. The ion energies varied in the range of 0.5 – 3 keV with the doses of $5{\cdot}10^{15}$ cm^{-2} for Ar^+ and $2{\cdot}10^{16}$ cm^{-2} for He^+ pre-amorphization implants. A perturbation of Si lattice together with porosity was found. The utilization of heavy and light ions causes distinctive microstructural peculiarities. The presence of oxygen and its distribution near the surface layer was determined by electron energy loss spectroscopy.

1. Introduction

The downscaling of design rules for ultra-large scale integration circuits (ULSI) to sub-32 nm requires manufacturing extremely highly doped and thin source and drain regions of integrated Si MOSFETs. The thickness of doped Si layer should not exceed 10-15 nm. New technology of doping – the plasma immersion ion implantation (PIII) is a promising technique for that purposes [1]. Most difficulties appear with p-type of doping where B has no alternatives due to high solubility threshold in Si. However, the small B atoms implanted in Si undergo well-known channeling effect and to prevent this, the preliminary amorphization of thin Si layer is required. Conventional pre-amorphizing implants (PAI) are Si^+ or non-active ions like Ar^+ or other inert gases. The use of heavy ions and high dose PAI results in the surface sputtering together with Si amorphization. The sputtering is very critical for shallow junctions [2]: the surface recess limits for sub-32 nm MOSFETS is about 1 nm.

Therefore, the PAI processes of Si with light ions, especially with He^+, is of great interest and has been extensively studied recently [3, 4]. The mechanisms of the amorphization and sputtering of solids by high energy ions are different for heavy and light ions [5]. The microstructure of amorphous layers could be different as well. That could cause peculiarities in diffusion and electrical activation behaviour of B dopant during rapid thermal annealing (RTA) process of implanted Si layers. The anomalous transient retrograde diffusion of B in Si during RTA after PIII-process with He^+ PAI found in [6] was not observed in Ar^+ PAI [7]. It was proposed that specific defects of an amorphous silicon layer caused by He^+ implantation and its behaviour during subsequent annealing was responsible for difference in B diffusion [6]. However, no systematic research of the Ar^+ and He^+ implanted Si with low ion energy $E < 3$ keV was done. In this study a thorough analysis of the microstructure of Si after 0.5-3 keV Ar^+ and He^+ PAI was performed.

2. Experiment.

Ions of He^+ and Ar^+ were implanted in wafers of n-type Si(100) (4.5 Ohm·cm). The implantation process was carried out using experimental setup in plasma immersion mode by switching the wafer on negative square shaped pulsed accelerating potential with duration of 10 μsec. The repetition rate of accelerating pulses was 1 kHz. The ion current density measured during pulses of implantation was in the range of 10 – 20 mA/cm^2. The wafer was placed on a water cooled chuck with helium feeding between the chuck and the sample to prevent overheating and the wafer temperatures at implantation process did not exceed 100°C. The experimental conditions for all samples are described in Table 1.

Table 1: The ion energies and the exposure dose used.

	Group I – Ar^+ PAI			Group II – He^+ PAI		
Sample #	1	2	3	4	5	6
Energy, keV	0.5	1	3	0.5	1	3
Dose, cm^{-2}	5 x 10^{15}	5 x 10^{15}	5 x 10^{15}	2 x 10^{16}	2 x 10^{16}	2 x 10^{16}

The cross-sections of all samples for TEM were prepared by conventional way: mechanical thinning of glued slabs to the thickness of 10-20 μm followed by ion milling in Gatan PIPS (Gatan, USA) by Ar^+ ions at an angle of 6° and an energy of 4 keV from the both sides of the sample. After the perforation of the sample the final cleaning with Ar^+ ions with an energy of 100 eV was carried out for 5 minutes.

The microstructural study was performed in a Titan 80-300 TEM/STEM (FEI, USA) with an accelerating voltage of 300 kV. The electron microscope was equipped with probe C_s corrector, high angle annular dark field detector (HAADF, Fischione, USA), Gatan image filter (GIF) and windowless energy dispersive X-ray spectrometer (EDXS, EDAX, USA). The bright field (BF) TEM, high resolution (HRTEM) and HAADF STEM operation modes together with 3-window elemental mapping technique were used in our investigations.

3. Results and discussion.

3.1. The microstructure of Si after Ar^+ PAI.

The BF HRTEM images obtained from cross-sectional specimens after Ar^+ PAI at E_{Ar} = 0.5 keV, 1 keV, 3 keV are presented in Figure 1 (a-c) correspondingly. Several layers having different contrast were found and their presence and thickness depended on ion energy. One or two amorphous layers were observed near the specimen surface. They could be resolved by slightly different contrast and Fast Fourier Transform (FFT) power spectra which were obtained from these parts of the image to prove the difference. Energy filtering (EF) of the images was used to determine O distribution in the amorphous layers. The sample surface after Ar^+ PAI with E_{Ar} = 0.5 keV looked wavy and the amorphous layer thickness determined from HREM image was 5±1 nm. The thickness of the layer was close to native SiO_x (3-5 nm) with similarly cleaned (001) Si surface. In that sample no difference in the contrast across the amorphous layer was observed and the small thickness of the layer prohibited the application of FFT analysis. The EF indicated that O was present in the amorphous layer from the top to the bottom of the layer. The attempts to obtain distinct Ar distribution failed and that could be due to additional Ar^+ implantation during Ar^+ milling of the sample. Below SiO_x/Si interface there was a crystalline layer exhibiting darker contrast, probably due to the presence of Ar^+ ions in the end of range. The bottom interface of that area was not flat and displayed faceting parallel to {111} Si planes. Close inspection of Si crystal planes in the adjacent area demonstrated distortion of planes, however, defects like dislocations or stacking faults were not found.

The HRTEM image of the sample after Ar^+ PAI at E_{Ar}=1 keV is shown in Figure 1 b. The overall thickness of the amorphous layer was 8 nm. The surface of the sample was not flat revealing an unevenness up to 1 nm height. The EF imaging indicated that 6-7 nm thick SiO_x layer was formed near the surface and the adjacent a-Si layer was about 1-2 nm. The crystalline Si (c-Si) surface was wavy with unevenness height of 0.5-1 nm. Again, the layer exhibiting darker contrast was below the a-Si/c-

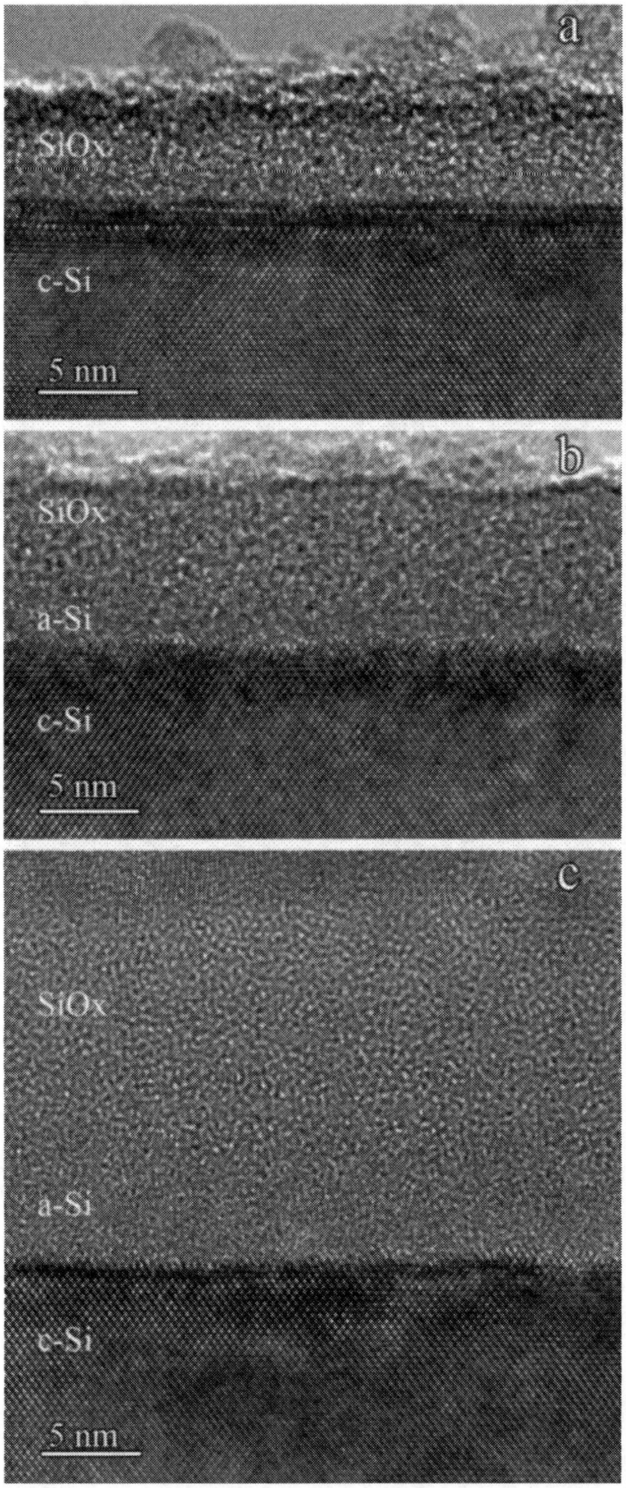

Figure 1. Cross-sectional BF TEM images of Ar^+ PAI sample with energies of 0.5 (a), 1 (b) and 3 keV (c).

Si interface and defects were found in the wavy (top) part which could be due to the overlapping of a-Si and c-Si.

The amorphous layer in the sample after Ar^+ PAI at E_{Ar}=3 keV was 17 nm thick (Figure 1 c) and, according to the EF data, the top part was O enriched. The thickness of that SiO_x layer was 8-11 nm, which is much thicker than typical native SiO_2. The FFT spectra analysis demonstrated a single maximum for the SiO_x layer and double maxima for the a-Si layer with peak positions close to the previous sample. The upper layers of c-Si had a microstructure similar to the sample after PAI at E_{Ar}=1kV. The study of the reason for the formation of this SiO_2 layer is on the way.

3.2. The microstructure of Si after He^+ PAI

After He^+ PAI with E_{He}=0.5 keV only one amorphous layer with thickness of 2 nm was observed (Figure 2 a). According to the EF study, the layer contained O. On several images we found a visibly distorted layer (0.5 nm) on the top of c-Si.

Increasing the He+ ions energy to 1 keV resulted in severe changes of overall microstructure. Three amorphous layers were found on the Si surface and they are shown in Figure 2 b:

1. a-SiO_x layer with thickness of 5-6 nm. The thickness of the layer was determined with the EF imaging.
2. a-Si layer. The thickness of the layer was estimated from BF and HAADF images and was 5-6 nm .
3. 10 nm thick a-Si layer with pores. The pore dimensions were up to 10 nm.
4. c-Si with high density of structural defects. HREM study revealed terminated crystal planes and black-white contrast typical for dislocations. The dark contrast on HAADF images indicated the presence of smaller pores (< 4 nm) in the top part of the layer. Bottom part of the layer exhibited darker contrast in HREM image (Fig. 2 b), probably due to absence of pores.

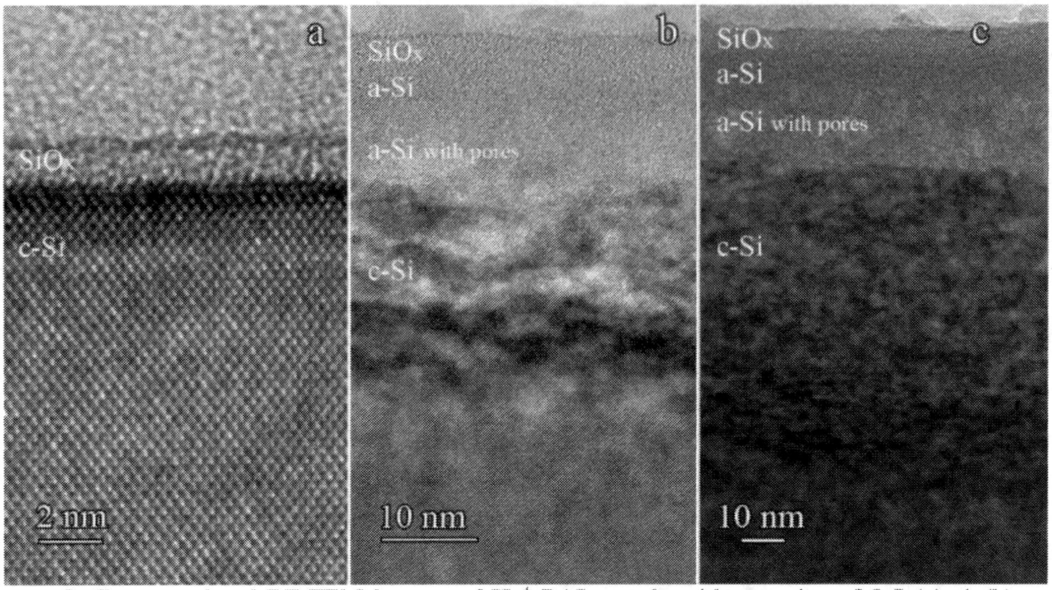

Figure 2. Cross-sectional BF TEM images of He^+ PAI sample with energies of 0.5 (a), 1 (b) and 3 keV (c).

These pores could be the consequence of helium bubbles formed in the upper layers. After He^+ PAI with E_{He}=3 keV the number of the layers and their morphology were similar to the He^+ PAI at E_{He} =1 keV. The thickness of a-SiO_x was 7 nm and that of a-Si 5-6 nm. The larger pores (up to 10 nm) were in the a-Si layer and smaller (<6 nm) in the c-Si. Again, the pore density was much higher in the top part of c-Si layer and that area appeared brighter in the HREM image (Figure 2 c).

4. Conclusion

Different microstructure was found near Si surface after Ar^+ and He^+ low energy plasma immersion ion implantation. The Ar^+ PAI results in formation of only amorphous layers while He^+ PAI leads to the amorphous together with highly defective crystalline layer formation. The presence of oxygen and its distribution near the surface layer was determined by electron energy loss spectroscopy.

References

[1] Anders A 2000 *Handbook of Plasma Immersion Ion Implantation and Deposition.* New-York, John Wiley & Sons.
[2] International Technology Roadmap for Semiconductors. 2011 Edition (online at http://www.itrs.net/Links/2011ITRS/Home2011.htm).
[3] Sasaki Y, Jin CG, Tamura H and Mizuno B 2004 *Proc. IEEE Symposium on VLSI Technology,* 180.
[4] Mizuno B 2008 *Electronic Device Architectures for the Nano-CMOS Era,* Pan Stanford Publ. 141.
[5] Falcone G 1990 *La Revista del Nuovo Cimento.* **13** 1.
[6] Lukichev V, Rudenko K, Orlikovsky A, Pustovit A and Vyatkin A 2008 *17th Int. Conf. on Ion Implantation Technol. AIP Conference Proceedings.* **1066** 481.
[7] Rudenko K, Averkin S, Lukichev V, Orlikovsky A, Pustovit A and Vyatkin A 2006 *Proc. SPIE* **6260** 626003.

Journal of Physics: Conference Series **471** (2013) 011004 doi:10.1088/1742-6596/471/1/011004

Subject Index

Journal of Physics: Conference Series **471** (2013) 011004 doi:10.1088/1742-6596/471/1/011004

Journal of Physics: Conference Series **471** (2013) 011003 doi:10.1088/1742-6596/471/1/011003

Author Index

Journal of Physics: Conference Series **471** (2013) 011003 doi:10.1088/1742-6596/471/1/011003

Dimroth F
Dobbie A
Durose K

Eliseev AA
Epicier T

Favia P
Favre L
Fischer AM
Fischione PE
Foxon CT
France RM
Franzo G
Furdyna JK

Galli M
Gass MH
Geelhaar L
Geisz JF
Gendry M
Gerace D
Geypen J
Goff LE
Golberg D
Goodman MCJ
Gould C
Grauer S
Griffiths IJ
Guo W
Gustafsson A

Hagedorn S
Halliday DP
Hangleiter A
Henneberger F
Hentschel T
Herfort J
Heurlin M
Hirohata A
Hoffmann L
Honda K
Hughes G
Humam NS
Humphreys CJ

Iacona F
Ihle S

Jäger W

Jenichen B
Jönen H
Jublot M

Kalmykov AE
Kappers MJ
Karczewski G
Kasama T
Kawaguchi K
Kelly TF
Kirmse H
Kiselev NA
Knauer A
Kong X
Kovács A
Kovalchuk MV
Krause FF
Krauss TF
Krylyuk O
Krylyuk S
Kueller V
Kukushkin A
Kumskov AS

Lábadi Z
Lagrain P
Lam T
Lari L
Larson DJ
Lawrence D
Lazarov VK
Lefebvre W
Lepinay K
Lezec HJ
Lindgren D
Liu X
Lo Savio R
Lodié D
Lorut F
Lukichev VF

Major JD
Marsh LM
Massabuau FC-P
Matesa JM
McCartney MR
McMahon WE
Mehrtens T
Mendis BG
Merckling C

Journal of Physics: Conference Series **471** (2013) 011003 doi:10.1088/1742-6596/471/1/011003

Journal of Physics: Conference Series **471** (2013) 011003 doi:10.1088/1742-6596/471/1/011003

Riechert H
Ripalda JM
Robins AC
Roddatis VV
Romero MJ
Rosenauer A
Ross IM
Rossow U
Rouvière J-L
Rudenko KV
Ruzmetov D
Ryll H

Sahonta S-L
Saiki T
Samuelson L
Sánchez AM
Schöne J
Schowalter M
Schreyeck S
Schumacher C
Servanton G
Shakoor A
Shannon MD
Shiraishi M
Smith DJ
Smith GXR
Solomkin FY
Soltau H
Songmuang R
Sorokin LM
Sparenberg M
Spiecker E
Strüder L
Strunk HP

Taboada AG
Takahashi M
Talin AA
Tarakina NV
Taylor AA
Tewes M
Texier M
Thagi Khani A
Thrush EJ
Trampert A
Tränkle G
Trinh-Xuan L
Tromas C
Tsumori N

Journal of Physics: Conference Series **471** (2013) 011003 doi:10.1088/1742-6596/471/1/011003